ANTIBODY FUSION PROTEINS

ANTIBODY FUSION PROTEINS

Edited by

STEVEN M. CHAMOW

and

AVI ASHKENAZI

WILEY-LISS

A John Wiley & Sons, Inc., Publication

New York · Chichester · Weinheim · Brisbane · Singapore · Toronto

This book is printed on acid-free paper. ∞

Printed simultaneously in Canada.

Library of Congress Cataloging-in-Publication Data:

Antibody fusion proteins / edited by Steven M. Chamow, Avi Ashkenazi.
p. cm.
Includes index.
ISBN 0-471-18358-X (cloth : alk. paper)
1. Antibody-drug conjugates. 2. Antibody-toxin conjugates.
I. Chamow, Steven Mark. II. Ashkenazi, Avi.
RS201.A56A665 1999
616.07'98--dc21 98-36471

Printed in the United States of America.

10 9 8 7 6 5 4 3 2 1

CONTENTS

PART II Fc FUSION PROTEINS

FOREWORD

The generation of specific antibodies in the immune system has been the prototype for protein engineering. Antibody molecules can be made to specifically bind just about anything, and for the most part, they have quite reasonable physical and chemical properties. Yet, the success of natural monoclonal antibodies in different fields of therapy has been far less than spectacular.

In this book, a number of leading scientists in the field outline that man-made molecules will be required not only to overcome the limitations of monoclonal antibodies, but also to extend the principle of selective targeting. The antibody molecule is designed primarily to fight viruses and bacteria, and does so quite effectively. The immunoglobulin is actually a specific adapter molecule. It mediates the contact between a target surface and an effector mechanism, which may be based on either activating whole cells or a cascade of enzymes, such as the complement system. It follows that specific binding to the target is essential, but is only half the story.

Taking the natural antibody as a source of inspiration, a variety of biological responses can be elicited by fusing other types of molecules to the antigen-binding site of the antibody. Many different and elegant ideas are discussed in the first part of this book, such as making fusion proteins with toxins, cytokines, or enzymes, which can activate a prodrug. A further strategy is to create a bispecific molecule by linking a second antigen binding site to the first one in order to activate an effector mechanism.

Another set of fusion proteins has been constructed which retain only the Fc part of the antibody. The Fc portion has two important features; not only does it dimerize the proteins fused to it, but it also extends their serum half-life.

The exploitation of this principle, giving rise to so-called immunoadhesins or Fc-fusion proteins, has led to a wealth of unique molecules with extremely interesting therapeutic potential, which is summarized in the second part of the book.

As elegant and promising as all of these approaches are, they are far from perfect. Studying the detailed in vitro and in vivo mechanisms of action of a number of these molecules, especially those that perform differently than expected while designed according to a plausible idea, will help develop the knowledge necessary for making even better second generation molecules. Such studies will advance the whole field of immunotargeting much more than serendipitous success.

Rapid progress in this field has become possible since many production issues have been solved, using either bacterial or eukaryotic expression systems. For example, Chapter 10 gives an outline of the state of the art in producing clincial grade material of immunoadhesins. Furthermore, the generation of specific anti-binding sites with reasonable affinity is a problem that is now largely solved through the use of antibody libraries and elegant selection tools such as phage display, to which another chapter is devoted. Whereas the tools for selection of binding molecules have come a long way, the biophysical understanding of what exactly makes a stable, high-affinity molecule is lagging far behind. Folding, expression, and stability remain challenging issues for the future and may ultimately decide the utility of a particular molecule in practical applications.

The great successes that have already been achieved with the strategies described in this book invite speculations on how the field might move ahead. In the beginning, the recombinant antibody technology strove to imitate nature in its combinatorial diversity and its pragmatic selection principles. At the next stage, the adapter principle of the antibody was taken to more abstract terms, leading to the variety of fusion proteins described in this volume. The third stage of abstraction may then be to combine the two, and fuse unnatural, optimized targeting domains to equally reengineered artificial effector functions in an approach merely inspired by nature's concepts. Although we now have many of the tools to achieve this, we still lack much of the required biological and molecular understanding.

This, however, is delightful for the researcher.

ANDREAS PLÜCKTHUN

ACKNOWLEDGMENTS

This book represents the culmination of more than a decade of work, during which the tools of protein engineering forged a new field in antibody fusion proteins. The work on immunoadhesins was begun at Genentech in the 1980s. We wish to acknowledge Dan Capon, who initiated this work, and Doug Smith, who helped Dan with the early experiments. In addition, we are grateful to Larry Lasky, Scot Marsters, and David Peers for their help in bringing the immunoadhesin technology to fruition. Moreover, we thank our wives, Judy and Chris, for their continued support and encouragement.

CONTRIBUTORS

Alejandro Aruffo, Department of Inflammation, Bristol Squibb Pharmaceutical Research Institute, P.O. Box 4000, Princeton, NJ 08543

Avi Ashkenazi, Department of Molecular Oncology, Genentech, Inc., 1 DNA Way, South San Francisco, CA 94080.

Steven M. Chamow, Department of Protein Chemistry, Scios, Inc., 2450 Bayshore Parkway, Mountain View, CA 94043

Michael Fanger, Department of Microbiology, Dartmouth Medical School, One Medical Center Drive, Lebanon, NH 03756; Medarex, Inc., 1545 Route 22 East, Annandale, NJ 08801

David FitzGerald, Laboratory of Molecular Biology/NCI, National Cancer Institute, Bldg. 37/4B-03, Bethesda, MD 20892-0001

Joel Goldstein, Department of Microbiology, Dartmouth Medical School, One Medical Center Drive, Lebanon, NH 03756; Medarex, Inc., 1545 Route 22 East, Annandale, NJ 08801

Robert F. Graziano, Department of Microbiology, Dartmouth Medical School, One Medical Center Drive, Lebanon, NH 03756; Medarex, Inc., 1545 Route 22 East, Annandale, NJ 08801

Werner Lesslauer, Department PRPN-D, F. Hoffmann-La Roche, Ltd., CH-4070 Basel, Switzerland

James D. Marks, Department of Anesthesia, San Francisco General Hospital, Room 3C-38, San Francisco, CA 94110

Sherie L. Morrison, Department of Microbiology and Molecular Genetics, University of California, 405 Hilgard Ave., Los Angeles, CA 90095-1489

Dianne L. Newton, Intramural Research Support Program, SAIC Frederick, National Cancer Institute-Frederick Cancer Research and Development Center, Frederick, MD 21702-1201

Manuel L. Penichet, Department of Microbiology and Molecular Genetics, University of California, 405 Hilgard Ave., Los Angeles, CA 90095-1489

David B. Powers, Department of Anesthesia, San Francisco General Hospital, Room 3C-38, San Francisco, CA 94110

Susanna M. Rybak, Building 567, Room 152, Laboratory of Biochemical Physiology, Division of Basic Science, National Cancer Institute, Frederick Cancer Research and Development Center, Frederick, MD 21702-1201

Seung-Uon Shin, Institute of Environment and Life Science, The Hallym Academy of Science, Hallym University, Cluchon, Kangwon-Do 200-702, Korea

J. Yun Tso, Protein Design Labs, Inc. 34801 Campus Drive, Fremont, CA 94555

Florian W. Wurm, Department of Chemistry, Swiss Federal Institute of Technology (EPFL), CH-1015, Lausanne, Switzerland

1

OVERVIEW

STEVEN M. CHAMOW
Sciros, Inc. Mountain View, CA 94043
AVI ASHKENAZI
Genentech, Inc. South San Francisco, CA 94080

1.1 ANTIBODIES

Immunoglobulins are the most critical components of the immune system. As proteins that bind to preselected molecular targets, immunoglobulins arise in response to foreign substances introduced into the body. The immunoglobulins comprise a heterogeneous group of proteins that account for approximately 20% of the total plasma protein in humans. In addition, different populations of immunoglobulins are found in extravascular fluids, in exocrine secretions, and on the surface of some lymphocytes. The biologic activity of immunoglobulins is best understood in the context of their structure.

The basic three-dimensional structure of antibodies was first elucidated as early as 1973,[1] when the crystal structure of a Fab fragment was solved. An antibody is a Y-shaped molecule, composed of two identical light chains and two identical heavy chains (Fig. 1.1). Both light and heavy chains contain variable and constant regions. The four chains are held together by disulfide bonds, which are located in a flexible region of the heavy chain known as the hinge. Variable regions of both heavy and light chains combine to form two identical antigen-binding sites, one on each arm of the Y. Heavy chain constant regions define five classes of antibodies: IgA, IgD, IgE, IgG, and IgM, each with its own class of heavy chain— α, δ, ε, γ, and μ, respectively. Each antibody class (termed an *isotype*) has distinct structural and functional characteristics. In isotypes such as IgM or IgA, multimeric assemblies of four-chain units

Antibody Fusion Proteins, Edited by Steven M. Chamow and Avi Ashkenazi
ISBN 0471-18358-X

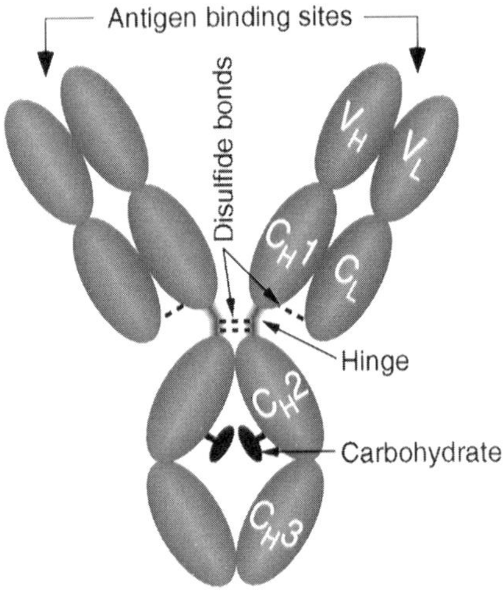

Figure 1.1 Structure of the antibody molecule. A schematic model of human IgG1, showing the basic four-chain structure and domains. V indicates variable region; C, constant region; V_L and C_L are domains of the light (L) chain; V_H, C_H1, C_H2, and C_H3 are domains of the heavy (H) chain. The hinge region, containing two inter-H chain disulfide bonds (dotted lines) and disulfide bonds connecting H and L chains, is indicated. The antigen-binding sites, of which there are two in each antibody, are comprised of unique segments contained within the V_H and V_L domains. A single asparagine-linked carbohydrate is encoded within the C_H2 domain of each H chain. X-ray structural studies confirm that these carbohydrates are sequestered within a pocket formed by the C_H2 and C_H3 domains of the Fc.

produce antibody molecules with ten or four antigen binding sites, respectively. In addition, there are a number of subclasses of IgG and IgA immunoglobulins; for example, within the human IgG isotype, there are four subclasses (IgG1, IgG2, IgG3, and IgG4) having $\gamma1$, $\gamma2$, $\gamma3$, and $\gamma4$ heavy chains, respectively. Effector functions of antibodies, such as complement activation, binding to phagocyte Fc receptors, antigen-dependent cellular cytotoxicity, and transport across the placenta, are mediated by structural determinants within the Fc region (the tail of the Y shape).

Two major discoveries in the 1960s ushered in the period of detailed structural study of antibodies. The first was the finding that enzymes and reducing agents could be used to digest or dissociate immunoglobulin molecules into smaller components. The second was the realization that the electrophoretically homogeneous proteins found abundantly in serum and urine of patients with a disease called multiple myeloma were related to normal immunoglobulins. These myeloma proteins were found to be structurally homogeneous. They are also called monoclonal proteins, since they are synthesized by single clones of malignant plasma cells.

To help the reader understand the information presented in this book, we have compiled a list of concepts and definitions that lay the groundwork for grasping the structural elements of immunoglobulins. Because most of the protein engineering approaches described in this book are based on the structural framework of human IgG, the list below defines concepts that apply specifically to the IgG molecule.

Structural Elements of Immunoglobulins

Basic unit (monomer): Each immunoglobulin contains at least one basic unit or monomer comprising four polypeptide chains (Fig. 1.1). The oligomeric structure is stabilized by interchain disulfide bonds that connect all four chains together.

H and L chains: Immunoglobulins contain two types of polypeptide chains. One pair of identical polypeptide chains contains approximately twice the number of amino acids, or is approximately twice the molecular weight, of the other pair of identical polypeptide chains. The chains of higher molecular weight are designated heavy (H) chains, and those of lower molecular weight, light (L) chains.

V and C regions: Each polypeptide contains an amino terminal portion, the variable (V) region; and a carboxyl terminal portion, the constant (C) region. These terms denote the considerable heterogeneity or variability in the amino acid residues in the V region compared to the C region. Heavy and light chains each have a single V region, and light chains possess a single C region. Heavy chains contain three C regions.

Antigen binding site: The part of the antibody molecule that binds to antigen is formed only by small numbers of amino acids in the V regions of H and L chains. These amino acids, contained within six complementarity-determining regions (CDRs), are brought into close proximity by the folding of the V regions.

Domains: The polypeptide chains are folded by disulfide bonds into globular regions called domains. The domains in H chains are designated as V_H, C_H1, C_H2 and C_H3; those in L chains are designated V_L and C_L. An IgG monomer contains two Asn-linked oligosaccharides, one attached to each H chain within the C_H2 domain. These oligosaccharides are oriented into a pocket formed by the bowing outward of the C_H2 and C_H3 domains of the Fc (Fig. 1.1).

Hinge region: The area of the H chains in the region between C_H1 and C_H2 domains. This region of approximately 12 amino acid residues is quite flexible and is more exposed than are other regions of the molecule to enzymatic and chemical cleavage. Thus, papain acts here to produce Fab and Fc fragments (see following).

1.1.1 Immunoglobulin Fragments

Fragmentation of antibodies to produce segments that retain biological function (Fig. 1.2) has been important to elucidation of antibody structure. Several examples of antibody fragments that are themselves useful reagents are presented in this book. Initially, these fragments were made by proteolytic treatment of purified whole antibodies. More recently, fragments have been produced directly by recombinant means. The possibility of generating antibody Fab fragments as recombinant proteins in *Escherichia coli*[2] has been a major breakthrough in the field of antibody engineering. Some definitions of fragments follow:

Fab and Fc fragments: Digestion of an IgG molecule by the enzyme papain cleaves the molecule within the hinge region at a site upstream of the inter-H chain disulfide bonds, producing two Fab (ab = antigen binding) fragments and one Fc (c = crystallizable) fragment[3] (Fig. 1.2A).

F(ab′)$_2$ fragment: Digestion of an IgG molecule by the enzyme pepsin cleaves the molecule also within the hinge region, but at a site downstream of the inter-H chain disulfide bonds (Fig. 1.2B). In addition, pepsin cleaves the Fc fragment into several peptide fragments. Thus, pepsin cleavage results in production of one F(ab′)$_2$ molecule.[4] The F(ab′)$_2$ molecule is composed of two Fab units and the hinge (hence the nomenclature "Fab′" to distinguish it from the "Fab" produced by papain). It contains two antigen-binding sites and can therefore bind bivalently to antigen.

Fv fragment: A fragment consisting only of the two V regions, V_L and V_H (Fig. 1.2C). Capable of binding to antigen, this fragment is unstable due to noncovalent association of its two polypeptide chains. Fv fragments were originally produced enzymatically by cleavage of IgG[5].

Single-chain Fv fragment (scFv): A stable variant Fv fragment in which the two V region polypeptides are covalently attached via a linker peptide (Fig. 1.2C).[6] This can be done in either orientation, so that the linker peptide attaches the C-terminus of V_H to the N-terminus of V_L, or vice versa. In either molecular construct, antigen-binding activity can be retained. Single-chain Fv fragments are most commonly produced by direct expression of recombinant fragments in bacteria.[7]

Fd fragment: The N-terminal half of the H chain comprising V_H and C_H1 (Fig. 1.2C).[8] The receptor domain of an Fc fusion protein is sometimes referred to as an "Fd-like" fragment, by analogy to this H chain nomenclature.

1.1.2 Genetically Engineered Immunoglobulins

For monoclonal antibodies to be maximally useful as human therapeutics, they must possess several qualities: (1) high affinity binding to antigen, (2) an ability

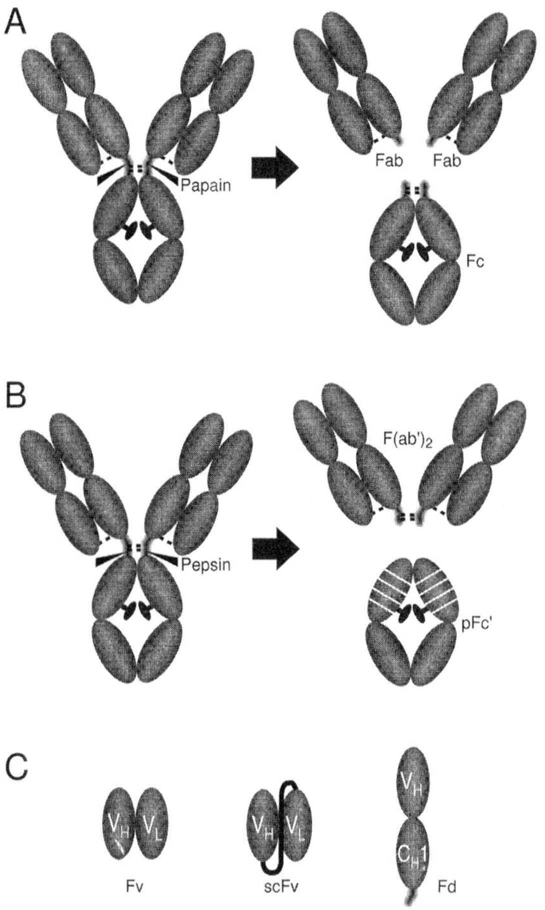

Figure 1.2 Immunoglobulins can be fragmented by partial digestion with proteases. The resulting fragments retain full antigen-binding activity. *A*. Treatment with papain cleaves the antibody molecule at a unique site in the hinge region, upstream of the interheavy chain disulfide bonds. This results in generation of three fragments: two identical Fab fragments that each bind monovalently to antigen, and the Fc fragment that retains its ability to bind to Fc receptors, including *Staphylococcus aureus* protein A. *B*. Treatment with pepsin cleaves the molecule at sites within the hinge region and also in the C_H2 domain, generating many peptides. The Fc (pFc′) is largely destroyed by this treatment. The cleavage site within the hinge region is downstream of the interheavy chains disulfide bonds, resulting in production of a bivalent antigen-binding fragment, $F(ab')_2$. *C*. Additional fragments to which the text refers are Fv, single-chain Fv (scFv), and Fd. Fv and scFv represent the smallest antigen-binding fragments yet produced. Fd fragments do not bind to antigen, but when this region is replaced in an Fc fusion protein, the ligand-binding domain is sometimes referred to as an "Fd-like" fragment. Fab, $F(ab')_2$, and scFv fragments are versatile alternatives to full-length antibodies that have been produced in recombinant form by direct expression in bacteria.

to neutralize antigen activity, (3) long serum half-life, and (4) low immunogenicity. Antibodies from different animal species can have some of these qualities, but only human antibodies are nonimmunogenic when injected into patients. Thus, the goal of therapeutic antibody research during the past 20 years has been to create specific antibodies that are increasingly human in sequence, in order to reduce immunogenicity of the molecule.

The application of genetic engineering to antibodies was not possible without creation of a source of a single, homogeneous antibody of defined antigen specificity. This was achieved in 1975 with the advent of hybridoma technology to produce murine monoclonal antibodies[9] (Fig. 1.3). Despite the success of this method for production of mouse antibodies, however, hybridoma technology has not been successful in production of human antibodies.

To overcome this limitation, investigators in the 1980s sought to use the emerging tools of protein engineering to convert murine antibodies into human forms. A first step toward this goal was the development of genetically engineered antibodies that contained some human constant region sequence but retained the mouse V regions—mouse–human chimeric antibodies.[10] Mouse–human chimeric antibodies are less immunogenic than are mouse antibodies in human patients. A further refinement was the construction of engineered antibodies with even more human sequence—humanized antibodies.[11] These were mouse antibodies in which all portions of the molecule were replaced by human sequence except the six CDRs within the V_H and V_L domains—the amino acid sequences responsible for antigen binding. With further advances that have been made more recently—specifically, the development of mice that have been genetically reconstituted with parts of the human immune system[12] and antibodies generated from phage libraries[13]—production of fully human antibodies is a goal that is just now beginning to be realized.

1.2 ANTIBODY FUSION PROTEINS

1.2.1 Fab Fusions

Antibody fusion proteins fall generally into two classes that are distinguished structurally by the molecular entity that provides the targeting function. Molecules in the first class are termed *Fab fusions* (Fig 1.4A–F): proteins in which the variable regions of the antibody molecule, which are responsible for antigen recognition, are retained. The non-immunoglobulin fusion partner is added to (Fig. 1.4A) or replaces (Fig. 1.4B–C), the Fc domain. Additionally, the nonimmunoglobulin fusion partner can be attached to the Fab portion of an intact IgG (Fig. 1.4D) or to a single chain Fv (Fig. 1.4E–F). Fab fusion proteins are the focus of Part I of this book.

Part I comprises Chapters 2 through 7. In Chapter 2, "Fab Fusion Proteins: Immunoligands," Manuel Penichet, Seung-Uon Shin, and Sherie Morrison of

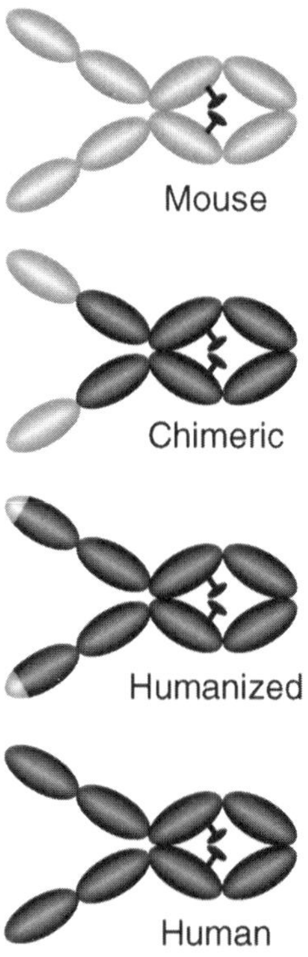

Figure 1.3 Routes to monoclonal antibody therapeutics. Hybridoma technology, which began the monoclonal antibody revolution, provided a means of producing mouse monoclonal antibodies. Although widely useful for research purposes in vitro, mouse monoclonal antibodies (top) have limited use as human therapeutics in vivo because they are immunogenic. Over the past 15 years, techniques of genetic engineering have been applied to address this problem, introducing human immunoglobulin sequences to replace mouse domains. In "chimeric" antibodies, all constant regions are human; the variable regions (V_H and V_L) are mouse. A further refinement is antibodies that are "humanized." In this case, the entire mouse molecule is replaced by human sequence, except the CDRs of the variable regions. Finally, recent progress in antibody phage display technology and in mice genetically reconstituted with human immune systems has made possible the production of therapeutic antibodies that are fully human in sequence.

the University of California, Los Angeles, describe Fab fusion proteins in which the fusion partner is a nonimmunoglobulin-derived binding protein such as a cytokine or growth factor. These authors describe in vitro and in vivo properties of several Fab immunoligand fusions, including antibodies fused to interleukin-2, insulinlike growth factors 1 and 2, and transferrin. This latter construct has exciting therapeutic potential, as it is able to cross the blood–brain barrier by receptor-mediated transfer.

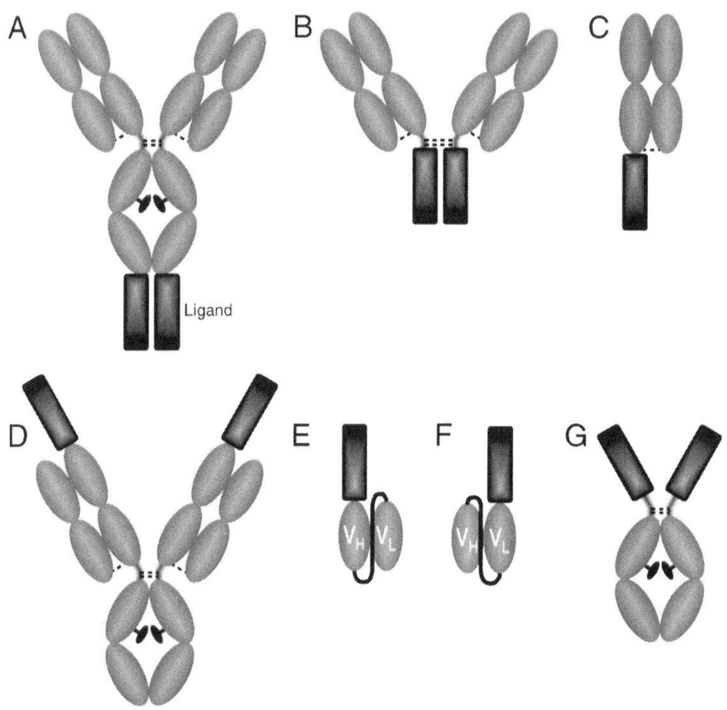

Figure 1.4 Antibody fusion proteins. Antibody fusion proteins of various structural designs are described in this book. Most of the molecules that are found in Chapters 2–10 fall into one of the structural paradigms shown here. *A–F* are different types of Fab fusion proteins (Part I: Chapters 2–7); *G* is an Fc fusion protein (Part II: Chapters 8–10). The structural paradigms include: *A*. a full-length immunoglobulin molecule fused at the carboxyl-terminus of each H chain to a nonimmunoglobulin ligand; *B*. a bivalent $F(ab')_2$ fused at the carboxyl-terminus of each H chain to a nonimmunoglobulin ligand; *C*. a monovalent Fab fused at the carboxyl-terminus of the H chain to a nonimmunoglobulin ligand; *D*. a full-length immunoglobulin molecule fused at the amino-terminus of each H chain to a non-immunoglobulin ligand; *E*. a single chain Fv fused at the amino-terminus or *F*. at the carboxyl terminus to a nonimmunoglobulin ligand; and *G*. an Fc fusion protein, in which the antigen-binding region of the immunoglobulin is replaced by a ligand-binding portion of a receptor.

Another type of Fab fusion protein, "Immunoenzymes," is featured in Chapter 3. Susanna Rybak and Dianne Newton of the Frederick Cancer Research Center in Frederick, Maryland, focus on antibody–enzyme fusions for the treatment of cancer and for thrombolysis. Two targeting strategies are discussed. In direct targeting, the enzyme itself is the drug, and it is transported to the target by the antibody. Nucleases can be used for this purpose. Antibody–nuclease fusion proteins can be specifically delivered and internalized to kill cancer cells that bear the appropriate antigen on the target cell surface. In indirect targeting, a product resulting from the activity of the enzyme is the drug. Indirect targeting, also known as antibody-dependent enzyme prodrug therapy (ADEPT), is a two-step approach. The antibody-enzyme is first administered so that it binds to the target; unbound fusion protein is cleared from circulation. A prodrug is then administered, which is converted to drug product by the enzyme only at the site of the target. β-Glucuronidase and β-lactamase are enzymes that have been used for this purpose. The indirect strategy can be significantly more efficient than the direct strategy, since the fusion protein does not have to translocate into the cell.

Incorporation of a toxin into a recombinant Fab-containing construct is outlined in Chapter 4, "Recombinant Immunotoxins," by David FitzGerald of the National Cancer Institute in Bethesda, Maryland. FitzGerald describes the construction of recombinant immunotoxins based upon bacterial (pseudomonas and diphtheria) and plant (ricin) toxins. Immunotoxins bind to receptors and antigens that are present on the surface of diseased cells and enter those cells by endocytosis. Toxins are processed intracellularly to produce active toxin fragments, which then kill the target cells. Immunotoxins targeted to cancer cells have demonstrated antitumor activities in animal models and are currently being evaluated in clinical trials.

$F(ab')_2$ fragments have also been produced in a recombinant form, and J. Yun Tso of Protein Design Labs in Fremont, California, discusses them in detail in Chapter 5, "$F(ab')_2$ Fusion Proteins and Bispecific $F(ab')_2$." $F(ab')_2$ is the largest proteolytic fragment that retains the bivalent binding sites of an antibody. Similar to Fab, it can serve as a targeting molecule to bring toxins, enzymes, or effector cells to antigen-bearing cells. A novel property of $F(ab')_2$ is its ability to create a bispecific targeting molecule; that is, when the two Fab′s contained in $F(ab')_2$ are derived from antibodies specific for different antigens, then a bispecific molecule results. Although Fab′ can be expressed in *E. coli*, conversion of Fab′ to $F(ab')_2$ via this route is inefficient.

Design of a fusion protein makes practical the bacterial production of recombinant $F(ab')_2$ (Fig. 1.4B). As Dr. Tso explains, by creating a fusion of Fab′ with a 30 amino acid segment of a leucine zipper-forming DNA binding protein,[14] a bispecific molecule can be efficiently produced. The leucine zipper segments of Fos and Jun,[15] each fused to a different Fab′ and purified separately, direct heterodimerization of the Fab′s, resulting in a bispecific $F(ab')_2$.

Chapter 6 illustrates an application of fusion protein technology with far-reaching implications—to select high-affinity antibodies from in vitro phage libraries. In "Monovalent Phage Display of Fab and scFv Fusions," David Powers and James Marks of the University of California, San Francisco, outline the principles and methods of antibody phage display. This technology has resulted from concurrent progress in several areas, including prokaryotic expression of antibody fragments, PCR cloning of antibody gene repertoires, and display of peptides and proteins as fusion proteins on filamentous bacteriophages.[16] Antibody fragments have been expressed as fusions with either of two bacteriophage-encoded proteins, pIII or pVIII (Fig. 1.4C and F). This technology is one route by which human antibodies, selected for high affinity binding to a particular antigen, have now been produced.[17]

In the final chapter of Part I, strategies for bispecific molecules are reviewed. In Chapter 7, "Bispecific Fusion Proteins," Joel Goldstein, Robert Graziano, and Michael Fanger of Dartmouth Medical School in Lebanon, New Hampshire, and Medarex, Inc., focus on an alternative approach to bispecifics, single-chain bispecific antibodies. Concentrating on anti-CD64, these investigators show that scFv fusion constructs can be made in which a fusion partner, such as epidermal growth factor, heregulin, or bombesin, can be fused to the amino or carboxyl terminus of the scFv (Fig. 1.4E–F), and the scFv remains functional.

1.2.2 Fc Fusions

The second class of antibody fusion proteins are Fc fusions, molecules also known as immunoadhesins. In an Fc fusion, the variable regions of the antibody molecule, which are responsible for antigen recognition, are replaced by the ligand-binding region of a receptor, while the antibody Fc region is retained (Fig. 1.4G). Depending on the Ig isotype, the Fc region can confer a long half-life in circulation, as well as immune effector functions. In addition, the hinge region is retained to provide conformational flexibility that can allow the Fc and receptor regions to function independently. Immunoadhesins of IgG, IgM, and IgE isotypes have been described.

Fc fusion proteins are the focus of Part II of this book, comprising Chapters 8–10. Chapter 8, "Immunoglobulin Fusion Proteins," written by Alejandro Aruffo of the Bristol-Myers Squibb Pharmaceutical Research Institute in Princeton, New Jersey, provides an overview of Fc fusion proteins. Types of Fc fusions, their characteristics, and application as research reagents are the focus of this chapter.

Chapter 9 discusses one particular Fc fusion protein that has been developed for clinical evaluation. In "TNF Receptor IgG Fusion Protein: Principles, Design, and Activities," Werner Lesslauer of Hoffmann-La Roche, Ltd., in Basel, Switzerland, details the biological rationale, molecular construction, preclinical and clinical development of an Fc fusion protein that incorporates, in place of an antibody Fab, the ectodomain of the p55 TNF receptor as the targeting moiety. This immunoadhesin binds to TNF with high affinity and effectively neutralizes the activity of TNF. Since TNF is a central mediator of inflammation, a molecule of this or similar design could potentially be useful in severe sepsis[18] and has been demonstrated to be efficacious for rheumatoid arthritis.[19]

In the final chapter, Florian Wurm of the Swiss Federal Institute of Technology in Lausanne, Switzerland, Avi Ashkenazi of Genentech, Inc., in South San Francisco, California, and Steven Chamow of Scios, Inc., in Mountain View, California outline methods for optimizing production and recovery of immunoadhesins. This chapter focuses exclusively on methods that have been developed for large-scale, clinical production of Fc fusion proteins. The authors draw from experience gained in developing large scale clinical processes for two immunoadhesins, CD4-IgG[20] and p55TNF receptor-IgG.[21]

This book represents the emergence of a new field within the broader discipline of antibody engineering. We have sought, for the first time, to compile progress made in a number of related scientific areas in which the common theme is design and use of antibody fusion proteins. These types of genetically engineered molecules have found broad application, in both research laboratories and in hospital clinics, and their use will no doubt continue to grow.

REFERENCES

1. Poljak, R. J., L. M. Amzel, H. Avery, B. L. Chen, R. P. Phizackerley, F. Saul. 1973. Three-dimensional structure of the Fab-fragment of a human immunoglobulin at 2.8Å resolution. *Proc. Natl. Acad. Sci. USA* 70: 3305–3310.
2. Skerra, A., A. Pluckthun. 1988. Assembly of a functional immunoglobulin Fv fragment in *Escherichia coli. Science* 240: 1038–1043.
3. Porter, R. R. 1959. The hydrolysis of rabbit g–globulin and antibodies with crystalline papain. *Biochem. J.* 73: 119–126.
4. Nisonoff, A., F. C. Wissler, L. N. Lipman, D. L. Woernley. 1960. Separation of univalent fragments from the bivalent rabbit antibody molecule by reduction of disulfide bonds. *Arch. Biochem. Biophys.* 89: 230–244.
5. Silvestris, F., R. C. Williams, Jr., R. P. Searles. 1986. Human anti-$F(ab')_2$ antibodies and pepsin agglutinators react with Fv determinants. *Scand. J. Immunol.* 23: 499–508.
6. Huston, J. S., D. Levinson, M. Mudgett–Hunter, M. S. Tai, J. Novotnyu, M. N. Margolies, R. J. Ridge, R. E. Bruccoleri, E. Haber, R. Crea. 1988. Protein engineering of antibody binding sites: recovery of specific activity in an anti-digoxin single-chain Fv analogue produced in *Escherichia coli. Proc. Natl. Acad. Sci. USA* 85: 5879–5883.
7. Pluckthun, A. 1991. Antibody engineering: advances from the use of *Escherichia coli* expression systems. *Bio/Technology* 9: 545–551.
8. Bigelow, C. C., B. R. Smith, K. J. Dorrington. 1974. Equilibrium and kinetic aspects of subunit association in immunoglobulin G. *Biochemistry* 13: 4602–4608.
9. Kohler, G., C. Milstein. 1975. Continuous cultures of fused cells secreting antibody of predefined specificity. *Nature* 256: 495–497.
10. Morrison, S. L., M. J. Johnson, L. H. Herzenberg, V. T. Oi. 1984. Chimeric human antibody molecules: mouse antigen-binding domains with human constant region domains. *Proc. Natl. Acad. Sci. USA* 81: 6851–6855.
11. Riechmann, L., M. Clark, H. Waldmann, G. Winter. 1988. Reshaping human antibodies for therapy. *Nature* 332: 323–327.
12. Green, L. L., M. C. Hardy, C. E. Maynard-Currie, H. Tsuda, D. M. Louie, M. J. Mendez, H. Abderrahim, M. Noguchi, D. H. Smith, Y. Zeng, N. E. David, H. Sasai, D. Garza, D. G. Brenner, J. F. Hales, R. P. McGuinness, D. J. Capon, S. Klapholz,

A. Jakobovits. 1994. Antigen-specific human monoclonal antibodies from mice engineered with human Ig heavy and light chain YACs. *Nature Genetics* 7: 13–21.

13. McCafferty, J., A. D. Griffiths, G. Winter, D. J. Chiswell. 1990. Phage antibodies: filamentous phage displaying antibody variable domains. *Nature* 348: 552–554.
14. Landschulz, W. H., P. F. Johnson, S. L. McKnight. 1988. The leucine zipper: a hypothetical structure common to a new class of DNA binding proteins. *Science* 240: 1759–1764.
15. O'Shea, E. K., R. Rutkowski, P. S. Kim. 1989. Preferential heterodimer formation by isolated leucine zippers from Fos and Jun. *Science* 245: 646–648.
16. Smith, G. P. 1985. Filamentous fusion phage: novel expression vectors that display cloned antigens on the virion surface. *Science* 228: 1315–1317.
17. Vaughan, T. J., A. J. Williams, K. Pritchard, J. K. Osbourn, A. R. Pope, J. C. Earnshaw, J. McCafferty, R. A. Hodits, J. Wilton, K. S. Johnson. 1996. Human antibodies with subnanomolar affinities isolated from a large non-immunized phage display library. *Nature Biotechnol.* 14: 309–314.
18. Abraham, E., M. P. Glauser, T. Butler, J. Garbino, D. Gelmont, P. F. Laterre, K. Kudsk, H. A. Bruining, C. Otto, E. Tobin, C. Zwingelstein, W. Lesslauer, A. Leighton. 1997. p55 Tumor necrosis factor receptor fusion protein in the treatment of patients with severe sepsis and septic shock. *JAMA* 277: 1531–1538.
19. Moreland, L. W., S. W. Baumgartner, M. H. Schiff, E. A. Tindall, R. M. Fleischmann, A. L. Weaver, R. E. Ettlinger, S. Cohen, W. J. Koopman, K. Mohler, M. B. Widmer, C. M. Blosch. 1997. Treatment of rheumatoid arthritis with a recombinant human tumor necrosis factor receptor (p75)-Fc fusion protein. *New Engl. J. Med.* 337: 141–147.
20. Byrn, R. A., J. Mordenti, C. Lucas, D. Smith, S. A. Marsters, J. S. Johnson, P. Cossum, S. M. Chamow, F. M. Wurm, T. Gregory, J. E. Groopman, D. J. Capon. 1990. Biological properties of a CD4 immunoadhesin. *Nature* 344: 667–670.
21. Ashkenazi, A., S. A. Marsters, D. J. Capon, S. M. Chamow, I. S. Figari, D. Pennica, D. V. Goeddel, M. A. Palladino, D. H. Smith. 1991. Protection against endotoxic shock by a tumor necrosis factor receptor immunoadhesin. *Proc. Natl. Acad. Sci. USA* 88: 10535–10539.

PART I

Fab FUSION PROTEINS

2

Fab FUSION PROTEINS: IMMUNOLIGANDS

MANUEL L. PENICHET

University of California, Los Angeles, CA 90095

SEUNG-UON SHIN

Hallym University, Kangwon-Do, Korea

SHERIE L. MORRISON

University of California, Los Angeles, CA 90095

2.1 STRUCTURE OF IMMUNOGLOBULINS AND IMMUNOLIGANDS

2.1.1 The Basic Structure of IgG

The basic structure of all immunoglobulin molecules (antibodies) is a unit consisting of two identical light chain polypeptide chains and two identical heavy polypeptide chains linked together by disulphide bonds (Fig. 2.1A). The amino-terminal end with the antigen binding site is characterized by sequence variability (V) in both the heavy and light chains, referred to as the V_H and V_L regions respectively. The rest of the molecule has a relatively constant (C) structure. The constant portion of the light chain is termed the C_L region. The constant portion of the heavy chain is further divided into three structurally discrete globular domains stabilized by intrachain disulfide bonds: C_H1, C_H2 and C_H3. The hinge region, a segment of heavy chain between the C_H1 and C_H2 domains, provides flexibility in the molecule. The constant region of the heavy

Antibody Fusion Proteins, Edited by Steven M. Chamow and Avi Ashkenazi
ISBN 0471-18358-X

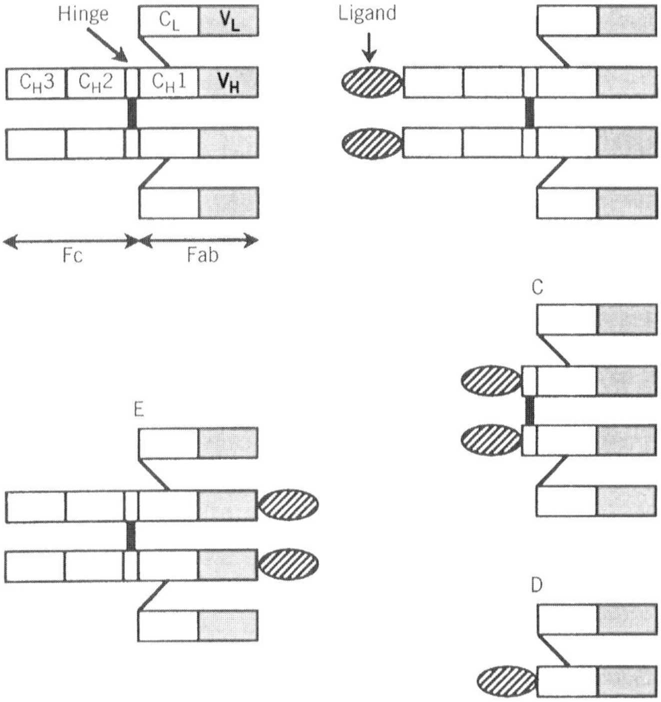

Figure 2.1 The structure of IgG and different Fab fusion proteins. *A*. The basic structure of IgG. The amino-terminal end, which binds antigen, is characterized by sequence variability (V) in both the heavy and light chains, referred to as the V_H and V_L regions respectively. The rest of the molecule has a relatively constant (C) sequence. The constant portion of the light chain is termed the C_L region. The constant portion of the heavy chain is further divided into three structurally discrete globular regions called *domains*, which are stabilized by intrachain disulphide bonds: C_H1, C_H2, and C_H3. The hinge region of the heavy chain between the C_H1 and C_H2 domains provides the molecule with flexibility. Papain digestion produces two functionally distinct fragments: The Fab consisting of the L chain and V_H and C_H1 and the Fc consisting of the hinge C_H2 and C_H3. *B*. Schematic diagram of an antibody fusion protein with the ligand positioned at the end of the C_H3 domain (C_H3-ligand). *C*. Schematic diagram of an antibody fusion protein with the ligand positioned immediately after hinge (H-ligand) *D*. Schematic diagram of an antibody fusion protein with the ligand positioned after the C_H1 domain (C_H1-ligand). *E*. Schematic diagram of an antibody fusion protein in which the ligand has been joined to the N-terminus of the heavy chain.

chain determines the isotype or class of the antibody (Ab). There are four human IgG subclasses: IgG1, IgG2, IgG3, and IgG4. Papain digestion of IgG yields two Fab and one Fc fragment. The Fab region binds antigen, while the Fc region mediates effector functions such as complement activation, monocyte binding, and placental transmission.

2.1.2 Structure of the IgG Immunoligands

Advances in genetic engineering and expression systems have led to rapid progress in the development of immunoglobulins with defined or novel functional properties. The domain structure of the antibody facilitates protein engineering. Functional domains carrying antigen binding activities (Fabs) or effector functions (Fc) can be exchanged between antibodies, and novel nonimmunoglobulin proteins can be added to create antibody fusion proteins. Antibody fusion proteins that contain an intact Fab (Fab fusion protein) should retain the ability to bind antigen while the attached ligand should be able to bind its respective binding partner. Figure 2.1B, 2.1C, and 2.1D show Fab fusion proteins with the ligand fused to the C-terminus of the heavy chain. Molecules engineered in this way should retain the binding specificity of the antibody and, depending on the position of the substitution, may retain antibody-related effector functions and biological properties. When the ligand is fused to the end of the C_H3 domain (C_H3-ligand) (Fig. 2.1B) the antibody combining specificity can be used to provide specific delivery of an associated biological activity as well as antibody-related effector functions. Alternatively, fusion of the ligand immediately after hinge (H-ligand) (Fig.2.1C) or to the C_H1 domain (C_H1-ligand)(Fig. 2.1D) may be useful when the antibody-related effector functions are unnecessary or harmful. In addition, for many applications the small size of H-ligand and C_H1-ligand may be an advantage over the larger C_H3-ligand.

An alternative approach is to construct Fab fusion proteins with the ligand fused to the N-terminus of the heavy chain (Fig. 2.1E). This may be necessary for proteins that require N-terminal processing or folding for activity. An example is nerve growth factor antibody fusion protein in which nerve growth factor (NGF) is attached to the amino terminus of the heavy chain. This protein retains both the ability to bind antigen as well as NGF activity.[1] NGF fused to the C terminus of the antibody did not retain activity. Similar results have been obtained with an antibody B7-1 fusion protein.[2]

Although Figure 2.1 shows the ligand fused to the heavy chain, it should be appreciated that the ligand can also be fused to the light chain. In addition, it is also possible to construct Fab fusion proteins that combine more than one kind of ligand at the amino or carboxy termini of the heavy and/or light chains.

2.2 PRODUCTION

2.2.1 Expression System and Vector Design

Although several expression systems are available for antibody production including bacteria, yeast, plants, baculovirus, and mammalian cells, intact, fully functional monoclonal antibodies have been most successfully expressed in mammalian cells, as these cells possess the mechanisms required for correct immunoglobulin assembly, posttranslational modification, and secretion. Posttranslational modifications can influence biologic properties and effector functions,[3–5] important considerations especially when the antibody is to be used for diagnosis and therapy.

Cloning cassettes have been constructed to allow proteins to be joined at various positions to human IgG3. Human IgG3 was chosen because its extended hinge region provides spacing and flexibility, thereby facilitating simultaneous antigen and receptor binding (Box 2.1). In fact, IgG3 is the most flexible human IgG.[6,7] In addition, IgG3 is the most effective of the human isotypes in complement activation[3] and, like IgG1, shows strong binding to Fc gamma receptors (FcγR).[8] Unique restriction sites were generated at the 3′ end of the C_H1 exon, immediately after the hinge at the 5′ end of the C_H2 exon, or at the 3′ end of the C_H3 exon using site directed mutagenesis.[9] In most cases polymerase chain reaction (PCR) is used to clone the product to be fused in a processed form (i.e., lacking any sequences that are removed posttranslationally). The variable region can be of any desired combining specificity; the heavy chain fusion genes are expressed with the light chain of the corresponding specificity in the appropriate recipient cell.

2.2.2 Purification and Storage Strategies.

For many experiments it is necessary to obtain pure fusion proteins. The fusion protein can be purified from culture supernatants by affinity column chromatography[10] using the binding to antigen or to protein A or G depending on the IgG subclass. After purification, the fusion proteins can be stored in phosphate buffered saline (PBS) (or other compatible buffer) at 4°C if they will be used in the short term (i.e., less than 6 months). For longer storage the fusion proteins can be aliquoted, snap frozen, and stored at -70°C. In both cases (short and long term), a concentration of $\geqslant 1$ mg/ml is recommended. Since freezing the fusion protein may be harmful for some ligands, it is recommended that the ligand-specific activity of each new fusion protein be determined before and after freezing.

2.3 PROPERTIES OF IMMUNOLIGANDS AND THEIR APPLICATIONS

2.3.1 Antibody-IL2 Fusion Proteins

2.3.1.1 Efficacy of Recombinant Human Interleukin-2 (rhIL-2) is Limited by Toxicity. The rapid development of molecular biology and immunological knowledge has greatly expanded the possible use of immunotherapy in the treatment of cancer. In particular, attention has focused on the potential use of interleukin-2 (IL-2) as a therapeutic agent.

IL-2 is a 15 kDa lymphokine produced by T-helper cells that stimulates T cells to proliferate and become cytotoxic.[11] IL-2 can stimulate resident, inactive

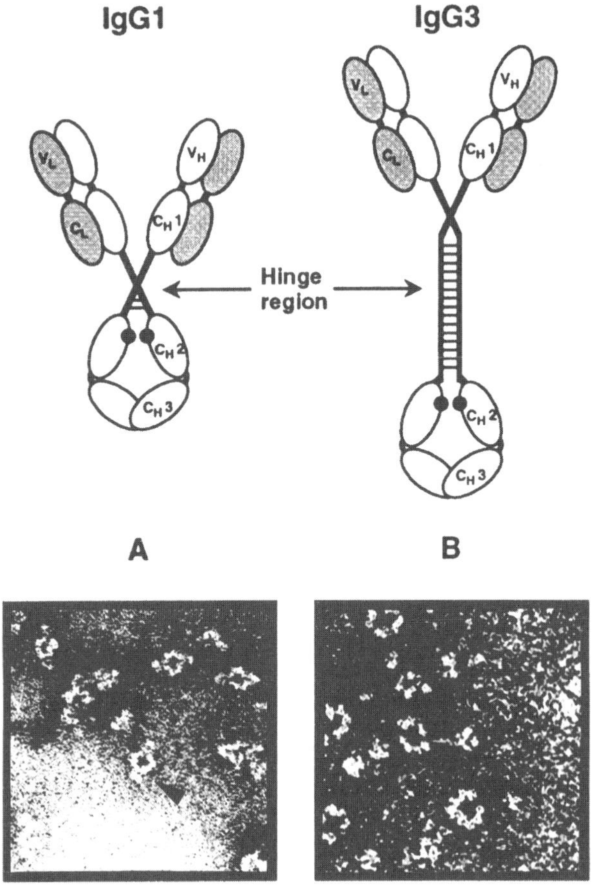

Box 2.1 Comparison of Human IgG1 and IgG3.
Human IgG is grouped into four subclasses termed IgG1, IgG2, IgG3, and IgG4. Although the $\gamma 1$, $\gamma 2$, $\gamma 3$, and $\gamma 4$, heavy chains are very similar in structure, each has characteristic differences that determine their unique biological properties. A distinctive characteristic of human IgG3 is its extended hinge region of 62 amino acids, which makes it somewhat larger than the other three subclasses. IgG3 is the most flexible human IgG. This extended hinge provides spacing and flexibility to fusion proteins, thereby facilitating simultaneous binding of different ligands by the amino and carboxy terminus of the protein. The models of human IgG1 and IgG3 illustrate the globular domains of heavy (H) and light (L) chains. The carbohydrate units, encoded within C_H2 domains, are directed inward. Note the differences in the length of the hinge regions and the number of inter-H chain disulphide bonds (in this figure the interchain disulphide bonds between H and L chains are not shown).

Box 2.1 ***continued.*** Panels A and B show electron migrographs of immune complexes formed between the symmetrical, bivalent hapten *bis*-dansyl cadaverine and chimeric recombinant human anti-dansyl IgG1 and IgG3, respectively. The complexes are composed of two antibodies forming a ring-shaped complex, presumably through the binding of bivalent haptens. The hinge regions of IgG1 and IgG3 are indicated by arrowheads. Notice the very long hinge region of IgG3. It exhibits considerable flexibility, allowing the Fc to be displaced far toward either side.

The electron micrograph was adapted from Figure 2, pages 1205 and 1206 of M. L. Phillips, M. H. Tao, S. L. Morrison, V. N. Schumaker, 1994. Human/Mouse chimeric monoclonal antibodies with human IgG1, IgG2, IgG3, and IgG4 constant domains: Electron microscopy and hydrodynamic characterization. *Mol. Immunol.* Vol. 31: 1201–1210. Copyright 1994, with kind permission of Elsevier Science Ltd., Netherlands.

tumor-infiltrating lymphocytes to proliferate and become cytotoxic for tumors.[11–14] IL-2 also stimulates natural killer (NK) cells, which respond with increased cytotoxicity toward tumor cells.[13] These properties suggest that IL-2 could be effective in the treatment of cancer.

With systemically administered rhIL-2, it was possible to stimulate an antitumor response only using high doses, and the treatment was only effective in a limited number of patients with renal cancer or melanoma.[15–17] Furthermore, high-dose IL-2 mainly stimulates nonspecific lymphokine-activated killer (LAK) activity through low-affinity IL-2 receptors (IL-2R) and does not lead to systemic immunity.[15,18] Moreover, the systemic administration of high-dose IL-2 had severe toxic side effects including increased vascular permeability that resulted in vascular leak syndrome, edema, anemia, fevers and chills, nausea, and hypotension.[15,16,19–22] The extremely short in vivo half-life of IL-2[23] [resulting at least in part from its small size (15 kDa)], which allows it to be eliminated rapidly by the kidney, also limits its utility. Due to these pharmacokinetic properties, either multiple injections or continuous infusion is required to maintain an effective concentration,[24] making the use of high-dose IL-2 very expensive. Although high dose, systemically applied IL-2 has had some success, the overall results are not very promising.

Several groups have reported that polyethylene glycol-modified IL-2 (PEG-IL2), which exhibits longer circulating half-life than IL-2, in some cases exhibits superior antitumor efficacy compared with rhIL-2. However, its use is also limited by its toxicity.[25–28]

The experiences just described suggest that more effective treatment with IL-2 could be achieved if methods could be developed to increase the effective local concentration and limit the generalized toxicity of the cytokine by selectively delivering it to the site of the tumor.[15,16] In fact, the first demonstration of an in vivo antitumor effect of IL-2 was obtained after local therapy with IL-2-containing preparations.[29] Indeed, studies have shown that intratumoral low doses of IL-2 can be highly effective against cancer without toxic side effects. Local low-dose rIL-2 treatment resulted in the eradication of tumor loads constituting up to 6% of the total body weight of a mouse. Moreover, elimination of the injected tumor and induction of a strong systemic antitumor reaction led to the elimination of tumors at distant sites.[15,29–31] This therapeutic approach has proved effective in many systems, including different tumor types in mice, hepatocellular carcinoma in guinea pigs, and vulval papilloma and ocular carcinoma in cattle.[15,30–34] Low doses of IL-2 also preferentially stimulate the specific antitumor immune response,[15,18] suggesting that low-dose IL-2 should be more effective in generating generalized tumor specific immunity. However, local injection has several problems: First, not all tumors are easily accessible for injection due to their size and/or anatomical localization; second, the procedure can disrupt the tumor and contribute to the systemic dissemination of tumor cells; and third, the IL-2 remains at the site of injection for only a short time.

Several studies have demonstrated that potent antitumor immunity can be induced using cytokine gene transfer, a strategy termed *transgenic immunotherapy*. The objective of this approach is to express the cytokine genes in the vicinity of tumor cells, either by transducing tumor cells themselves, or by delivering cytokine-expressing endothelial cells to tumor sites. As a result of the local concentration of IL-2, the tumor cells become highly immunogenic and elicit an immune response that provides protection against later challenge with the parental, nonimmunogenic tumor.[35,36] These data demonstrate that sufficient concentrations of IL-2 at the site of a tumor can effectively stimulate the host immune system to identify and destroy the tumor cells; however, transfection and reintroduction of cells is difficult to apply broadly in the clinical setting. The intratumoral injection of an adenovirus expressing IL-2 has also been shown to induce regression and immunity in a murine cancer model[37]; however this approach suffers from the shortcomings previously listed for intratumoral injection such as tumor accessibility, potential tumor disruption and dissemination of tumor cells, and diffusion of drug from the site of injection.

2.3.1.2 An Antibody-IL2 Fusion Protein Increases the Therapeutic Index of IL-2. Thus, the challenge is to develop an alternative approach for achieving effective local concentrations of IL-2. Indeed, antibody-IL2 (Ab-IL2) fusion proteins possess a combination of properties that make them candidates for providing effective immune stimulation at the site of the tumor.

Although monoclonal antibodies have a number of properties that make them potentially useful tools in the diagnosis and treatment of cancer, antibody-based therapeutic regimens face several problems. A combination of factors including poor vascularization, low flow rate, and low transvascular transport makes it difficult for the antibody to reach all the regions of a solid tumor in adequate quantities, especially when the tumor mass is large.[38,39] Another problem is that tumor cells mutate quickly and frequently lose expression of surface antigens that are the target of the therapeutic antibody. Antibodies alone kill tumor cells only inefficiently, and toxins or radionuclides conjugated to the antibody can cause collateral damage to normal tissues.[40] Coadministration of high doses of IL-2[41–43] has been shown to enhance the antitumor activity of monoclonal antibodies, but is limited by the toxicity associated with IL-2.

One approach to solving the problems described above is to create Ab-IL2 fusion proteins. To date several groups have produced Ab-IL2 fusion proteins that combine the Ab-related functions such as antigen binding with the immune stimulatory activity of IL-2.[44–51] However, the most complete studies of the efficacy of the fusion proteins in providing protection against malignancy has been done by the groups of Reisfeld and Gillies.[52–59]

By design, Ab-IL2 fusion proteins should have many advantages in anticancer therapy. First, the antibody variable region should provide specific targeting to tumor-associated antigens (TAAs),[60] thereby increasing the effective concentration of IL-2 in the tumor microenvironment. Second, IL-2 should activate the immune system to recognize not only the epitope targeted by the antibody but also a broad range of neoepitopes present on the cancer cells, thereby generating effective immunity in the face of antigenic modulation. Weakly immunogenic tumor cells[61] should be rendered highly immunogenic by the local presence of IL-2. Third, both the increased size and inherent stability of the Ab should extend the short half-life of the IL-2.[49,53] In addition, due to the binding of the Ab to the TAA, the IL-2 will remain at the site of the tumor. Fourth, the Ab-IL2 fusion protein should enhance access to the tumor by increasing vascular permeability, as has been observed for IL-2/Ab combination therapy, Ab-IL2 chemical conjugates, and Ab-IL2 fusion proteins.[41,42,51,62] Fifth, the simultaneous triggering of T-cell function via IL-2 and Fc-mediated effector functions [including the antibody-dependent cell-mediated cytotoxicity (ADCC) and complement (C′) activation] could synergize to further improve the immune stimulation. IL-2 is known to increase ADCC,[14,43] and C′ activation releases factors that are chemotactic for immune cells that could subsequently respond to IL-2 and become cytotoxic. In addition, by simultaneously binding to tumor cells and IL-2R-bearing cells, Ab-IL2 may crosslink the tumor with effector cells to further facilitate tumor killing. For all these reasons, Ab-IL2 fusion proteins should be more effective than either antibody or IL-2-based therapeutics in activation of the immune system to kill tumours locally as well as to attack distant metastases.

2.3.1.3 Properties of a Human IgG3-C$_H$3-IL2 Fusion Protein.

In vitro Properties. To produce a human IgG3-C$_H$3-IL2 fusion protein similar to the molecule described in Figure 2.1B, we used the vector described above with unique restriction sites at the 3′ end of the C$_H$3 exon.[9] Reverse transcription (RT)-PCR was used to clone a fully processed form (lacking any sequences that are removed posttranslationally) of IL-2.[48] This fusion protein, linking the N-terminus of human IL-2 to the C-terminus of human IgG3, was expressed in myeloma cells to yield a full length IgG3-IL2. This fusion protein combines the antigen specificity and Fc effector functions of the human IgG3 with the immune stimulatory activities of IL-2, providing a therapeutic agent with an improved repertoire of properties and activities.

Although our long-term goal was to provide tumor specific therapeutics, the first chimeric IgG3-IL2 antibody that we made was specific for the hapten dansyl (DNS).[48] In myeloma cells, the IgG3-IL2 fusion protein was properly assembled and secreted as an H_2L_2 tetramer. When analyzed by sodium dodecylsulfate polyacrylamide gel electrophoresis (SDS-PAGE) the fusion protein migrated with the expected molecular weight under both reducing and nonreducing conditions. Treatment of cells with tunicamycin, an inhibitor of *N*-glycosylation, causes a decrease in the apparent molecular weight of the heavy chain, showing that the Fc region contains the N-linked carbohydrate necessary for both Fc gamma receptor (FcγR) binding and C′ activation.[4,63] The anti-DNS IgG3-IL2 fusion protein binds to antigen as demonstrated both by enzyme-linked immunosorbent assay (ELISA) and the ability to purify the fusion protein using an antigen affinity column. The pure protein was tested for its ability to perform both IL-2 and Ab-related activities. Studies by others have shown that an IgG1-IL2 fusion protein of similar structure binds to antigen, generates ADCC, stimulates T-cell proliferation, and increases the cytotoxicity of various effector cell types.[44,52,54,64]

Our anti-DNS IgG3-C_H3-IL2 fusion protein was able to stimulate proliferation of the IL2-dependent murine cell lines CTLL-2 (Fig. 2.2A) or HT-2. Maximal proliferation levels achieved with IgG3-IL2 were at least as high as those achieved with rhIL-2. Half maximal proliferation was achieved with IgG3-IL2 concentrations (200 to 400 pM) that were only slightly higher than those required for rhIL-2 alone (100 to 200 pM). Both rhIL-2 and IgG3-IL2 activities in this assay were abrogated by anti-IL2 neutralizing antibodies. When the ability of the fusion protein and rhIL-2 to generate LAK cells against ^{51}Cr-loaded target cells was determined (Fig. 2.2B), rhIL-2 was found to generate maximal killing at 1000 IU/ml and an effector-to-target (E/T) ratio of 50. In contrast IgG3-IL2 achieved maximal killing at 10-fold lower concentration (100 IU/ml) with the same E/T ratio. This trend was consistent; at significantly lower concentration, IgG3-IL2 generated more efficient killing than rhIL-2. Even at low concentrations (10 IU/ml) IgG3-IL2 continued to be active whereas very little activity was observed with rhIL-2.

Therefore these studies showed that while IgG3-IL2 and rhIL-2 showed similar stimulation of CTLL-2 proliferation, IgG3-IL2 was much more effective than rhIL-2 in stimulating cytotoxicity. This difference could reflect enhanced stability resulting from the antibody portion of the fusion molecule, which may enable the fused protein to remain active for a longer period of time during the 3- to 5-days culture involved in the LAK assay. Alternatively, the bivalency or altered conformation of IL-2 in IgG3-IL2 could affect the signaling of the

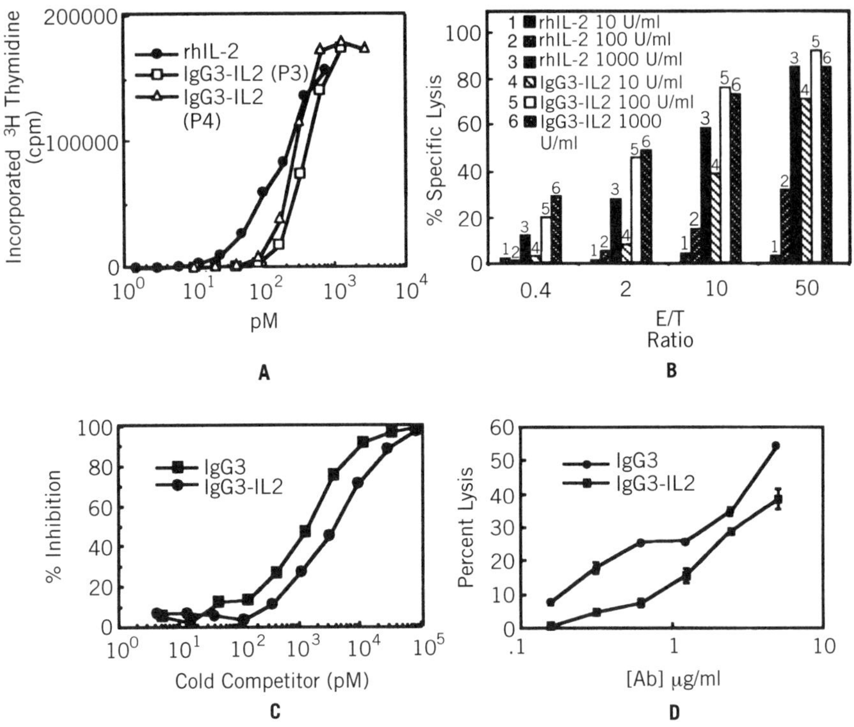

Figure 2.2 In vitro activities of IgG3-C_H3-IL2 fusion protein. *A*. Stimulation of CTLL-2 cell proliferation by rhIL-2 and IgG3-C_H3-IL2. Comparison of the ability of rhIL-2 and two separate preparations of IgG3-C_H3-IL2 to stimulate the proliferation of the IL-2 dependent cell line CTLL-2. Cell proliferation was measured by incorporation of [^{3}H] thymidine into newly synthesized DNA. *B*. Generation of LAK activity by various concentrations of rhIL-2 or IgG3-C_H3-IL2. PBL cultured for 3 days with rhIL-2 or IgG3-C_H3-IL2 at the indicated doses were tested for their cytotoxicity against ^{51}Cr-loaded Raji target cells. Results are expressed as additional percentage of total ^{51}Cr specifically released by LAK cells compared to that of unstimulated PBL. Results are expressed as the average of three samples. *C*. FcγRI binding by IgG3 and IgG3-C_H3-IL2. U937 cells stimulated with gamma interferon were incubated with ^{125}I-labeled IgG3 and varying concentrations of unlabeled IgG3 or IgG3-C_H3-IL2 as competitor. Following 3 h incubation, receptor- bound radioactivity was quantitated. Values are expressed as percentage of maximum inhibition obtained with a 200-fold excess of IgG3. *D*. Complement-mediated hemolysis by IgG3 and IgG3-C_H3-IL2. Anti-DNS IgG3-C_H3-IL2 and anti-DNS IgG3 were assayed for their ability to effect complement-mediated hemolysis of ^{51}Cr-loaded dansylated sheep red blood cells. Results are expressed as the percentage of the total available counts released. (Reprinted from E. T. Harvill and S. L. Morrison. 1995. An IgG3-IL-2 fusion protein activates complement, binds Fc gamma RI, generates LAK activity and shows enhanced binding to the high affinity IL-2R. *Immunotechnology* 1:95–105. Copyright 1995. With kind permission of Elsevier Science — NL.)

molecule or the internalization and degradation process that eventually leads to cessation of signaling. The bivalent nature of IgG3-IL2 could also lead to an increased overall affinity for its membrane-bound receptor, leading to greater receptor occupancy and greater signaling.

The IL-2R is composed of three subunits with varying individual and combined affinities for IL-2 (Box 2.2).[65] The α subunit is referred to as the low affinity receptor (IL-2Rα, $K_D = 1 \times 10^{-8}$ M). The β and γ_c subunits comprise the intermediate affinity receptor (IL-2R$\beta\gamma_c$, $K_D = 1 \times 10^{-9}$ M). Activation of T cells induces the expression of the α subunit, which combines with the β and γ subunits of the IL-2R$\beta\gamma_c$ to form the high affinity receptor (IL-2R$\alpha\beta\gamma_c$, $K_D = 1 \times 10^{-11}$ M).

To further investigate the enhanced cytotoxic activity of the IgG3-IL2 fusion protein,[13] binding of rhIL-2 and IgG3-IL2 to the IL-2R was examined.[48] When cells expressing the IL-2R$\beta\gamma_c$ on their surface were used, half-maximal inhibition of the binding of ^{125}I-labeled IL-2 occurred at 3 to 4 nM for rhIL-2

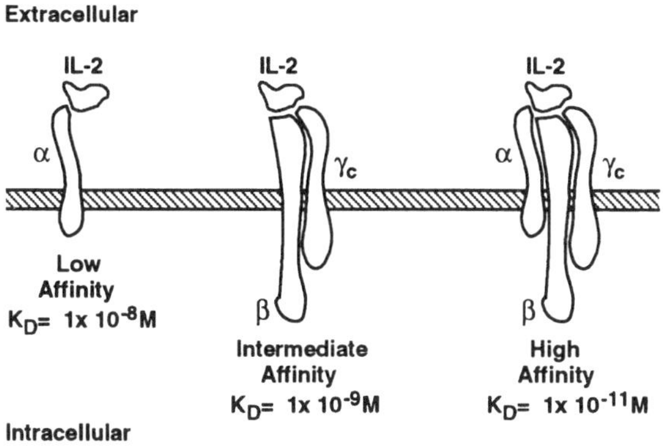

Box 2.2 Interleukin-2 and the Three Classes of Interleukin-2 Receptors. IL-2, a single chain 15 kDa lymphokine produced by T-helper cells, stimulates T cells to proliferate and become cytotoxic. The diagram illustrates IL-2 interacting with the three different forms of IL-2R. IL-2 rapidly associates and dissociates with the low affinity receptor (IL-2Rα) comprised only of the 55 kDa α subunit. An intermediate affinity receptor (IL-2R$\beta\gamma_c$) comprised of the β (75 kDa) and γ_c (64 kDa) subunits is constitutively expressed on many cell types including T cells. Activation of T cells induces the expression of the α subunit, which combines with the β and γ_c subunits to form the high affinity receptor (IL-2R$\alpha\beta\gamma_c$). T-cell expansion following activation is critically dependent upon the interaction of IL-2 with the high affinity receptor.

and at 7 to 8 nM for IgG3-IL2. In contrast, when cells expressing the IL-2R$\alpha\beta\gamma_c$ were used, half-maximal inhibition occurred for rhIL-2 at 2 to 3 nM and for IgG3-IL2 at 0.7 to 0.8 nM. Similarly, when unlabeled rhIL-2 or IgG3-IL2 were used to inhibit ^{125}I-labeled IgG3-IL2 binding to IL-2R$\alpha\beta\gamma_c$-bearing cells, half maximal inhibition occurred for rhIL2 at 10–15 nM and for IgG3-IL2 at only 0.9–1 nM. Unlabeled IgG3 did not affect binding to IL-2R in these assays.

The foregoing data suggested that while IgG3-IL2 and rhIL-2 have similar affinities for the IL-2R$\beta\gamma_c$, IgG3-IL2 has a significantly higher affinity than rhIL-2 for the IL-2R$\alpha\beta\gamma_c$. These two forms of the receptor differ by the presence or absence of the α subunit.[65] To directly determine binding of both rhIL-2 and IgG3-IL2 to the α subunit, a soluble form of the α subunit was immobilized on the surface of an IAsys optical biosensor cuvette (Fison Applied Sensor Technology). Analysis of the rhIL-2 binding to α subunit yielded an association rate constant (k_a) of $7.0 \times 10^4\,M^{-1}\,sec^{-1}$ and a dissociation rate constant (k_d) of 0.013 sec^{-1}, giving an equilibrium dissociation constant (K_D) of 1.9×10^{-7} M (Table 2.1). Measurement of IgG3-IL2 binding to sIL-2Rα yielded a k_a of 7.5 x $10^4\,M^{-1}\,sec^{-1}$ similar to that of rhIL-2. However the dissociation of IgG3-IL2 differed significantly from that of rhIL-2 with greater than 60% of the bound IgG3-IL2 remaining bound even after repeated buffer washes over several hours. The fraction of the fusion protein that did dissociate exhibited a k_d of 0.004 sec^{-1}. These k_a and k_d values were used to calculate a K_D of 5.3×10^{-8} M. Therefore under these conditions, IgG3-IL2 exhibits at least a 20-fold higher affinity for the α subunit than native IL-2.

Several characteristics of IgG3-IL2 could influence its binding to the IL-2R. Separate regions of IL-2 bind to the three subunits of the IL-2 receptor.[66] The fusion of IgG3 to IL-2 may alter the structure of IL-2 or its accessibility for binding. In addition, IgG3-IL2 contains two IL-2 molecules; bivalency might render it more efficient in binding the α subunit. The higher affinity shown by

TABLE 2.1 Rate Constants

Ligand	k_a	k_d	K_D
rhIL-2	$7.0 \times 10^4\,M^{-1}\,sec^{-1}$	$0.013\,sec^{-1}$	1.9×10^{-7} M
IgG3-IL2	$7.5 \times 10^4\,M^{-1}\,sec^{-1}$	$0.004\,sec^{-1}$	5.3×10^{-8} M

Note: A soluble form of α subunit was immobilized on the surface of an IAsys optical biosensor. Association at various concentration of both rhIL-2 and IgG3-IL2 to the α subunit was monitored over time. Following association, the cuvette was washed with PBS/Tween, and dissociation was followed. k_{on} values calculated using the FastFit program were plotted against concentration. The slope of the best fit line yielding k_a. k_d was determined both from the *y* intercept of the graph and from direct analysis of the dissociation curves using the FastFit program. K_D was calculated as k_d/k_a.

the IgG3-IL2 fusion protein for the IL-2R$\alpha\beta\gamma_c$ may be critical in cancer therapy and may contribute to the activation of cytotoxic T cells recognizing the tumor.

Because effector functions such as C′ activation[4,5,67,68] and FcγR binding[63,69,70] are associated with the C_H2 of the IgG molecule, they should be present in the IgG3-IL2 fusion protein. Indeed, competition studies of ^{125}I-IgG3 binding to FcγRI (Fig. 2.2C) showed that half-maximal inhibition of binding was obtained with 1–2 nM IgG3 or with 3–4 nM IgG3-IL2, indicating that IgG3-IL2 binds FcγRI with slightly decreased affinity compared to native IgG3. Although IgG3-IL2 was able to direct C′-mediated lysis of antigen-coated sheep red blood cells in a dose-dependent manner (Fig. 2.2D), it was somewhat less effective than IgG3. These results demonstrate that the Fc region of IgG3-IL2 retains its appropriate structure and effector functions.[48]

In vivo Properties. The ultimate utility of an Ab-IL2 fusion protein depends on its pharmacokinetic properties. The biological half-life of ^{125}I-labeled IgG3-IL2 injected intraperitoneally into BALB/c mice was approximately 8 h,[49] somewhat shorter than the 3–5 day half-life of IgG3, but considerably longer than the 25 min half-life reported for IL-2.[18,24] IgG3 and IgG3-IL2 showed different biodistribution patterns. Four hours after injection, the highest concentration of IgG3 was found in the blood, whereas the highest concentration of IgG3-IL2 was in the thymus, spleen, and lymph nodes; all of which are organs known to possess IL-2Rs. SDS-PAGE analysis of ^{125}I-labeled IgG3-IL2 recovered from blood 4 hr following injection showed that it remained intact. Even though the fusion protein had an increased in vivo half-life, it did not cause serious toxicity with intraperitoneal injection of 100 μg into mice, resulting only in transient weight gain.

Early in vivo studies with antitumor forms of an IgG1-IL2 fusion protein showed that this protein can prevent the spread of human neuroblastoma and melanoma in human LAK-reconstituted severe combined immunodeficiency disease (SCID) mice.[54,59] However, a xenografted immune system may not be able to differentiate between cancerous and normal cells in the same way that a normal immune system would. Additionally, generating LAK cells involves several days of in vitro treatment with high levels of IL-2[14,54,59] and the nonspecific LAK activity through low-affinity IL-2 receptors does not lead to systemic immunity.[15,18]

To assess the antitumor activity of the IgG3-IL2 fusion protein using a syngeneic tumor in an animal with an intact immune system,[71] we created an IgG3-IL2 fusion protein containing variable regions specific for the idiotype of the antibody expressed on the surface of the B-cell lymphoma, 38C13, that arose in a carcinogen (7,12-dimethylbenz(a)anthracene) treated C3H/HeN mouse.[72,73] Anti-idiotypic (anti-Id) antibodies bind the surface immunoglobulin on the B-cell lymphoma specifically, providing a model system for evaluating the therapeutic potential of antibodies directed at TAAs.[74] This tumor model system has previously been used to demonstrate that antibodies of different isotypes differ in their ability to affect tumor growth and that antibody and IL-2 administered together provide a more effective tumor treatment than does either administered alone.[41,42,75]

Gamma camera imaging of C3H/HeN mice bearing subcutaneous tumors on the right flank and injected with ^{131}I-anti-Id IgG3-IL2 showed specific targeting of the fusion protein, with the majority of the ^{131}I-anti-Id IgG3-IL2 present at 24 hr localized at the site of the tumor. In contrast, a non-tumor-specific fusion protein (^{131}I-anti-DNS IgG3-IL2) localized to the vicinity of the spleen and liver. In tumor-bearing mice anti-Id IgG3-IL2 has a half-life of approximately 7 hr with about 12% of the injected radioactivity remaining at 24 hr, the time of the gamma camera image.

To determine the effect of anti-Id IgG3-IL2 on tumor growth, C3H/HeN mice were injected intraperitoneally (i.p.) with 1000 38C13 cells, a 10-fold higher dose than that necessary to give rise to tumors in 100% of injected mice.[42] After injection, the mice were randomly segregated into 5 cages. The following day, groups of 6 mice each received different single i.p. injections of PBS; 10 μg of anti-Id IgG3; 30,000 IU of rhIL-2; both 10 μg anti-Id IgG3 and 30,000 IU of rhIL-2; or 10 μg of anti-Id IgG3-IL2. For animals in each group, survival was followed over time. Three mice from the group treated with anti-Id IgG3-IL2 and one treated with anti-Id IgG3 did not develop any signs of tumor (Table 2.2, Expt. 1). All of the other mice died of the tumor by day 35 with a mean survival of approximately 25 days. The injection of anti-Id IgG3-IL2 resulted in 50% disease-free survivors, producing a statistically significant improvement in survival compared with Ab alone ($p = 0.05$), IL-2 alone ($p = 0.01$), or their combination ($p = 0.01$).

TABLE 2.2 Results of In Vivo Therapy Experiments

		Disease-Free Survivors[a]	
Group	Treatment	Experiment 1[b]	Experiment 2[c]
1	PBS	0/6 (0%)	0/8 (0%)
2	anti-Id IgG3	1/6 (16.7%)	2/8 (25%)
3	IL-2	0/6 (0%)	0/8 (0%)
4	anti-Id IgG3 + IL-2	0/6 (0%)	4/8 (50%)
5	anti-Id IgG3-IL2	3/6 (50%)	7/8 (87.5%)
6	control IgG3-IL2	not done	2/8 (25%)

[a]Animals surviving 60 days without evidence of tumor were considered to be tumor free.

[b]Groups of 6 C3H/HeN mice were injected i.p. with 1000 38C13 cells. The following day each group received single i.p. injections of PBS, 10 μg of anti-Id IgG3, 30,000 IU of rhIL-2, both 10 μg anti-Id IgG3 and 30,000 IU of rhIL-2, or 10 μg of anti-Id IgG3-IL2.

[c]Groups of 8 C3H/HeN mice were injected s.c. with 1000 38C13 cells. The following day each group received the first of five daily i.p. injections. Treatments included PBS, 10 μg of anti-Id IgG3, 30,000 IU of rhIL-2, both 10 μg anti-Id IgG3 and 30,000 IU of rhIL-2, 10 μg of anti-Id IgG3-IL2 and 10 μg of an irrelevant human IgG3-IL2 fusion protein without binding specificity for 38C13 (control IgG3-IL2).

In order to investigate the importance of T cells in the antitumor effects of anti-Id IgG3-IL2, a similar experiment was performed in nude mice that were severely depleted of T cells. None of the treatments significantly increased the survival of the nude mice.

To determine if the four surviving mice from the first experiments were protected against subsequent tumor growth, they were rechallenged with 10,000 38C13 cells by subcutaneous (s.c.) injection, a dose that resulted in the death of the 30 naive control mice between days 19 and 28 (average 23 days). All of the mice surviving the initial exposure to tumor showed some resistance to tumor growth. The mouse previously treated with anti-Id IgG3 died 37 days after injection of the tumor cells. One of the mice previously treated with anti-Id IgG3-IL2 succumbed to tumor on day 39. A second, designated E2, showed very slow tumor growth and survived for much longer, finally becoming moribund at 60 days, at which time it was sacrificed and tissue was removed for analysis. The third mouse previously treated with anti-Id IgG3-IL2 did not show any signs of tumor growth following this tenfold larger inoculum of cells.

Immunohistochemical analysis of the tumor from E2 showed extensive infiltration of $CD8^+$ cells. $CD4^+$ cells were also noted at relatively high concentrations in some regions of the E2 tumor with $NK1.1^+$ cells observed occasionally. Similar analysis of the tumor from a naive mouse showed no $CD8^+$, $CD4^+$, or $NK1.1^+$ cells in multiple sections from various locations. Therefore in mouse E2, long-term survival and a decreased rate of tumor growth is associated with extensive immune infiltration of the tumor, suggesting that memory T cells contributed to the tumoricidal activity.

Experiments were next performed to determine if anti-Id IgG3-IL2 could also affect the growth of tumors distant from the site of injection, and if repeated treatment with the fusion protein would be more effective. One thousand tumor cells were injected s.c. in 48 mice that were randomly segregated into six cages. The following day each group of eight mice received the first of five daily i.p. injections. Treatments included PBS; 10 μg of anti-Id IgG3; 30,000 IU of rhIL-2; both 10 μg anti-Id IgG3 and 30,000 IU of rhIL-2; 10 μg of anti-Id IgG3-IL2; and 10 μg of an irrelevant human IgG3-IL2 fusion protein (lacking binding specificity for 38C13). This new control served to evaluate the importance of tumor targeting versus increased half-life in the antitumor immune response. Tumors in mice treated with PBS or rhIL-2 grew rapidly, and all of the mice in these two groups were dead by day 38 with a mean survival of approximately 33 days. Two of the eight mice treated with anti-Id IgG3 or the irrelevant human IgG3-IL2 fusion protein remained disease free (25%). All but one of the mice treated with the combination of anti-Id IgG3 and IL-2 showed an extended survival with four of eight mice remaining disease free (50%). The best result was obtained with the group treated with anti-Id IgG3-IL2 in which all mice showed extended survival and seven of eight remained disease free (87.5%). Thus, four treatments showed some efficacy, but the best was the treatment with the anti-Id IgG3-C_H3-IL2

(Table 2.2, Expt. 2). Statistical analysis showed it to be superior to the combination of anti-Id IgG3 and IL-2 ($p = 0.09$), anti-Id IgG3 ($p = 0.007$) and the irrelevant human IgG3-IL2 fusion protein ($p = 0.01$).

The surviving mice were then rechallenged with a s.c. injection of 10,000 38C13 cells. The majority of the mice showed no or a minimal increase in survival compared to the untreated control group.There were two disease-free survivors, one previously treated with the combination of anti-Id IgG3 and IL-2 and the other previously treated with the irrelevant human IgG3-IL2 fusion protein. Although no disease-free survivors were seen in the group previously treated with the anti-Id IgG3-IL2, this group showed a mean of survival of 34.4 days compared to 28.5 days for naive mice with two mice dying at day 41 and 56, respectively. The results of the rechallenge of the surviving mice provided little evidence for immunologic memory following the treatment with the specific Ab-IL2 fusion protcin, suggesting that specific cytotoxic T cells did not play a significant role in providing protection. This result is consistent with previous reports demonstrating that IL-2 can augment anti-Id therapy of a murine B-cell lymphoma by increasing the number and activity of cells, such as NK, that mediate ADCC.[41,42] The dose dependence of the type of response elicited by IL-2 may also help explain the different results obtained upon rechallenge of the survivors from the first and second trial. Low doses of IL-2 have been shown to activate T cells, while higher doses lead to NK activation.[15,18] In fact, using IL-2 expressing transfected cells as cancer vaccines, it has been observed that high levels of IL-2 production is associated with a failure to generate tumor-specific CTLs.[36]

Although our studies have clearly shown the utility of Ab-IL2 fusion protein in cancer immunotherapy, they have not been as dramatically successful as those reported by the Reisfeld/Gillies group.[56–59] Using an Ab-IL2 fusion protein, their studies showed virtually complete regression of metastases in immunocompetent mice, with approximately 50% of the mice resistant to tumor on rechallenge due to specific activation of $CD8^+$ CTLs. Several factors could explain these differences. The tumor they used was a transfectant of the B16 melanoma expressing the GD2 antigen. We used a lymphoma, and clearly, differences in immunogenicity and response to cytokine therapy are seen with different tumors.[15,76] Several differences observed with IL-2 expressing tumor cells are as follows: (i) IL-2 producing murine colon adenocarcinoma CT-26 and the murine mastocytoma cells P815 elicit tumor-specific $CD8^+$ CTLs[77,78] while protection against the IL-2 producing murine lymphoma EL4 is mainly mediated by NK cells.[79] Factors produced by tumor cells could explain these differences.[80] (ii) The melanoma B16 is syngeneic with C57BL/6 mice while the lymphoma was derived from C3H/HeN mice and strain-specific differences in immune responsiveness are well known.[81–83] (iii) The isotype and binding specificity of the fusion proteins also differ, being an IgG1 anti-GD2 in one case and an IgG3 anti-Id in the other. (iv) Different doses as well as different routes of administration were used ($10\,\mu g \times 5$ i.p. in one anti-Id treatment versus 8 or $16\,\mu g \times 7$ i.v. in an anti-GD2 treatment). (v) The different pharmacokinetic properties of the Abs could result in the delivery of vastly different effective doses. These findings underscore the importance of further defining the functional properties of the Ab-IL2 fusion proteins so that proteins with the optimal combination of properties for each application can be prepared.

The experiments described here have demonstrated the ability of the anti-Id IgG3-IL2 fusion protein to localize to the tumor, producing both outstanding tumor imaging and enhancement of antitumor activity. However, our results suggest that, in the tumor model we used, the dose of the fusion protein has a significant effect on whether nonspecific killers or CTLs with immunologic memory are generated. These studies emphasize the complexity of this treatment approach and indicate that further studies involving both different types of tumors and different treatment regimens will be required before a complete understanding is achieved.

Ab-IL2 fusion proteins can be used to potentiate the immune response not only against tumors, but also against any associated antigen. Indeed, direct fusion of IL-2 to candidate antigens has been used in vaccines to increase the immune response to attached antigens.[84–88] Although the mechanism is not well understood, IL-2 may target the molecule to IL-2R-bearing cells for more effective antigen presentation, and/or may deliver a stimulatory signal that directly or indirectly enhances the immune response to the attached antigen. Physically linking the antigen and IL-2 has been shown to be critical to maximizing the immune response to the antigen,[84–88] but these experiments have required considerable effort to construct, express, purify, and characterize each fusion protein individually. Moreover in this approach, it is probable that some IL-2 fusion proteins will be expressed only poorly and that others, while well expressed, might show a decrease or lack of critical IL-2 activity.

The anti-DNS IgG3-IL2 that we constructed can be easily targeted to any protein antigen simply by dansylating the protein antigen of interest.[48] As discussed previously, this fusion protein retains IL-2 activity and has a greatly improved binding ability for the high affinity IL-2R.[48–50] The hapten DNS can be easily linked to primary amino groups present on the antigen so that the resulting dansylated antigen is recognized by the anti-DNS fusion protein, which can then potentiate the immune response to the dansylated antigen. Indeed, when Sepharose beads coated with DNS-BSA were incubated with either PBS, anti-DNS IgG3, or anti-DNS IgG3-IL2 and injected intraperitoneally into mice, an enhanced antibody response to both DNS and BSA was observed when the antigen was bound by anti-DNS IgG3-IL2, but not by anti-DNS IgG3. A particularly dramatic increase (especially at early time points) in antigen-specific antibodies of all isotypes was noted with the IgG2 response (Fig. 2.3), in agreement with previous reports that IL-2 increases production of this isotype.[89] Binding of antigen to fusion protein was critical for the enhancement of the immune response, since coinjection of anti-idiotype IgG3-IL2 generated only a modest increase in antibody response. Anti-DNS IgG3-IL2 also increased the secondary antibody response to soluble DNS-BSA.[49]

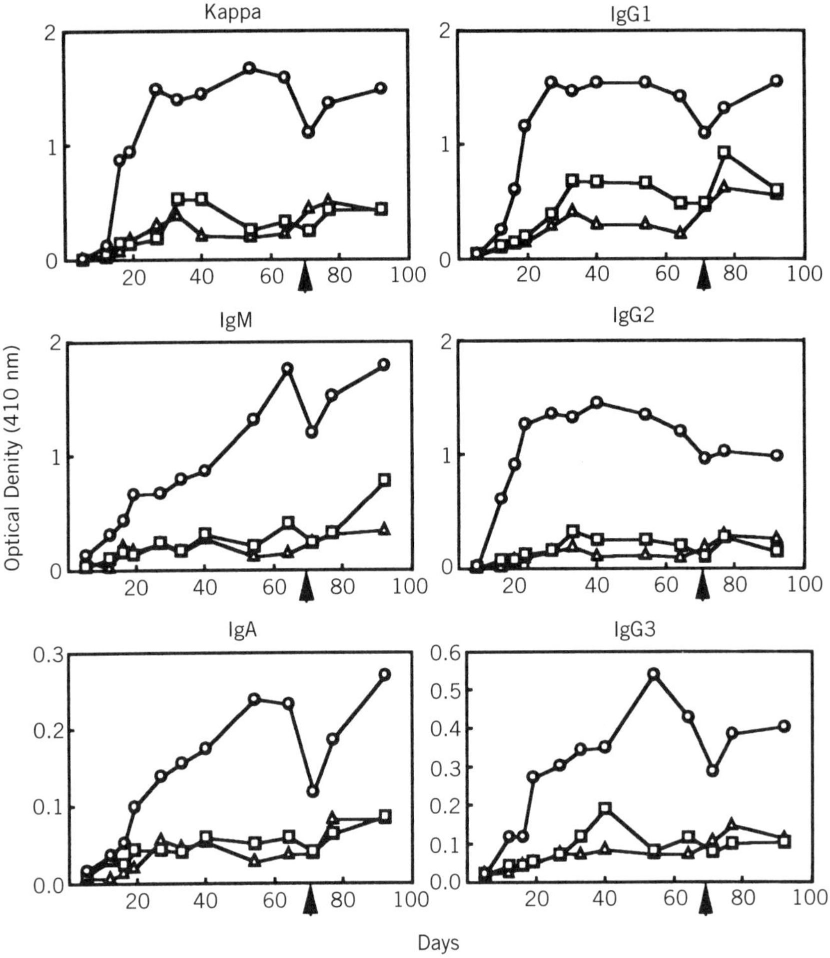

Figure 2.3 Immunostimulation due to anti-DNS IgG3-IL2 injection. Five mice per group were injected with DNS-BSA-Sepharose alone (open triangles), DNS-BSA-Sepharose bound to anti-DNS IgG3 (open squares), or DNS-BSA-Sepharose bound to anti-DNS IgG3-IL2 (open circles). Identical booster injections were delivered on day 70, as indicated by the arrow. Mice were bled at the various times, the sera pooled and assayed for DNS specific antibodies using DNS-BSA coated ELISA plates and alkaline phosphastase-labeled detecting antisera specific for the different mouse isotypes. The results are plotted as the optical density at 410 nm versus the time the mice were bled following the initial antigen injection. (Reprinted from E. T. Harvill, J. M. Fleming, and S. L. Morrison. 1996. In vivo properties of an IgG3-IL-2 fusion protein: a general strategy for immune potentiation. *J. Immunol.* 157:3165–3170. Copyright 1996. With kind permission of The American Association of Immunologists.)

The results obtained suggest that this approach to vaccination may be successful with many antigens, with the versatility of the DNS system providing a novel tool for the generation of IL2-potentiated immune responses. Using this approach, a panel of dansylated antigens can be rapidly evaluated and the protein, or peptides, that would be most useful in a therapeutic application identified. In addition, the anti-DNS Ab-IL2 fusion proteins could be used to deliver dansylated toxins to IL-2R-bearing cells for the therapy of virally infected T cells, T-cell leukemias, or autoimmune diseases. Of particular interest would be the ability of this kind of fusion protein to stimulate an immune response in immune compromised individuals, such as acquired immunodeficiency syndrome (AIDS) patients, who may lack effective cytokine secreting T-helper cells. There is evidence that IL-2 may improve the immune status of such patients.[90,91]

2.3.1.4 Potential Drawbacks of the Use of Ab-IL2 Fusion Proteins. Even though Ab-IL2 fusion proteins show potential antitumor effects, problems remain. The effect elicited by IL-2 depends on its dose. At low doses, IL-2 has cytotoxic adjuvant activity mediated via helper cells. At intermediate doses, suppressor cell activity is supported or induced,[15,18,36] and the use of IL-2 might enhance tumor growth. However, at still higher IL-2 doses (near the maximal tolerable doses), tumor regression can be seen but this time it is due to the stimulation of nonspecific antitumor activity (LAK activity). Effective immunotherapy with IL-2 should use IL-2 doses that give optimal stimulation of a specific immune reaction. The precise contributions of dose and pharmacokinetic properties to the effectiveness of the Ab-IL2 fusion proteins remain to be determined.

Another potential problem is that under some conditions the presence of a tumor-specific immune reaction appears to aid tumor growth. The immunofacilitation theory attempts to explain this paradoxical phenomenon and postulates that there exists an optimal level of immune reaction peculiar to each tumor that facilitates its growth. Although it is unclear if this is true generally, it has been shown to be the case for several tumors that an intermediate level of antitumor immune reaction is conducive to oncogenesis while a greater or a lesser level is not.[92–94] This problem can, of course, be overcome if an antitumor immune reaction greater than the intermediate level is elicited.

In these experiments, IL-2 was fused to the C_H3 domain of the IgG3 in order to keep intact the Fc-associated effector functions that might contribute to an enhanced tumoricidal activity. However, we can not exclude the possibility that this bispecific fusion protein may be cytotoxic to IL-2R-bearing cells. In several studies in which IL-2 replaced the variable region of the antibody, the chimeric protein mediated the specific lysis of IL-2 receptor-bearing cells[95] or abolished the cell-mediated immunity in vivo.[96] Although in some cases the protein elicits cytotoxicity against IL-2 receptor-bearing cells, in other cases this inhibition of immune function occurred despite a profound proliferation and accumulation of splenic T cells. These T cells appeared to be retained in the reticuloendothelial system of the spleen and liver and were unable to take part in the cellular immune response.[96] Therefore, although the IgG3-IL2 fusion proteins show some immunostimulating and tumoricidal properties, it is not clear if the presence of the Fc leads to some immunosuppressive properties. If this is indeed the case, perhaps the tumoricidal activity can be improved by mutating the Fc region to eliminate effector functions, changing the isotype, or changing the fusion construction such that IL-2 is fused after the hinge or C_H1 (Fig. 2.1C and 2.1D).

2.3.2 Antibody Fusion Proteins for Receptor-Mediated Transport across the Blood-Brain Barrier

2.3.2.1 General Strategies for Drug Delivery into the Central Nervous System. The efficient and specific targeting of the active agent to the desired site can be critical for the successful diagnosis or therapy of many diseases. One region of the body particularly difficult to target is the central nervous system (CNS), due to the presence of the blood-brain barrier (BBB). This high resistance barrier, which maintains homeostasis within the CNS, is formed by tightly joined capillary endothelial cell membranes.[97–101] The BBB effectively restricts transport from the blood of certain molecules, especially those that are water soluble and larger than several hundred daltons.[102] In fact the clinical utility of many proteins of therapeutic interest for the CNS, such as the neurotrophic agent nerve growth factor (NGF), is limited by their inability to cross the BBB. In some cases neurotrophic factors have been administered to the CNS by invasive neurosurgical procedures[103,104] or grafting neurotrophin-producing cells into CNS sites.[105,106] An alternative strategy allowing transient access to the CNS is the disruption of the BBB by hyperosmolar shock[107] or by the administration of vasoactive substances such as leukotrienes.[108] However, disruption of the BBB allows uncontrolled access of blood-borne compounds to the brain, and when the BBB is repetitively opened, neuropathologic changes take place in the brain.[109] Peptide lipidization[110] does not appear to be useful for peptides of more than five amino acids.[99] Moreover liposomes are too large to traverse the BBB and have not proved effective as drug delivery vehicles for the brain.[99]

Because the CNS requires many blood-borne compounds, the brain capillary endothelial cells possess an array of systems for transport across the BBB (Box 2.3).[99,101,111] In fact, the BBB has been shown to have specific receptors that allow the transport of several macromolecules from the blood to the CNS including insulin,[112] iron attached to transferrin (Tf),[113] and insulin-like growth factors 1 and 2 (IGF1 and IGF2).[112,114] Therefore, one approach to the delivery of drugs to the CNS is to attach the drug of interest to one of these molecules with receptors on the BBB, which would then serve as a vector for transport of the drug–protein congugate across the brain capillary endothelial cells and its release into the brain interstitial space.

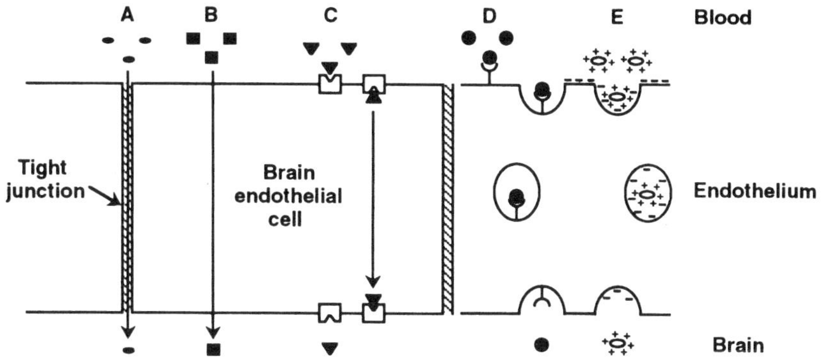

BOX 2.3. The Blood-brain Barrier, Showing Routes Across the Brain Endothelium.

In contrast to the blood vessels of in the systemic (peripheral) circulation, which are relatively open, the vessels of the central nervous system (CNS) have tight junctions between endothelial cells. As a consequence, a high-resistance barrier called the blood-brain barrier (BBB) is formed by the endothelium lining the CNS microvessels. The BBB permits the rigorous control of the CNS microenvironment that is necessary for complex neural signaling. Different routes across the brain endothelium are illustrated. *A*. Paracellular aqueous pathway for water-soluble agents; *B*. Transcellular lipophilic pathway for lipid soluble agents; *C*. Transport proteins for molecules such as glucose, amino acids, and purines; *D*. Specific receptor-mediated endocytosis for proteins such as insulin, IGF-1, IGF-2, and transferrin; *E*. Absorptive endocytosis for cationized proteins. Most plasma proteins are poorly transported; however, cationization can increase their transport through the BBB. (Illustration adapted from N. J. Abbott and I. A. Romero. 1996. Transporting therapeutics across the blood-brain barrier. *Mol. Med. Today* **2**:106–113.)

The endogenous transport systems present at the BBB can be exploited for noninvasive delivery of drugs into the CNS. Indeed, both NGF and CD4 will cross the BBB when chemically conjugated to an antibody directed against the transferrin receptor (TfR).[115–117] Therefore, notwithstanding exclusion from the brain normally,[118] Abs can be an effective vehicle for the delivery of other molecules into the brain parenchyma, if the Abs have specificity for receptors of the BBB. An Ab-NGF conjugate retained NGF activity, and its intravenous injection prevented the loss of striatal choline acetyltransferase-immunoreactive neurons in a rat model of Huntington's disease.[119] Recently NGF has been expressed in a recombinant form fused to the N-terminus of an antibody directed against human TfR (Fig. 2.1E).[1] This fusion protein showed both antigen binding and NGF activity suggesting its therapeutic utility. A monoclonal antibody against the human insulin receptor[120], with high affinity binding to human brain capillaries in vitro and an unexpectedly high degree of transcytosis through the BBB in vivo in primates, is another promising delivery vehicle.

2.3.2.2 The Use of Antibody Fusion Proteins for Drug Delivery into the CNS. An alternative approach to targeting BBB receptor-bearing cells is to use the ligand itself as a targeting moiety as a means of delivering an antibody of therapeutic interest. For example, in this approach an antibody fused to a growth factor can be transported across the BBB by binding to a growth factor receptor (Fig. 2.4). In these fusion constructs, the antibody-combining site could be used for transport of a drug or for secondary targeting within the CNS.

In an initial attempt to use this approach, we positioned rat insulin-like growth factor 1 (IGF1, 7.7 kDa), after the hinge of a chimeric mouse–human IgG3 anti-dansyl (anti-DNS) antibody (Fig. 2.1C).[121] Human IgG3 was used because its extended hinge region provides spacing and flexibility (Box 2.1) optimizing simultaneous binding to the BBB receptor and antigen. Indeed, this fusion protein was assembled and secreted as an H_2L_2 tetramer, which bound to both antigen and the IGF1 receptor and exhibited functions of IGF1 [e.g., increasing uptake of α-aminoisobutyric acid and 2-deoxy-D-glucose by receptor-bearing cells[122]].

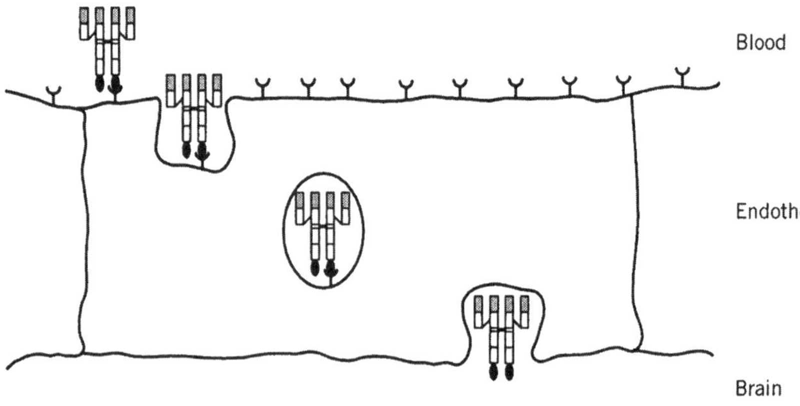

Figure 2.4 Schematic representation of the transport of Fab fusion proteins that contain ligands specific for receptors on the BBB. The antibody conjugates bind to the BBB receptors present on the luminal membrane of brain capillary endothelial cells. Through the process of receptor-mediated endocytosis, the fusion protein is internalized into vesicular structures within the endothelial cells. Then, the fusion protein is transported to and released from the abluminal surface of the capillary endothelial cell and, once released into the brain, diffuses into the parenchyma.

We expanded our studies by creating a family of fusion proteins capable of targeting different specific BBB receptors[9,123] using, in addition to IGF1, human insulin-like growth factor 2 (IGF2, 7.4 kDa) and transferrin (Tf, 180 kDa). Fusion proteins were produced with the three ligands joined individually to an antibody-binding site after C_H1 (producing Fab) (Fig. 2.1D), after the hinge (producing $F(ab)'_2$) (Fig. 2.1C) and after C_H3 (producing full length IgG) (Fig. 2.1B). SDS-PAGE analysis of proteins purified from culture supernatants showed that all of the fusion proteins were properly assembled and had the expected size corresponding to the molecular weight of the immunoglobulin moiety plus that of the attached non-IgG partner. All retained their ability to bind antigen (Ag). Ag binding is an extremely important feature of these proteins and can potentially be utilized for secondary targeting within the brain, or in the case of antidrug antibodies, for noncovalent attachment of a therapeutic molecule to the carrier. Although in our initial experiments we used an anti-DNS antibody, other variable regions can be substituted to generate fusion proteins with virtually any specificity.

2.3.2.3 Properties of Human IgG3-ligand Fusion Proteins for Receptor Mediated Transport across the Blood-Brain Barrier

In vitro Properties. A critical attribute of fusion proteins is that they retain the ability to bind to their respective receptors. Therefore competition assays with ^{125}I-ligand were used to compare the relative affinities of the fusion proteins and the native ligands for their receptors.[9]

In competitive binding studies with the IGF-1 receptor, IgG3-H-IGF1 and IgG3-C_H3-IGF1 showed approximately a 1000-fold decrease in affinity compared with IGF-1. Approximately 1 μM of either IgG3-H-IGF1 or IgG3-C_H3-IGF1 was required for 50% inhibition (IC_{50}) while recombinant human IGF-1 had a IC_{50} of 1 nM. Similarly, an IGF1-PE40 (*Pseudomonas* exotoxin) fusion protein showed a 250-fold decrease in binding efficiency to the IGF1 receptor[124]. Another construct, the IgG3-C_H1-IGF1 Fab molecule, was significantly impaired in its binding ability, and only achieved 30% inhibition at the highest concentration tested (1.5 μM). Wild-type chimeric IgG3 did not show any inhibition of ^{125}I-IGF1 binding. Taken together, these data confirm that competition by the chimeric fusion proteins occurred as a consequence of the presence of the IGF-1 moiety in each molecule.

The same analysis performed with IGF-2 and the IGF-2 fusion proteins showed an approximately 10- to 50-fold higher IC_{50} (e.g., reduced affinity) for the three IGF-2 fusion proteins (30–90 nM) than for recombinant IGF-2 (2 nM). Surprisingly, wild-type chimeric IgG3 showed inhibition of ^{125}I-IGF2 binding, but the IC_{50}(2 μM) was approximately 100 times that of the fusion proteins. These results indicate that the IGF2 moiety in the fusion protein enables it to specifically bind the IGF2 receptor, albeit with decreased affinity compared to IGF2. The difference in inhibition by IgG3 seen with IGF-2 compared to IGF-1 also suggests that the two IGFs are binding to different receptors on IM-9 cells, as would be expected.[9]

Competition by each of the IgG3-Tf fusion proteins for binding to TfR showed a IC_{50} of 3.1 nM for IgG3-H-Tf and 2.6 nM for IgG3-C_H3-Tf, only slightly higher than that of Tf (1.2 nM). In contrast the IC_{50} for IgG3-C_H1-Tf (19.3 nM) was 15-fold higher than that for Tf. It should be noted that there are two Tf molecules in IgG3-H-Tf and IgG3-C_H3-Tf, which may facilitate their binding to the dimeric TfR,[125,126] while the IgG3-C_H1-Tf contains only a single Tf molecule. Wild-type chimeric IgG3 did not show any inhibition of ^{125}I-Tf binding, even at a concentration as high as 6 μM.

Taken together, the foregoing results indicate that the growth factor moiety in the fusion proteins enables them to specifically bind their respective receptors, although in some cases with decreased affinity compared to unconjugated protein.

Effector functions such as complement (C′) activation[4,5,67,68] and Fc gamma receptor (FcγRI) binding[63,69,70] are associated with the C_H2 of the IgG molecule. Therefore the IgG3-C_H3-ligand fusion protein should be capable of these effector functions, while the IgG3-H-ligand and IgG3-C_H1-ligand should not. We examined the ability of the various IgG3-C_H3-ligands both to activate C′ and to bind FcγRI.

IgG3-C_H3-IGF1 was able to carry out C′ mediated lysis of Ag-coated (DNS) target ^{51}Cr-loaded sheep red blood cell (SRBC), but it was much less effective than unconjugated IgG3. However, surprisingly IgG3-C_H3-IGF2 showed an approximately 50-fold increase in its ability to induce C′ mediated cell lysis compared to IgG3, the most effective of the human isotypes.[3] This therefore represents an unusual situation in which an engineered antibody is more effective in carrying out an effector function than is the wild-type protein. This enhancement was Ag-specific, since no lysis was seen with SRBC not coated with antigen. In addition, specific binding of the IgG3-C_H3-IGF1 and IgG3-C_H3-IGF2 fusion proteins to the antigen (DNS) was analyzed, and it was not significantly different from that of IgG3. Therefore, the observed differences in antigen-dependent complement activation were not the result of differences in binding by fusion proteins.

Both the IgG3-C_H3-IGF1 and IgG3-C_H3-IGF2 were similar to wild-type IgG3 in their ability to bind C1q, the first component of the complement cascade, even though they were extremely different in their ability to carry out complement-mediated hemolysis. The classical pathway of complement activation requires both Ca^{++}, and Mg^{++} while the alternative pathway requires only Mg^{++}. The IgG3-C_H3-IGF2 fusion protein failed to carry out complement-mediated cytolysis in the presence of EGTA/Mg^{++} (which chelates Ca^{++} but not Mg^{++}), indicating that the enhancement of SRBC lysis with IgG3-C_H3-IGF2 takes place via the classical pathway.

In contrast to the IgG3-C_H3-IGF, the IgG3-C_H3-Tf was unable to carry out complement-mediated hemolysis of Ag-coated SRBC. However, IgG3-C_H3-Tf did consume complement in an antigen-dependent manner but not as effectively as chimeric IgG3. Since IgG3-C_H3-Tf binds C1q as well as wild-type IgG3, it is possible that the fusion of Tf to IgG3 affects subsequent steps in the activation cascade. Thus, despite similarities in their ability to bind C1q, the IgG3-C_H3-IGF1 and IgG3-C_H3-Tf fusion proteins exhibit impairment, whereas the IgG3-C_H3-IGF2 fusion protein exhibits enhancement of their ability to induce C′-mediated lysis.

The high affinity IgG-specific Fc receptors (FcγRI) expressed on a variety of lymphoid cell types are important for host defense. When the IgG3-C_H3-ligand fusion proteins were used to compete with ^{125}I-IgG3 for binding to FcγRI, IC_{50} values of 7 nM were obtained for IgG3-C_H3-IGF1, 5 nM for IgG3-C_H3-IGF2, and 3.5 nM for IgG3-C_H3-Tf; all similar to the value of 10 nM obtained for IgG3. Therefore the fusion proteins resemble IgG3 in their ability to bind FcγRI, indicating that the attached ligand moieties did not affect accessibility of the hinge-proximal FcγRI binding site.[63,69,70]

The Fc associated effector functions retained by the IgG3-C_H3-ligand fusion proteins may be useful properties even inside the CNS. Microglia cells have been implicated as central mediators of immune regulatory/effector responses in the CNS[127] and can function as antigen-presenting cells.[127–131] In addition to FcγR mediated uptake of opsonized antigens,[132] ligation of FcγR on microglia can also activate microglia resulting in upregulated cytokine production.[130,131] Moreover, there are indications that the astrocytes (and perhaps microglia) in the CNS express a complete, functional complement system,[133] which might play an important role in protection and pathogenesis within the brain.[134] All of these data indicate the potential importance of the Fc effector functions present in the IgG3-C_H3-ligand fusion proteins.

In vivo Properties. The experiment just described demonstrated that the IgG3 fusion proteins retained their Ab-associated activities. To examine the abilities of these proteins to cross the BBB, purified proteins were labeled using ^{3}H-succinimidyl propionate and injected into the tail vein of female Sprague-Dawley rats. Animals were sacrificed and the brains removed at various times postinjection, the brain tissue homogenates depleted of capillaries by density gradient centrifugation, and the radioactivity in the brain parenchyma (post-capillary supernatant) and the brain capillary pellet determined. Comparison of the level of radioactivity in the different brain fractions as a function of time indicates whether material has crossed the BBB and been taken up into brain. In all cases, ^{14}C-labeled IgG3, which shows no detectable targeting to the brain, was used as an internal control to correct for blood contamination.

All three IGF-1 fusion proteins were able to cross the BBB (Fig. 2.5A) twenty-four hr following injection, similar amounts of IgG3-C_H1-IGF1 (0.15% ID), IgG3-H-IGF1 (0.15% ID), and IgG3-C_H3-IGF1 (0.10% ID) were found in the brain parenchyma. However, the kinetics of uptake differed for each protein, with IgG3-C_H1-IGF1 initially taken up more rapidly than IgG3-H-IGF1, but with similar uptake by 4 hr postinjection. The IgG3-C_H3-IGF1 fusion protein is taken up most rapidly reaching the peak parenchymal

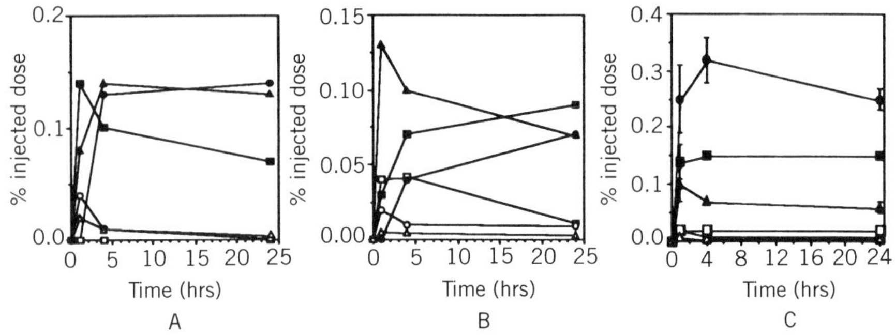

Figure 2.5 Uptake of Fab fusion proteins into intact rat brain analyzed by the capillary depletion method.[^{3}H]-labeled fusion proteins were injected into rats, and the brains harvested at varying times following injection. The capillaries were separated from the brain parenchyma and the radioactivity in each fraction determined. Solid symbols represent the parenchyma fractions containing proteins transcytosed into the brain, open symbols are the capillary fractions that contain nontransfered proteins. C_H1-ligand (triangles), H-ligand (circles) and C_H3-ligand (squares) were analyzed at 1, 4, and 24 h. *A*. IgG3-IGF1 fusion proteins. *B*. IgG3-IGF2 fusion proteins. *C*. IgG3-Tf fusion proteins. (*A* and *B* reprinted from S. U. Shin, P. Friden, M. Moran, and S. L. Morrison. 1994. Functional properties of antibody insulin-like growth factor fusion proteins. *J. Biol. Chem.* 269: 4979–4985. Copyright 1994. With kind permission of The American Society for Biochemistry and Molecular Biology. *C* reprinted from S. U. Shin, P. Friden, M. Moran, T. Olson, Y. S. Kang, W. M. Pardridge, and S. L. Morrison. 1995. Transferrin-antibody fusion proteins are affective in brain targeting. *Proc. Natl. Acad. Sci. USA*. 92: 2820–2824. Copyright 1995. With kind permission of The National Academy of Sciences, USA.)

levels 1 h after injection; this fusion protein is never found associated with the capillary fraction while significant amounts of both IgG3-C_H1-IGF1 and IgG3-H-IGF1 can be recovered from the capillary fraction between 0 and 5 hr postinjection. IGF1 showed very little brain targeting, with only 0.02% of the ID found in the brain parenchyma after 30 min and decreasing to less than 0.005% at 4 hr (data not shown).

The brain uptake of the IGF2 fusion proteins also depended on their structure (Fig. 2.5B). IgG3-C_H1-IGF2 was rapidly taken up, reaching peak parenchymal levels 1 hr after injection while the IgG3-H-IGF2 was delayed in its uptake and 1 hr postinjection was found associated with the capillary pellet but not in the brain parenchyma. However, by 4 hr postinjection some uptake (0.04% ID) of IgG3-H-IGF2 had occurred. IgG3-C_H3-IGF2 was intermediate in its kinetics of uptake: at 1 hr postinjection the protein (approximately 0.025% ID) was associated with both the capillary pellet and the brain parenchyma; however by 4 hr postinjection it (approximately 0.07% ID) was detected only in the brain parenchyma. By 24 hr, similar amounts of all three fusion proteins (IgG3-C_H1-IGF1, 0.06% ID; IgG3-H-IGF1, 0.067 ID; and

IgG3-C_H3-IGF1, 0.08%ID) were found in the brain parenchyma.

Among the fusion proteins, IgG3-Tf showed the most promise for use in brain targeting (Fig. 2.5C). The initial kinetics of uptake of IgG3-C_H3-Tf and IgG3-C_H1-Tf were similar with approximately 0.10 to 0.15%ID in brain parenchyma and no uptake in the brain capillaries 1 h postinjection. However, by 24 hr postinjection parenchymal levels of IgG3-C_H3-Tf remained at 0.15%ID, whereas levels of IgG3-C_H1-Tf had fallen to approximately 0.07%ID. IgG-H-Tf achieved the highest level in the brain with 0.3% of the ID appearing in the parenchyma 15 min after injection; this level decreased only slightly by 24 hr postinjection (0.25%ID). Transferrin targeted to the brain less efficiently than the fusion proteins, with less than 0.1% of the injected dose recovered in the brain parenchyma. To confirm that the radioactivity measured in the brain parenchyma represented intact fusion protein, ^{125}I-IgG3-C_H3-Tf iodinated to high specific activity was infused through the right internal carotid artery, and brain fractions were prepared. Radioactivity from either the perfusate or the parenchyma was immunoprecipitated with either DNS-Sepharose or protein G and analyzed by nonreducing SDS-PAGE. These procedures demonstrated the presence of a 330 kDa protein, representing the tetrameric form of IgG3-C_H3-Tf in the brain parenchyma, indicating that this fusion protein crosses the BBB as an intact molecule.

The transport of IgG3-Tf fusion proteins presumably occurs through an interaction with the TfRs, which are abundant on the brain endothelial cells[135] that make up the BBB.[136] At the earliest times tested, the majority of the recovered protein was in the brain parenchyma, suggesting that these molecules have a very rapid transit time across BBB. The efficient uptake of Tf fusion proteins is somewhat surprising, given the fact that the concentration of Tf in plasma[137] is higher than the dissociation constant of the TfR.[138] Anticipating a problem as a consequence of the Tf concentration in plasma, early studies used an anti-TfR antibody instead of Tf as a carrier molecule for drug delivery into the brain.[139] It is possible that the increased brain uptake of the IgG3-H-Tf fusion protein relative to Tf reflects differences in bindings to TfR, perhaps a result of bivalency in the fusion protein. After binding to its receptor on the cell surface, Tf is internalized into an acidic compartment. Within this intracellular compartment, the iron dissociates, and the apo-TfR is returned to the cell surface where ligand-receptor dissociation occurs. If the ability of the Ab-Tf fusion proteins to release iron were altered or if the affinity of the apo-Tf fusion proteins for the receptor were modified, mistrafficking of the fusion protein and transport across the BBB may result.

Antibodies to the TfR have been used[115–117,140,141] to deliver drugs (methotrexate), peptides (vasoactive intestinal peptide), and proteins (NGF and CD4) to the brain. With anti-TfR antibodies, 0.4–0.5% of the injected dose accumulated in the brain parenchyma. Independent of the molecule being delivered, these values are only slightly higher than that observed with IgG3-H-Tf (0.3%) but are significantly greater than that observed with either IgG3-C_H3-Tf or IgG3-C_H1-Tf. In contrast to the Tf fusion proteins which do not accumulate significantly in the capillary pellet, anti-TfR antibodies were recovered primarily in the capillary pellet at early times after administration (0–4 hr after injection) and reached a maximum concentration in the brain parenchyma between 4 and 24 hr after injection. The difference in the kinetics of uptake most likely reflects different mechanisms by which these proteins interact with the receptor: The Tf fusion proteins bind to the ligand binding site on the receptor and, like Tf, are internalized via the receptor soon after binding,[142,143] whereas the anti-TfR antibodies interact with an epitope on the TfR distinct from the ligand binding site.

The foregoing data indicate that fusion proteins, especially those containing Tf, can be used for CNS targeting with potential applications in the diagnosis and/or therapy of neurological disorders.

2.3.2.4 Multipurpose Delivery Vehicles. The ideal brain delivery vector should be able to deliver many different compounds. It is therefore desirable to develop a universal delivery vector that eliminates the need to make an antibody-ligand fusion for each individual case. As described previously, a family of anti-DNS growth factor fusion proteins is available, the members of which retain the ability to bind DNS and to cross the BBB. These anti-DNS antibodies can serve potentially as "universal vectors" for the delivery of any dansylated moiety.

Another strategy for developing a universal delivery vehicle is to exploit the broadly used avidin-biotin technology. An anti-rat-TfR antibody-avidin chemical conjugate has been used to deliver biotinylated vasoactive intestinal peptide through the BBB in rats, with CNS effects induced by intracarotid infusion of this conjugate.[141] However, the cationic nature of avidin leads to its rapid clearance from the systemic circulation by the liver and kidneys,[144] reducing the effective delivery of biotin to the brain. [This despite a comparable BBB permeability coefficient for the unconjugated antibody and the avidin-antibody conjugate.[145]] Use of a neutral form of avidin can solve this problem.[146] For transport across the BBB, the biotinylated therapeutic is coinjected with the avidin-vector conjugate. Therefore, it is necessary to use monobiotinylated compounds to prevent the formation of high molecular weight aggregates that are selectively cleared by the liver.[141,147]

An important drawback of the chemical coupling procedure is the difficulty in producing a reproducible, homogeneous product. Genetic engineering of Ab-avidin fusion proteins provides an alternative approach. Recently, chicken avidin has been expressed genetically fused to an anti-DNS mouse-human IgG3 heavy chain in several constructs: at the end of C_H1(IgG3- C_H1-Av), after the hinge (IgG3-H-Av), and at the end of C_H3 (IgG3-C_H3-Av).[148] The resulting molecules bound biotinylated albumin. Importantly the IgG3-Av fusion proteins have longer serum half-life in rats than avidin alone. These results suggest that Ab-Av fusion proteins can be used effectively to deliver biotinylated ligands such as drugs and peptides to locales expressing any antigen recognized by the associated antibody.

REFERENCES

1. McGrath, J. P., X. Cao, A. Schutz, P. Lynch, T. Ebendal, M. J. Coloma, S. L. Morrison, S. D. Putney. 1997. Bifunctional fusion between nerve growth factor and a transferrin receptor antibody. *J Neurosci. Res.* 47: 123–133.
2. Challita, P. M., M. L. Penichet, S. U. Shin, N. Mosammaparast, T. M. Poles, K. Mahmood, D. J. Slamon, S. L. Morrison, J. D. Rosenblatt. 1998. A B7.1-antibody fusion protein retains antibody specificity and ability to activate via the T cell costimulatory pathway. *J Immunol.* 160: 3419–3426.
3. Tao, M. H., S. L. Morrison. 1989. Studies of aglycosylated chimeric mouse-human IgG. Role of carbohydrate in the structure and effector functions mediated by the human IgG constant region. *J. Immunol.* 143: 2595–2601.
4. Tao, M. H., S. M. Canfield, S. L. Morrison. 1991. The differential ability of human IgG1 and IgG4 to activate complement is determined by the COOH-terminal sequence of the CH2 domain. *J Exp. Med.* 173: 1025–1028
5. Tao, M. H., R. I. Smith, S. L. Morrison. 1993. Structural features of human immunoglobulin G that determine isotype-specific differences in complement activation. *J Exp Med.* 178: 661–667.
6. Dangl, J. L., T. G. Wensel, S. L. Morrison, L. Stryer, L. A. Herzenberg, V. T. Oi. 1988. Segmental flexibility and complement fixation of genetically engineered chimeric human, rabbit and mouse antibodies. *Embo J.* 7: 1989–1994.
7. Phillips, M. L., M. H. Tao, S. L. Morrison, V. N. Schumaker. 1994. Human/mouse chimeric monoclonal antibodies with human IgG1, IgG2, IgG3 and IgG4 constant domains: electron microscopic and hydrodynamic characterization. *Mol. Immunol.* 31: 1201–1210.
8. Roitt, I. M., J. Brostoff, K. Male, eds. 1989. *Immunology*, 2d ed. London: Gower Medical Publishing, pp. 5.7–5.8.
9. Shin, S. U., P. Friden, M. Moran, S. L. Morrison. 1994. Functional properties of antibody insulin-like growth factor fusion proteins. *J Biol Chem.* 269: 4979–4985.
10. Morrison, S. L., S. Canfield, S. Porter, L. K. Tan, M. H. Tao, L. A. Wims. 1988. Production and characterization of genetically engineered antibody molecules. *Clinical Chemistry* 34: 1668–1675.
11. Lotze, M. T., E. A. Grimm, A. Mazumder, J. L. Strausser, S. A. Rosenberg. 1981. Lysis of fresh and cultured autologous tumor by human lymphocytes cultured in T-cell growth factor. *Cancer Res.* 41: 4420–4425.
12. Yron, I., T. Wood, Jr., P. J. Spiess, S. A. Rosenberg. 1980. In vitro growth of murine T cells. V. The isolation and growth of lymphoid cells infiltrating syngeneic solid tumors. *J Immunol.* 125: 238–245.
13. Grimm, E. A., A. Mazumder, H. Z. Zhang, S. A. Rosenberg. 1982. Lymphokine-activated killer cell phenomenon. Lysis of natural killer-resistant fresh solid tumor cells by interleukin 2-activated autologous human peripheral blood lymphocytes. *J. Exp. Med.* 155: 1823–1841.
14. Hank, J. A., R. R. Robinson, J. Surfus, B. M. Mueller, R. A. Reisfeld, N. K. Cheung, P. M. Sondel. 1990. Augmentation of antibody dependent cell mediated cytotoxicity following in vivo therapy with recombinant interleukin 2. *Cancer Res.* 50: 5234–5239.

15. Maas, R. A., H. F. Dullens, W. Den Otter. 1993. Interleukin-2 in cancer treatment: disappointing or (still) promising? A review. *Cancer Immunol. Immunother.* 36: 141–148.
16. Rosenberg, S. A., J. R. Yannelli, J. C. Yang, S. L. Topalian, D. J. Schwartzentruber, J. S. Weber, D. R. Parkinson, C. A. Seipp, J. H. Einhorn, D. E. White. 1994. Treatment of patients with metastatic melanoma with autologous tumor-infiltrating lymphocytes and interleukin 2. *J. Natl. Cancer Inst.* 86: 1159–1166.
17. Rubin, J. T., L. J. Elwood, S. A. Rosenberg, M. T. Lotze. 1989. Immunohistochemical correlates of response to recombinant interleukin-2-based immunotherapy in humans. *Cancer Res.* 49: 7086–7092.
18. Talmadge, J. E., H. Phillips, J. Schindler, H. Tribble, R. Pennington. 1987. Systematic preclinical study on the therapeutic properties of recombinant human interleukin 2 for the treatment of metastatic disease. *Cancer Res.* 47: 5725–5732.
19. Vial, T., J. Descotes. 1992. Clinical toxicity of interleukin-2. *Drug Saf.* 7: 417–433.
20. Siegel, J. P., R. K. Puri. 1991. Interleukin-2 toxicity. *J Clin Oncol.* 9: 694–704.
21. Ponce, P., J. Cruz, J. Travassos, P. Moreira, J. Oliveira, G. E. Melo, J. Gouveia. 1993. Renal toxicity mediated by continuous infusion of recombinant interleukin-2. *Nephron.* 64: 114–118.
22. Harada, Y., I. Yahara. 1993. Pathogenesis of toxicity with human–derived interleukin-2 in experimental animals. *Int. Rev. Exp. Pathol.* 34 Pt A: 37–55.
23. Lotze, M. T., Y. L. Matory, S. E. Ettinghausen, A. A. Rayner, S. O. Sharrow, C. A. Seipp, M. C. Custer, S. A. Rosenberg. 1985. In vivo administration of purified human interleukin 2. II. Half life, immunologic effects, and expansion of peripheral lymphoid cells in vivo with recombinant IL 2. *J. Immunol.* 135: 2865–2875.
24. Donohue, J. H., S. A. Rosenberg. 1983. The fate of interleukin-2 after in vivo administration. *J. Immunol.* 130: 2203–2208.
25. Katre, N. V., M. J. Knauf, W. J. Laird. 1987. Chemical modification of recombinant interleukin 2 by polyethylene glycol increases its potency in the murine Meth A sarcoma model. *Proc. Natl. Acad. Sci. USA* 84: 1487–1491.
26. Zimmerman, R. J., S. L. Aukerman, N. V. Katre, J. L. Winkelhake, J. D. Young. 1989. Schedule dependency of the antitumor activity and toxicity of polyethylene glycol-modified interleukin 2 in murine tumor models. *Cancer Res.* 49: 6521–6528.
27. Mattijssen, V., L. T. M. Balemans, P. A. Steerenberg, P. H. M. Demulder. 1992. Polyethylene-glycol-modified interleukin-2 is superior to interleukin-2 in locoregional immunotherapy of established guinea-pig tumors. *Int. J. Cancer.* 51: 812–817.
28. Bernsen, M. R., H. F. Dullens, W. Den Otter, P. M. Heintz. 1995. Reevaluation of the superiority of polyethylene glycol-modified interleukin-2 over regular recombinant interleukin-2. *J. Interferon Cytokine Res.* 15: 641–645.
29. Bubeník, J., P. Perlmann, M. Indrov, J. Símov, T. Jandlov, J. Neuwirt. 1983. Growth inhibition of an MC-induced mouse sarcoma by TCGF (IL 2)-containing preparations. Preliminary report. *Cancer Immunol. Immunother.* 14: 205–206.
30. Maas, R. A., H. F. Dullens, W. H. De Jong, W. Den Otter. 1989. Immunotherapy of mice with a large burden of disseminated lymphoma with low-dose interleukin 2. *Cancer Res.* 49: 7037–7040.
31. Maas, R. A., D. H. Van Weering, H. F. Dullens, W. Den Otter. 1991. Intratumoral

low-dose interleukin-2 induces rejection of distant solid tumour. *Cancer Immunol. Immunother.* 33: 389–394.

32. Cortesina, G., A. De Stefani, M. Giovarelli, M. G. Barioglio, G. P. Cavallo, C. Jemma, G. Forni. 1988. Treatment of recurrent squamous cell carcinoma of the head and neck with low doses of interleukin-2 injected perilymphatically. *Cancer* 62: 2482–2485.
33. Pizza, G., G. Severini, D. Menniti, C. De Vinci, F. Corrado. 1984. Tumour regression after intralesional injection of interleukin 2 (IL-2) in bladder cancer. Preliminary report. *Int. J. Cancer* 34: 359–367.
34. Rutten, V. P., W. R. Klein, W. A. De Jong, W. Misdorp, W. Den Otter, P. A. Steerenberg, W. H. De Jong, E. J. Ruitenberg. 1989. Local interleukin-2 therapy in bovine ocular squamous cell carcinoma. A pilot study. *Cancer Immunol. Immunother.* 30: 165–169.
35. Su, N., J. O. Ojeifo, A. MacPherson, J. A. Zwiebel. 1994. Breast cancer gene therapy: transgenic immunotherapy. *Breast Cancer Res. Treat.* 31: 349–356.
36. Schmidt, W., T. Schweighoffer, E. Herbst, G. Maass, M. Berger, F. Schilcher, G. Schaffner, M. L. Birnstiel. 1995. Cancer vaccines: the interleukin 2 dosage effect. *Proc. Natl. Acad. Sci. USA* 92: 4711–4714.
37. Addison, C. L., T. Braciak, R. Ralston, W. J. Muller, J. Gauldie, F. L. Graham. 1995. Intratumoral injection of an adenovirus expressing interleukin 2 induces regression and immunity in a murine breast cancer model. *Proc. Natl. Acad. Sci. USA.* 92: 8522–8526.
38. Epenetos, A. A., D. Snook, H. Durbin, P. M. Johnson, P. J. Taylor. 1986. Limitations of radiolabeled monoclonal antibodies for localization of human neoplasms. *Cancer Res.* 46: 3183–3191.
39. Jain, R. K. 1990. Physiological barriers to delivery of monoclonal antibodies and other macromolecules in tumors. *Cancer Res.* 50: 8145–8195
40. Vitetta, E. S., P. E. Thorpe, J. W. Uhr. 1993. Immunotoxins: magic bullets or misguided missiles? *Immunol. Today.* 14: 252–259.
41. Berinstein, N., R. Levy. 1987. Treatment of a murine B cell lymphoma with monoclonal antibodies and IL 2. *J. Immunol.* 139: 971–976.
42. Berinstein, N., C. O. Starnes, R. Levy. 1988. Specific enhancement of the therapeutic effect of anti-idiotype antibodies on a murine B cell lymphoma by IL-2. *J. Immunol.* 140: 2839–2845.
43. Hank, J. A., J. Surfus, J. Gan, T. L. Chew, R. Hong, K. Tans, R. Reisfeld, R. C. Seeger, C. P. Reynolds, M. Bauer, et. al. 1994. Treatment of neuroblastoma patients with antiganglioside GD2 antibody plus interleukin-2 induces antibody-dependent cellular cytotoxicity against neuroblastoma detected in vitro. *J. Immunother.* 15: 29–37.
44. Fell, H. P., M. A. Gayle, L. Grosmaire, J. A. Ledbetter. 1991. Genetic construction and characterization of a fusion protein consisting of a chimeric F(ab') with specificity for carcinomas and human IL-2. *J. Immunol.* 146: 2446–2452.
45. Nicolet, C. M., J. K. Burkholder, J. Gan, J. Culp, S. V. Kashmiri, J. Schlom, N. S. Yang, P. M. Sondel. 1995. Expression of a tumor-reactive antibody-interleukin 2 fusion protein after in vivo particle-mediated gene delivery. *Cancer Gene Ther.* 2: 161–170.

46. Bei, R., J. Schlom, S. V. Kashmiri. 1995. Baculovirus expression of a functional single-chain immunoglobulin and its IL-2 fusion protein. *J. Immunol. Methods.* 186: 245–255.

47. Boleti, E., M. P. Deonarain, R. A. Spooner, A. J. Smith, A. A. Epenetos, A. J. George. 1995. Construction, expression and characterisation of a single chain anti-tumour antibody (scFv)-IL-2 fusion protein. *Ann. Oncol.* 6: 945–947.

48. Harvill, E. T., S. L. Morrion. 1995. An IgG3–IL2 fusion protein activates complement, binds Fc gamma RI, generates LAK activity and shows enhanced binding to the high affinity IL2-R. *Immunotechnology* 1: 95–105.

49. Harvill, E. T., J. M. Fleming, S. L. Morrison. 1996. In vivo properties of an IgG3-IL-2 fusion protein. A general strategy for immune potentiation. *J. Immunol.* 157: 3165–3170.

50. Harvill, E. T., S. L. Morrison. 1996. An IgG3-IL-2 fusion protein has higher affinity than hrIL-2 for the IL-2R alpha subunit: real time measurement of ligand binding. *Mol. Immunol.* 33: 1007–1014.

51. Hu, P., J. L. Hornick, M. S. Glasky, A. Yun, M. N. Milkie, L. A. Khawli, P. M. Anderson, A. L. Epstein. 1996. A chimeric Lym-1/interleukin 2 fusion protein for increasing tumor vascular permeability and enhancing antibody uptake. *Cancer Res.* 56: 4998–5004.

52. Gillies, S. D., E. B. Reilly, K. M. Lo, R. A. Reisfeld. 1992. Antibody-targeted interleukin 2 stimulates T-cell killing of autologous tumor cells. *Proc. Natl. Acad. Sci. USA.* 89: 1428–1432.

53. Gillies, S. D., D. Young, K. M. Lo, S. Roberts. 1993. Biological activity and in vivo clearance of antitumor antibody/cytokine fusion proteins. *Bioconjug. Chem.* 4: 230–235.

54. Sabzevari, H., S. D. Gillies, B. M. Mueller, J. D. Pancook, R. A. Reisfeld. 1994. A recombinant antibody-interleukin 2 fusion protein suppresses growth of hepatic human neuroblastoma metastases in severe combined immunodeficiency mice. *Proc. Natl. Acad. Sci. USA* 91: 9626–9630.

55. Pancook, J. D., J. C. Becker, S. D. Gillies, R. A. Reisfeld. 1996. Eradication of established hepatic human neuroblastoma metastases in mice with severe combined immunodeficiency by antibody-targeted interleukin-2. *Cancer Immunol. Immunother.* 42: 88–92.

56. Becker, J. C., N. Varki, S. D. Gillies, K. Furukawa, R. A. Reisfeld. 1996. Long-lived and transferable tumor immunity in mice after targeted interleukin-2 therapy. *J Clin. Invest.* 98: 2801–2804.

57. Becker, J. C., N. Varki, S. D. Gillies, K. Furukawa, R. A. Reisfeld. 1996. An antibody-interleukin 2 fusion protein overcomes tumor heterogeneity by induction of a cellular immune response. *Proc. Natl. Acad. Sci. USA.* 93: 7826–7831.

58. Becker, J. C., J. D. Pancook, S. D. Gillies, K. Furukawa, R. A. Reisfeld. 1996. T cell-mediated eradication of murine metastatic melanoma induced by targeted interleukin 2 therapy. *J. Exp. Med.* 183: 2361–2366.

59. Becker, J. C., J. D. Pancook, S. D. Gillies, J. Mendelsohn, R. A. Reisfeld. 1996. Eradication of human hepatic and pulmonary melanoma metastases in SCID mice by antibody-interleukin 2 fusion proteins. *Proc. Natl. Acad. Sci. USA.* 93: 2702–2707.

60. Riethmuller, G., G. E. Schneider, J. P. Johnson. 1993. Monoclonal antibodies in cancer therapy. *Curr. Opin. Immunol.* 5: 732–739.

61. Tsomides, T. J., H. N. Eisen. 1994. T-cell antigens in cancer. *Proc. Natl. Acad. Sci. USA.* 91: 3487–3489.

62. LeBerthon, B., L. A. Khawli, M. Alauddin, G. K. Miller, B. S. Charak, A. Mazumder, A. L. Epstein. 1991. Enhanced tumor uptake of macromolecules induced by a novel vasoactive interleukin 2 immunoconjugate. *Cancer Res.* 51: 2694–2698.

63. Canfield, S. M., S. L. Morrison. 1991. The binding affinity of human IgG for its high affinity Fc receptor is determined by multiple amino acids in the CH2 domain and is modulated by the hinge region. *J. Exper. Med.* 173: 1483–1491.

64. Naramura, M., S. D. Gillies, J. Mendelsohn, R. A. Reisfeld, B. M. Mueller. 1993. Mechanisms of cellular cytotoxicity mediated by a recombinant antibody-IL2 fusion protein against human melanoma cells. *Immunol. Lett.* 39: 91–99.

65. Minami, Y., T. Kono, T. Miyazaki, T. Taniguchi. 1993. The IL-2 receptor complex: its structure, function, and target genes. *Ann. Rev. Immunol.* 11: 245–268.

66. Zurawski, S. M., F. J. Vega, E. L. Doyle, B. Huyghe, K. Flaherty, D. B. McKay, G. Zurawski. 1993. Definition and spatial location of mouse interleukin-2 residues that interact with its heterotrimeric receptor. *EMBO. J.* 12: 5113–5119.

67. Duncan, A. R., G. Winter. 1988. The binding site for C1q on IgG. *Nature* 332: 738–40.

68. Tan, L. K., R. J. Shopes, V. T. Oi, S. L. Morrison. 1990. Influence of the hinge region on complement activation, C1q binding, and segmental flexibility in chimeric human immunoglobulins [published erratum appears in *Proc. Natl. Acad. Sci. USA* 88(11):5066]. *Proc. Natl. Acad. Sci. USA* 87: 162–166.

69. Duncan, A. R., J. M. Woof, L. J. Partridge, D. R. Burton, G. Winter. 1988. Localization of the binding site for the human high-affinity Fc receptor on IgG. *Nature* 332: 563–564.

70. Chappel, M. S., D. E. Isenman, M. Everett, Y. Y. Xu, K. J. Dorrington, M. H. Klein. 1991. Identification of the Fc gamma receptor class I binding site in human IgG through the use of recombinant IgG1/IgG2 hybrid and point-mutated antibodies. *Proc. Natl. Acad. Sci. USA* 88: 9036–9040.

71. Penichet, M. L., E. T., Harvill, S. L. Morrison. 1998. An IgG3-IL2 fusion protein recognizing a murine B cell lymphoma exhibits effective tumor imaging and anti-tumor activity. *J. Interferon Cytokine Res.* 18: 597–607.

72. Bergman, Y., J. Haimovich, F. Melchers. 1977. An IgM-producing tumor with biochemical characteristics of a small B lymphocyte. *Eur. J. Immunol.* 7: 574–579.

73. Bergman, Y., J. Haimovich. 1977. Characterization of a carcinogen-induced murine B lymphocyte cell line of C3H/eB origin. *Eur. J. Immunol.* 7: 413–417.

74. Maloney, D. G., M. S. Kaminski, D. Burowski, J. Haimovich, R. Levy. 1985. Monoclonal anti-idiotype antibodies against the murine B cell lymphoma 38C13: characterization and use as probes for the biology of the tumor in vivo and in vitro. *Hybridoma* 4: 191–209.

75. Kaminski, M. S., K. Kitamura, D. G. Maloney, M. J. Campbell, R. Levy. 1986. Importance of antibody isotype in monoclonal anti-idiotype therapy of a murine

B cell lymphoma. A study of hybridoma class switch variants. *J. Immunol.* 136: 1123–1130.

76. Whittington, R., D. Faulds. 1993. Interleukin-2. A review of its pharmacological properties and therapeutic use in patients with cancer. *Drugs* 46: 446–514.
77. Russell, S. J. 1990. Lymphokine gene therapy for cancer [see comments]. *Immunol. Today* 11: 196–200.
78. Gansbacher, B., K. Zier, B. Daniels, K. Cronin, R. Bannerji, E. Gilboa. 1990. Interleukin 2 gene transfer into tumor cells abrogates tumorigenicity and induces protective immunity. *J. Exp. Med.* 172: 1217–1224.
79. Visseren, M. J., M. Koot, E. I. van der Voort, L. A. Gravestein, H. J. Schoenmakers, W. M. Kast, M. Zijlstra, C. J. Melief. 1994. Production of interleukin-2 by EL4 tumor cells induces natural killer cell- and T-cell-mediated immunity. *J. Immunother. Emphasis Tumor Immunol.* 15: 119–128.
80. Mizoguchi, H., J. J. O'Shea, D. L. Longo, C. M. Loeffler, D. W. McVicar, A. C. Ochoa. 1992. Alterations in signal transduction molecules in T lymphocytes from tumor-bearing mice [see comments]. *Science* 258: 1795–1798.
81. Blach-Olszewska, Z., E. Zaczynska, E. Broniarek, A. D. Inglot. 1993. Production of cytokines by mouse peritoneal cells treated with Tolpa Torf Preparation (TTP): dependence on age and strain of mice. *Arch. Immunol. Ther. Exp. (Warsz).* 41: 81–85.
82. Ratajczak, H. V., P. T. Thomas, R. B. Sothern, T. Vollmuth, J. D. Heck. 1993. Evidence for genetic basis of seasonal differences in antibody formation between two mouse strains. *Chronobiol. Int.* 10: 383–394.
83. Kerékgyárt, C., L. Virág, L. Tank, G. Chihara, J. Fachet. 1996. Strain differences in the cytotoxic activity and TNF production of murine macrophages stimulated by lentinan. *Int. J. Immunopharmacol.* 18: 347–353.
84. Hinuma, S., M. Hazama, A. Mayumi, Y. Fujisawa. 1991. A novel strategy for converting recombinant viral protein into high immunogenic antigen. *FEBS Lett.* 288: 138–142.
85. Hazama, M., A. Mayumi-Aono, T. Miyazaki, S. Hinuma, Y. Fujisawa. 1993. Intranasal immunization against herpes simplex virus infection by using a recombinant glycoprotein D fused with immunomodulating proteins, the B subunit of Escherichia coli heat-labile enterotoxin and interleukin-2. *Immunology* 78: 643–649.
86. Hazama, M., A. Mayumi-Aono, N. Asakawa, S. Kuroda, S. Hinuma, Y. Fujisawa. 1993. Adjuvant-independent enhanced immune responses to recombinant herpes simplex virus type 1 glycoprotein D by fusion with biologically active interleukin-2. *Vaccine* 11: 629–636.
87. Chen, T. T., M. H. Tao, R. Levy. 1994. Idiotype–cytokine fusion proteins as cancer vaccines. Relative efficacy of IL–2, IL–4, and granulocyte–macrophage colony-stimulating factor. *J. Immunol.* 153: 4775–4787.
88. Nakao, M., M. Hazama, A. Mayumi–Aono, S. Hinuma, Y. Fujisawa. 1994. Immunotherapy of acute and recurrent herpes simplex virus type 2 infection with an adjuvant-free form of recombinant glycoprotein D-interleukin-2 fusion protein. *J. Infect. Dis.* 169: 787–791.
89. Amigorena, S., C. Bonnerot, W. H. Fridman, J. L. Teillaud. 1990. Recombinant

interleukin 2–activated natural killer cells regulate IgG2a production. *Eur. J. Immunol.* 20: 1781–1787.

90. Murray, H. W., K. Welte, J. L. Jacobs, B. Y. Rubin, R. Mertelsmann, R. B. Roberts. 1985. Production of an in vitro response to interleukin 2 in the acquired immunodeficiency syndrome. *J. Clin. Invest.* 76: 1959–1964.
91. Kovacs, J. A., M. Baseler, R. J. Dewar, S. Vogel, R. T. Davey, Jr., J. Falloon, M. A. Polis, R. E. Walker, R. Stevens, N. P. Salzman, et al. 1995. Increases in CD4 T lymphocytes with intermittent courses of interleukin-2 in patients with human immunodeficiency virus infection. A preliminary study [see comments]. *N. Engl. J. Med.* 332: 567–575.
92. Prehn, R. T., L. M. Prehn. 1987. The autoimmune nature of cancer. *Cancer Res.* 47: 927–932.
93. Prehn, R. T., L. M. Prehn. 1989. The flip side of tumor immunity. *Arch. Surg.* 124: 102–106.
94. Prehn, R. T. 1993. Tumor immunogenicity: how far can it be pushed? [comment]. *Proc. Natl. Acad. Sci. USA* 90: 4332–4333.
95. Landolfi, N. F. 1991. A chimeric IL-2/Ig molecule possesses the functional activity of both proteins. *J. Immunol.* 146: 915–919.
96. Kunzendorf, U., T. Pohl, S. Bulfone–Paus, H. Krause, M. Notter, A. Onu, G. Walz, T. Diamantstein. 1996. Suppression of cell-mediated and humoral immune responses by an interleukin-2-immunoglobulin fusion protein in mice. *J. Clin. Invest.* 97: 1204–1210.
97. Crone, C., S. P. Olesen. 1982. Electrical resistance of brain microvascular endothelium. *Brain Res.* 241: 49–55.
98. Butt, A. M., H. C. Jones, N. J. Abbott. 1990. Electrical resistance across the blood-brain barrier in anaesthetized rats: a developmental study. *J. Physiol. (Lond).* 429: 47–62.
99. Pardridge, W. M., ed. 1991. *Peptide Drug Delivery to the Brain.* New York: Raven Press, pp. 123–148.
100. Brightman, M. W., J. H. Tao-Cheng. 1993. Tight junctions of brain endothelium and epithelium. In *The Blood-Brain Barrier: Cellular and Molecular Biology,* ed. W. M. Pardridge, New York: Raven Press, pp. 107–125.
101. Abbott, N. J., I. A. Romero. 1996. Transporting therapeutics across the blood-brain barrier. *Mol. Med. Today* 2: 106–113.
102. Shapiro, W. R., J. R. Shapiro. 1986. Principles of brain tumor chemotherapy. *Semin. Oncol.* 13: 56–69.
103. Olson, L., E. O. Backlund, T. Ebendal, R. Freedman, B. Hamberger, P. Hansson, B. Hoffer, U. Lindblom, B. Meyerson, I. Strömberg, et al. 1991. Intraputaminal infusion of nerve growth factor to support adrenal medullary autografts in Parkinson's disease. One-year follow-up of first clinical trial. *Arch Neurol.* 48: 373–381.
104. Olson, L., A. Nordberg, H. Vonholst, L. Backman, T. Ebendal, I. Alafuzoff, K. Amberla, P. Hartvig, A. Herlitz, A. Lilja, H. Lundqvist, B. Langstrom, B. Meyerson, A. Persson, M. Viitanen, B. Winblad, A. Seiger. 1992. Nerve growth factor affects C-11-nicotine binding, blood flow, EEG, and verbal episodic memory in an Alzheimer patient. *J. Neur. TR-P* 4: 79–95.

105. Rosenberg, M. B., T. Friedmann, R. C. Robertson, M. Tuszynski, J. A. Wolff, X. O. Breakefield, F. H. Gage. 1988. Grafting genetically modified cells to the damaged brain: restorative effects of NGF expression. *Science* 242: 1575–1578.

106. Kordower, J. H., M. S. Fiandaca, M. F. Notter, J. T. Hansen, D. M. Gash. 1990. NGF-like trophic support from peripheral nerve for grafted rhesus adrenal chromaffin cells. *J. Neurosurg.* 73: 418–428.

107. Neuwelt, E. A., S. I. Rapoport. 1984. Modification of the blood-brain barrier in the chemotherapy of malignant brain tumors. *Fed. Proc.* 43: 214–219.

108. Baba, T., K. L. Black, K. Ikezaki, K. Chen, D. P. Becker. 1991. Intracarotid infusion of leukotriene-C4 selectively increases blood brain barrier permeability after focal ischemia in rats. *Cerebr. Blood Flow Metabol.* 11: 638–643.

109. Johansson, B. B., C. Owman, W. H., ed. 1990. *Pathophysiology of the Blood-Brain Barrier: Long Term Consequences of Barrier Dysfunction for the Brain.* Amsterdam: Elsevier, pp. 145–157.

110. Tsuzuki, N., T. Hama, T. Hibi, R. Konishi, S. Futaki, K. Kitagawa. 1991. Adamantane as a brain-directed drug carrier for poorly absorbed drug: antinociceptive effects of [D-Ala2]Leu-enkephalin derivatives conjugated with the 1–adamantane moiety. *Biochem. Pharmacol.* 41: R5–8.

111. Friden, P. M. 1993. Receptor-mediated transport of peptides and proteins across the blood-brain barrier. In *The Blood-Brain Barrier: Cellular and Molecular Biology*, W. M. Pardridge ed. New York: Raven Press, pp. 229–247. .

112. Duffy, K. R., W. M. Pardridge. 1987. Blood-brain barrier transcytosis of insulin in developing rabbits. *Brain Research* 420: 32–38.

113. Fishman, J. B., J. B. Rubin, J. V. Handrahan, J. R. Connor, R. E. Fine. 1987. Receptor-mediated transcytosis of transferrin across the blood-brain barrier. *J. Neurosci. Res.* 18: 299–304.

114. Rosenfeld, R. G., H. Pham, B. T. Keller, R. T. Borchardt, W. M. Pardridge. 1987. Demonstration and structural comparison of receptors for insulin-like growth factor-I and -II (IGF-I and -II) in brain and blood-brain barrier. *Biochem. Biophys. Res. Commun.* 149: 159–166.

115. Friden, P. M., L. R. Walus, G. F. Musso, M. A. Taylor, B. Malfroy, R. M. Starzyk. 1991. Anti-transferrin receptor antibody and antibody-drug conjugates cross the blood-brain barrier. *Proc. Natl. Acad. Sci. USA* 88: 4771–4775.

116. Friden, P. M., L. R. Walus, P. Watson, S. R. Doctrow, J. W. Kozarich, C. Beckman, H. Bergman, B. Hoffer, F. Bloom, A. C. Granholm. 1993. Blood-brain barrier penetration and in vivo activity of an NGF conjugate. *Science* 259: 373–377.

117. Walus, L. R., W. M. Pardridge, R. M. Starzyk, P. M. Friden. 1996. Enhanced uptake of rsCD4 across the rodent and primate blood-brain barrier after conjugation to anti-transferrin receptor antibodies. *J. Pharmacol. Exp. Ther.* 277: 1067–1075.

118. Bullard, D. E., M. Bourdon, D. D. Bigner. 1984. Comparison of various methods for delivering radiolabeled monoclonal antibody to normal rat brain. *J. Neurosurg.* 61: 901–911.

119. Kordower, J. H., V. Charles, R. Bayer, R. T. Bartus, S. Putney, L. R. Walus, P. M. Friden. 1994. Intravenous administration of a transferrin receptor antibody-nerve

growth factor conjugate prevents the degeneration of cholinergic striatal neurons in a model of Huntington disease. *Proc. Natl. Acad. Sci. USA* 91: 9077–9080.

120. Pardridge, W. M., Y. S. Kang, J. L. Buciak, J. Yang. 1995. Human insulin receptor monoclonal antibody undergoes high affinity binding to human brain capillaries in vitro and rapid transcytosis through the blood-brain barrier in vivo in the primate. *Pharm. Res.* 12: 807–816.

121. Shin, S. U., S. L. Morrison. 1990. Expression and characterization of an antibody binding specificity joined to insulin-like growth factor 1: potential applications for cellular targeting. *Proc. Natl. Acad. Sci. USA* 87: 5322–5326.

122. Beguinot, F., C. R. Kahn, A. C. Moses, R. J. Smith. 1985. Distinct biologically active receptors for insulin, insulin-like growth factor I, and insulin-like growth factor II in cultured skeletal muscle cells. *J. Biol. Chem.* 260: 15892–15898.

123. Shin, S. U., P. Friden, M. Moran, T. Olson, Y. S. Kang, W. M. Pardridge, S. L. Morrison. 1995. Transferrin-antibody fusion proteins are effective in brain targeting. *Proc. Natl. Acad. Sci. USA* 92: 2820–2824.

124. Prior, T. I., L. J. Helman, D. J. FitzGerald, I. Pastan. 1991. Cytotoxic activity of a recombinant fusion protein between insulin-like growth factor I and Pseudomonas exotoxin. *Cancer Res.* 51: 174–180.

125. Enns, C. A., H. H. Sussman. 1981. Similarities between the transferrin receptor proteins on human reticulocytes and human placentae. *J. Biol. Chem.* 256: 12620–12623.

126. Schneider, C., R. Sutherland, R. Newman, M. Greaves. 1982. Structural features of the cell surface receptor for transferrin that is recognized by the monoclonal antibody OKT9. *J. Biol. Chem.* 257: 8516–8522.

127. Williams, K., E. Ulvestad, J. Antel. 1994. Immune regulatory and effector properties of human adult microglia studies in vitro and in situ. *Adv. Neuroimmunol.* 4: 273–281.

128. Frei, K., C. Siepl, P. Groscurth, S. Bodmer, C. Schwerdel, A. Fontana. 1987. Antigen presentation and tumor cytotoxicity by interferon-gamma-treated microglial cells. *Eur. J. Immunol.* 17: 1271–1278.

129. Hickey, W. F., H. Kimura. 1988. Perivascular microglial cells of the CNS are bone marrow-derived and present antigen in vivo. *Science* 239: 290–292.

130. Ulvestad, E., K. Williams, S. Mork, J. Antel, H. Nyland. 1994. Phenotypic differences between human monocytes/macrophages and microglial cells studied in situ and in vitro. *J. Neuropathol. Exp. Neurol.* 53: 492–501.

131. Williams, K., E. Ulvestad, A. Waage, J. P. Antel, J. McLaurin. 1994. Activation of adult human derived microglia by myelin phagocytosis in vitro. *J. Neurosci. Res.* 38: 433–443.

132. Gosselin, E. J., K. Wardwell, D. R. Gosselin, N. Alter, J. L. Fisher, P. M. Guyre. 1992. Enhanced antigen presentation using human Fc-gamma receptor (monocyte macrophage)-specific immunogens. *J. Immunol.* 149: 3477–3481.

133. Gasque, P., M. Fontaine, B. P. Morgan. 1995. Complement expression in human brain. Biosynthesis of terminal pathway components and regulators in human glial cells and cell lines. *J. Immunol.* 154: 4726–4733.

134. Colten, H. R. 1992. Tissue-specific complement gene expression evidence for novel functions of the complement proteins. In *Progress in Immunology VII.* J. Gergerly,

A. Erdci, A. Falus, G. Fürst, G. Medgyesi, G. Petrányi, and E. Rajnavölgyi, eds. Springer, Budapest, Hungary, p. 483.

135. Pardridge, W. M. 1986. Receptor-mediated peptide transport through the blood-brain barrier. *Endocr. Rev.* 7: 314–330.

136. Goldstein, G. W., A. L. Betz. 1986. The blood-brain barrier. *Sci. Am.* 255: 74–83.

137. Huebers, H. A., C. A. Finch. 1987. The physiology of transferrin and transferrin receptors. *Physiol. Rev.* 67: 520–582.

138. Pardridge, W. M., J. Eisenberg, J. Yang. 1987. Human blood-brain barrier transferrin receptor. *Metabolism: Clin. Exper.* 36: 892–895.

139. Friden, P. M. 1994. Receptor-mediated transport of therapeutics across the blood-brain barrier. *Neurosurgery* 35: 294–298; discussion 298.

140. Pardridge, W. M., J. L. Buciak, P. M. Friden. 1991. Selective transport of an anti-transferrin receptor antibody through the blood-brain barrier in vivo. *J. Pharmacol. Exp. Ther.* 259: 66–70.

141. Bickel, U., T. Yoshikawa, E. M. Landaw, K. F. Faull, W. M. Pardridge. 1993. Pharmacologic effects in vivo in brain by vector-mediated peptide drug delivery. *Proc. Natl. Acad. Sci. USA* 90: 2618–2622.

142. Ciechanover, A., A. L. Schwartz, A. Dautry-Varsat, H. F. Lodish. 1983. Kinetics of internalization and recycling of transferrin and the transferrin receptor in a human hepatoma cell line. Effect of lysosomotropic agents. *J. Biol. Chem.* 258: 9681–9689.

143. Raub, T. J., C. R. Newton. 1991. Recycling kinetics and transcytosis of transferrin in primary cultures of bovine brain microvessel endothelial cells. *J. Cell. Physiol.* 149: 141–151.

144. Pardridge, W. M., R. J. Boado, J. L. Buciak. 1993. Drug delivery of antisense oligonucleotides or peptides to tissues in vivo using an avidin-biotin system. *Drug Target Del.* 1: 43–50.

145. Kang, Y. S., U. Bickel, W. M. Pardridge. 1994. Pharmacokinetics and saturable blood-brain barrier transport of biotin bound to a conjugate of avidin and a monoclonal antibody to the transferrin receptor. *Drug Metab. Disp.* 22: 99–105.

146. Kang, Y. S., W. M. Pardridge. 1994. Use of neutral avidin improves pharmacokinetics and brain delivery of biotin bound to an avidin-monoclonal antibody conjugate. *J. Pharmacol. Exper. Therapeut.* 269: 344–350.

147. Sinitsyn, V. V., A. G. Mamontova, Y. Y. Checkneva, A. A. Shnyra, S. P. Domogatsky. 1989. Rapid blood clearance of biotinylated IgG after infusion of avidin. *J. Nucl. Med.* 30: 66–69.

148. Shin, S. U., D. Wu, R. Ramanathan, W. M. Pardridge, S. L. Morrison. 1997. Functional and pharmacokinetic properties of antibody-avidin fusion proteins. *J. Immunol.* 158: 4797–4804.

3

IMMUNOENZYMES

Susanna M. Rybak and Dianne L. Newton
*National Cancer Institute, Frederick Cancer Research and Development Center
Frederick, MD 21702*

3.1 BACKGROUND AND HISTORY OF DIRECT AND INDIRECT TARGETING STRATEGIES

The specificity of the antibody-binding site is being exploited to deliver molecules with various effector functions to their molecular and cellular targets. Current work with recombinant antibody enzyme fusion proteins derives from observations that immunoglobulin DNA could be expressed in myeloma cells.[1–3] The possibility of producing sufficient amounts of recombinant antibodies for experimental use coupled with the feasibility of manipulating immunoglobulin gene DNA rapidly led to the report of recombinant antibodies with novel effector functions.[4] Particularly relevant to this chapter, one of the first antibody–enzyme fusion proteins was a Fab-nuclease in which the *S. aureus* nuclease gene was joined to the γ2b CH2 exon of the V_{NP} antibody.[4] The enzyme domain was able to refold and retain DNA degrading activity. Although the possibility of using recombinant antibody technology to "tag" enzymes for increased ease in purification was demonstrated,[5] the possibility of using this technology to make clinically useful reagents for the treatment of various diseases has aroused the interest of investigators and stimulated the development of this field.

Antibody Fusion Proteins, Edited by Steven M. Chamow and Avi Ashkenazi
ISBN 0471-18358-X

The concepts and results presented in this chapter recount the use of recombinant antibody–enzyme chimeras for the treatment of cancer and thrombolysis (Fig. 3.1). Two different strategies have been adopted for ultimately delivering the drug to the target. In direct targeting, the enzyme itself is the drug and is transported to the target by the antibody. Antibody nuclease fusion proteins are revisited in a new role as anticancer therapeutics. Both RNase and DNase can be specifically delivered and internalized to kill cancer

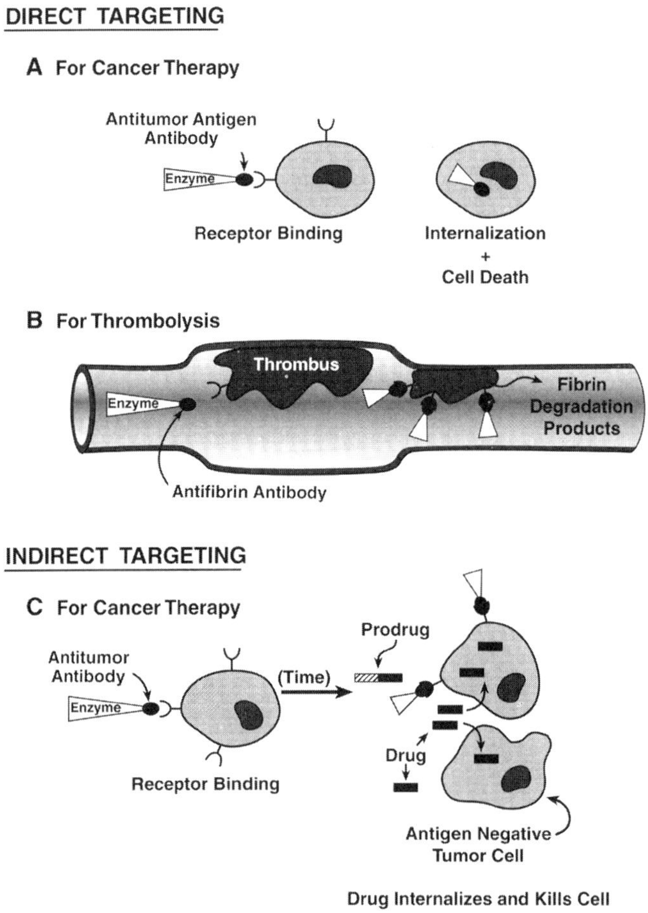

Figure 3.1 Two strategies for targeting immunoenzymes. Direct targeting: the effector agent (enzyme) is delivered to the target site by the antibody, the enzyme is internalized, and cell death occurs (*A*) or the enzyme directly functions at the target site (*B*). Indirect targeting: the effector agent (drug) is formed from an inactive prodrug at the target site by an activating enzyme previously delivered to the target site by an antibody (*C*). The released drug then diffuses to nearby antigen positive and negative cells to cause cell death.

cells that bear the appropriate antigen on the cell surface (Fig. 3.1A). The rationale for the direct targeting of nucleases, particularly human enzymes, is the ease of administration and the hope of reduced nonspecific toxic side effects and immunogenicity. This approach is similar to that associated with the use of immunotoxins based on plant and bacterial enzymes (discussed in Chapter 4). One strategy for the development of new thrombolytic agents exploits recombinant antibody technology to directly deliver plasminogen activators to the surface of a thrombus in an effort to increase both potency and selectivity (Fig. 3.1B). The target of this therapy is present in the vasculature, thus it is accessible to antibody-based reagents offering an excellent rationale for this form of therapy. Indirect targeting is another approach to increase the availability of the drug to its target. In the two-step approach shown in Figure 3.1C, the antibody–enzyme fusion protein is bound to the target antigen at the cell surface and time is allowed for circulating fusion protein to clear the vasculature. A relatively inactive prodrug is then administered, which becomes activated by the enzyme portion of the fusion protein only at the site of the tumor. A decided advantage of the indirect targeting approach for cancer is that the antibody fusion protein does not have to translocate into the cell, an inefficient process. Moreover, the activated drug can diffuse to nearby antigen negative tumor cells. Since prodrug therapy is designed to also kill antigen negative tumor cells, it circumvents the problem of antigen heterogeneity encountered in direct targeting strategies.

This chapter presents the progress in each of these three areas and cites pitfalls that are common as well as distinct to all of these strategies, along with their proposed solutions.

3.1.1 Targeting Members of the RNase Superfamily to Cancer Cells: Overview

In 1955, bovine pancreatic RNase A injected into tumor-bearing mice was reported to impede tumor growth.[6,7] Thus investigations into the clinical use of RNase A were stimulated and it was used in human clinical trials for the treatment of leukemia. Patients with chronic myelocytic leukemia were given daily s.c. injections of 0.5 to 1 mg of the bovine enzyme and were reported to have a decrease in spleen size and to show general improvement.[8] In another clinical study RNase A was given to 246 patients with tick-borne encephalitis and was reported to be tolerated well, generating no side effects with a clinical outcome superior to gamma globulin.[9] Another member of the RNase A superfamily (Onconase) is being evaluated for cancer therapy in ongoing clinical trials. Onconase was isolated from *Rana pipiens* oocytes by following cytotoxic activity against cancer cells in vitro[10] and in vivo.[11] Phase I and Phase I/II clinical trials of Onconase as a single therapeutic agent in patients with a variety of solid tumors[12] or combined with tamoxifen in patients with advanced pancreatic carcinoma have recently been completed and have progressed to Phase III clinical trials. Thus there is both historical and current precedence for the association of RNase A type enzymes with cancer therapy.

In this regard, the biology and history of members of the RNase A family is more extensively explored in Youle et al.[13]

RNase catalytic activity associated with fungal[14] or bacterial[15] enzymes was shown to effectively kill cancer cells. Yet, using members of the pancreatic RNase A superfamily as the toxic moiety of an immunoenzyme may be advantageous because they are homologous to human RNase A type enzymes (Box 3.1). For that reason they may not elicit a vigorous antienzyme immune

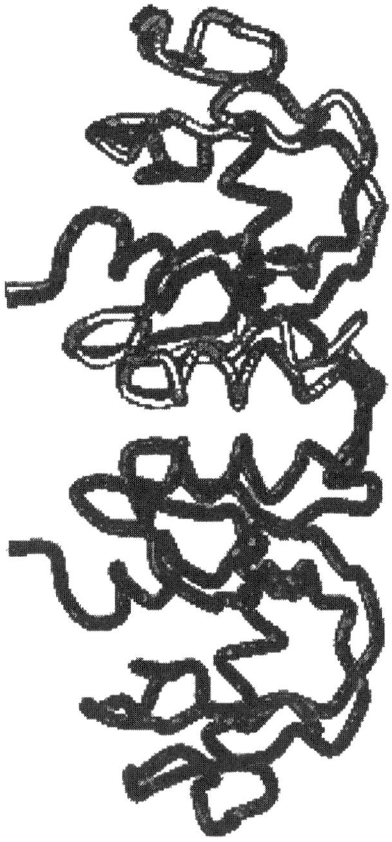

Box 3.1 Structure and Function of Pancreatic Type RNases.

Pancreatic type ribonucleases (RNases) typified by RNase A comprise a superfamily of proteins with similar catalytic activities and overall structure. They are small (M_r 12–14 kDa) basic proteins that cleave RNA endonucleolytically after pyrimidines to yield 3′-phosphomono- and oligonucleotides. Homologous proteins are found in the pancreas of mammals and some reptiles. Homologous proteins have also been identified in the fluids and tissues of other organs in mammals and amphibians. Overall pancreatic type

RNases share 60–70% homology and 25–35% identity in their primary structures. Importantly, all the amino acid residues required for catalysis are highly conserved thus conferring similar nuclease activity toward RNA substrates in enzymatic assays. Yet certain RNases also exhibit different functional activities. Angiogenin (ANG) originally isolated from human colon carcinoma cells is also found in human plasma and stimulates angiogenesis. Eosinophil-derived neurotoxin (EDN) and eosinophil cationic protein (ECP) were first isolated from human eosinophils and display antibacterial and antihelminthic activities. RNases from species such as *Rana pipiens* (Onconase) or *Rana catesbeiana* (frog lectin) are cytotoxic to mammalian cells because they internalize into cells, where they cause protein synthesis inhibition and cell death. Onconase is a chemosensitizer and is being evaluated as an anticancer therapeutic. Bovine seminal RNase (BSRNase) displays antitumor and immunosuppressive activities. It is a homologue of bovine RNase A sharing 80% identity in the primary structure but exists as a dimer cross-linked by disulfide bonds. The dimeric nature is required for the functional activities of BSRNase A since the monomeric forms are not active. In general, RNase enzymatic activity is necessary but not sufficient for the diverse functions displayed by this superfamily of proteins indicating that their mechanisms of action have not been fully elucidated. The structure of dimeric BSRNase is shown in shades of grey. The structure of monomeric RNase A (white) is superimposed with one of the subunits of BSRNase. (Diagram courtesy of Lluis Boque).

response. Figure 3.2 presents a sequence alignment of some of these RNases. The small size (12 kDa) as well as the homology (30% identity; 60–70% homology) between Onconase and the human RNases angiogenin (ANG),[16] and eosinophil-derived RNase (EDN),[17] that are present extracellularly in human plasma most probably explains the immunological tolerance of humans to this frog protein. It has been administered on a weekly basis for up to three years in some patients.[12] Although Onconase possesses inherent cytotoxic activity, its potency can be increased by chemically conjugating it to an antibody recognizing a tumor-associated antigen.[18] Similarly, a dimeric member of this protein superfamily, bovine seminal RNase (BSRNase) (reviewed in 19), possesses antitumor activity in vitro and in vivo.[20] The cytotoxic activity of this RNase is currently being exploited as the effector domain of an immunoenzyme.[21,22]

Conversely, human RNases such as EDN and ANG, like RNase A, are not cytotoxic to normal or tumor cells. Therefore, it is expected that they would elicit fewer toxic side effects in humans when used as the toxic component of a fusion protein. However, these nontoxic RNases can acquire cell-type-specific cytotoxic properties when chemically conjugated[23–26] or fused recombinantly

```
                               *                                 *
RNase A    ..KET.AAAK FERQHMDSST SAASSSNYCN QMMKSRNLTK DRCKPVNTFV
hpRNase    ..KET.AAAK FLTQHYD.AK PQGRDDRYCE SIMRRRGLTS P.CKDINTFI
ANG        ..QDNSRYTH FLTQHYD.AK PQGRDDRYCE SIMRRRGLTS P.CKDINTFI
EDN        KPPQFTWAQW FETQHINMTS QQ......CT NAMQVINNYQ RRCKNQNTFL
ECP        RPPQFTRAQW FAIQHISLNP PR......CT IAMRAINNYR WRCKNQNTFL
Onconase   ....E.DWLT FQKKHI.TNT RDVD....CD NIMSTNLFH. ..CKDKNTFI

RNase      HESLADVQAV CSQKNVACK. NGQT.NCYQS YSTMSITDCR ET..GSSKYP
hpRNase    HEPLVDVQNV CFQEKVTCK. NGQG.NCYQS YSTMSITDCR ET..GSSKYP
ANG        HGNKRSIKAI CENKNGNPH  RE...NLRIS KSSFQVTTCK LHGGSPWP..
EDN        LTTFANVVNV CGNPNMTCPS NKTRKNCHHS GSQVPLIHCN LTTPSPQNIS
ECP        RTTFANVVNV CGNQSIRCPH NRTLNNCHRS RFRVPLLHCD LINPGAQNIS
Onconase   YSRPEPVKAI CK.GIIA... ...S.KNVLT TSEFYLSDCN VT.....SRP

                                                *
RNase A    NCAYDTTQAN KHIIVACEGN .........P YVPVHFDASV ..
hpRNase    NCAYRTSPKE RHIIVACEGS .........P YVPVHFDATV
ANG        PCQYRATAGF RNVVVACENG .......... .LPVHLSQS  IFRRP
EDN        NCRYAQTPAN MFYIVACDNR DQRRDPPQYP VVPVHLDRII :.
ECP        NCRYADRPGR RFYVVACDNR DPR.DSPRYP VVPVHLDTTI ..
Onconase   .CKYKLKKST NKFCVTCEN. .......... QAPVHFVGVG SC
```

Figure 3.2 Sequence alignment of the amino acid sequences of several members of the RNase A superfamily. RNase A, bovine pancreatic RNase A;[178] hpRNase, human pancreatic RNase;[30] ANG, human angiogenin;[189] EDN, human eosinophil-derived neurotoxin;[190] ECP, human eosinophil cationic protein;[191] Onconase, frog ribonuclease isolated from the oocytes of *Rana pipiens*.[192] Cysteine residues are in shaded boxes, the putative catalytic histidine and lysine residues are marked by an asterisk, dots represent gaps introduced to align cysteine and catalytic residues.

to internalizing cell binding ligands.[27–31] For this reason they are being explored as a new type of specific cell-killing agent for cancer and other diseases.

3.1.2 Targeting Plasminogen Activators for Improved Thrombolytic Therapy: Overview

A thrombus is a clot that forms inside blood vessels (Fig. 3.1). It is a major medical problem, especially in developed societies, causing coronary occlusion that can trigger myocardial or cerebral infarction, disability, and death. The simplified scheme shown in Figure 3.3 portrays the major features of fibrinolysis referred to in this chapter when describing targeted plasminogen activators. A thrombus is fibrin rich due to the conversion of the soluble plasma protein fibrinogen to insoluble fibrin. During the polymerization of fibrin monomers, plasminogen and other blood proteins are incorporated into the fibrin clot. Plasminogen activators are thrombolytic agents that convert plasminogen to the active enzyme plasmin, which degrades fibrin into soluble degradation products. The plasminogen activators in turn can be cleaved; however the physiological relevance of this cleavage is uncertain.[32,33]

The two physiological plasminogen activators used to construct the anti-

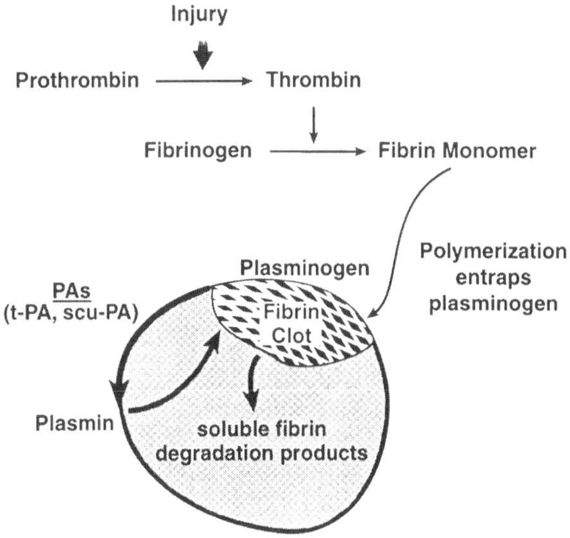

Figure 3.3 Schematic representation of the formation and lysis of a blood clot. A prothrombin activator formed in response to an injury catalyzes the conversion of prothrombin to thrombin. Thrombin in turn converts fibrinogen to fibrin monomers, which polymerize entrapping blood cells, platelets, plasma, and plasminogen. Plasminogen is converted to plasmin by the fibrin-specific plasminogen activators, tissue type (t-PA) or single chain urokinase-type (scu-PA) plasminogen activator. Plasmin degrades fibrin into soluble fibrin degradation products.

body–enzyme fusion proteins described herein are tissue-type plasminogen activator (t-PA) and single chain urokinase type plasminogen activator (scu-PA). They are among the thrombolytic agents currently approved for clinical use. Yet, new approaches to thrombolysis are being investigated due to limitations of the standard agents such as incomplete lysis of all thrombi, lysis that occurs too slowly, and reformation of the thrombus. In this regard the rationale for targeting plasminogen activators to the thrombus is similar to that hoped for in the treatment of cancer; that is, to widen the therapeutic index by increasing the potency and specificity of the drug.

The feasibility of the approach was pioneered by Bode et al.[34] who demonstrated that a murine monoclonal antibody chemically conjugated to urokinase retained the original binding specificity of the antibody and showed a 100-fold increase in fibrinolysis. Progress from chemical conjugates to engineered antibody plasminogen activator fusion proteins has issued mainly from the laboratories of Edgar Haber (Center for Prevention of Cardiovascular Disease, Harvard School of Public Health, Boston, Massachusetts) and later that of Desire Collen (Center for Molecular and Vascular Biology, University of Leuven, Leuven, Belgium). Recent excellent reviews from both laboratories are available for detailed reading in this area.[32,35–37]

3.1.3 Targeting Enzymes for Prodrug Therapy: Overview

In the prodrug strategy the targeted enzymes are not the drugs per se, rather they are used to activate the drug only in the vicinity of the tumor (Fig. 3.1C). The prodrug concept was first introduced in 1973 by Philpott et al.[38] with an antibody conjugated to glucose oxidase. The conjugate killed cells by iodination when it was combined with glucose, lactoperoxidase, and iodide. Subsequently, more direct enzyme systems were introduced independently by two laboratories (Kenneth Bagshawe, Medical Oncology, Charing Cross & Westminster Medical School, London, England[39] and Peter Senter, Bristol-Myers Squibb Pharmaceutical Research Institute, Seattle, Washington).[40]

The prodrug strategy was designed to incorporate the specificity of monoclonal antibodies (MAbs) into treatment with standard chemotherapeutic drugs or new versions thereof. Systemic administration of relatively nontoxic prodrugs decreases nonspecific toxic side effects associated with the parent drug. Moreover, problems experienced with direct targeting of tumor antigens such as antigen heterogeneity, drug distribution, and tumor penetration can be circumvented since the small M_r drug can enter antigen-free tumor cells and diffuse more readily throughout tumors than larger antibody–enzyme conjugates. For example a MAb–alkaline phosphatase conjugate was used to generate clinically approved etoposide from etoposide phosphate for the treatment of solid carcinomas.[40] The prodrug was less toxic systemically most likely because phosphorylation hindered cellular uptake. Other phosphorylated drug derivatives were also prepared, and drugs as mechanistically diverse as mitomycin[41] and doxorubicin[42] were generated. The versatility of the prodrug strategy is illustrated in Figure 3.4. Doxorubicin can be produced from doxorubicin glucuronide or cephalosporin doxorubicin by β-glucuronidase[43,44] or β-lactamase,[45] respectively (Fig. 3.4A). β-lactamase was also used to activate a cephalosporin based prodrug, demonstrating that a single monoclonal antibody–enzyme combination can activate a variety of prodrugs (Fig. 3.4B).

Preclinical and preliminary clinical work has been accomplished with a variety of prodrug–enzyme systems with chemical antibody–enzyme conjugates. The results of pharmacokinetic studies are promising because they show that the concentration of an activated prodrug achieved intratumorally is greater than that with the conventional administration of chemotherapeutic drugs.[46,47] Moreover, significant antitumor activity has been demonstrated in animal models of human cancer.[48–51] Clinical results using a benzoic acid mustard prodrug that was activated by an anti-CEA MAb-carboxypeptidase G2 (CPG2) conjugate have been obtained.[52–55] This pilot study demonstrated the feasibility of this approach in a clinical setting and is more fully discussed in section 3.5. For more extensive reading in this area, several reviews are available.[56–61]

To construct fusion proteins with more uniform properties, new approaches using recombinant DNA technology are being undertaken. Chemical conjugates are heterogeneous because the chemical cross-linking agents are not

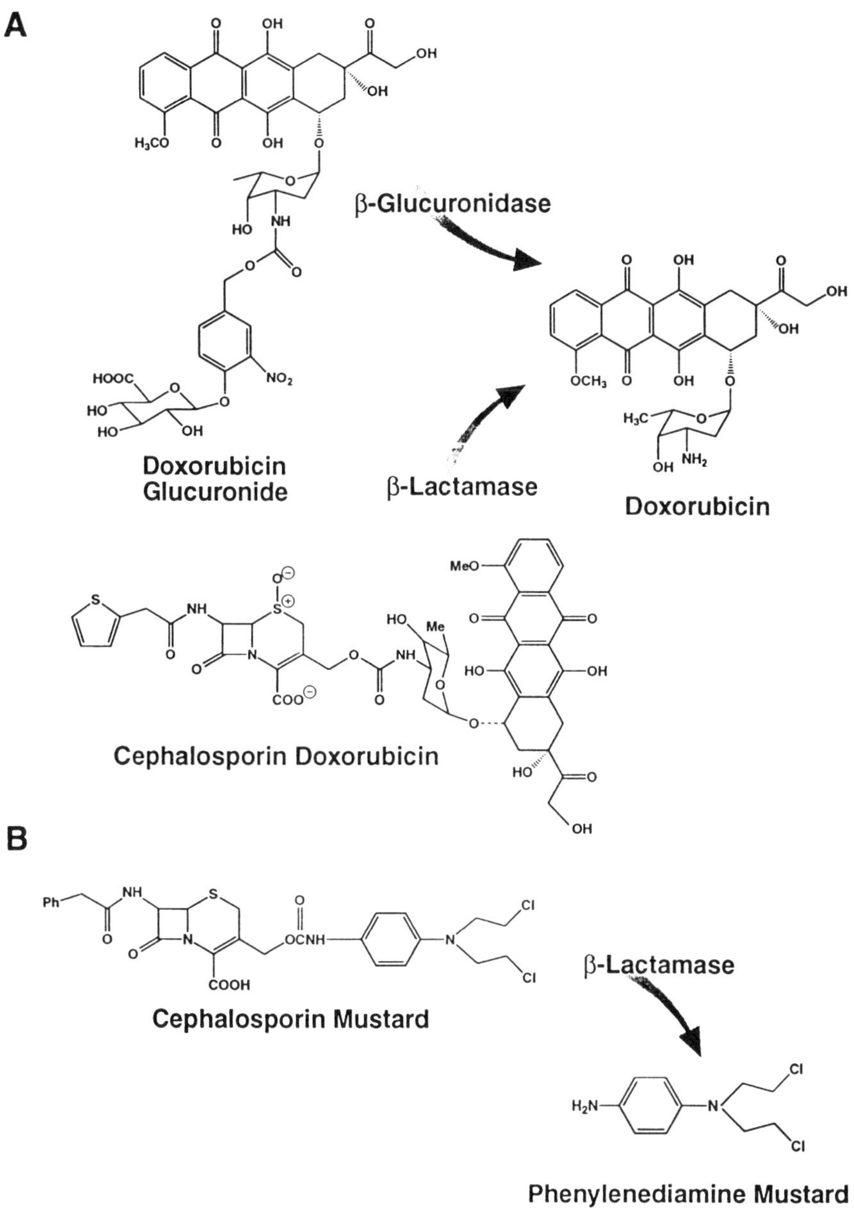

Figure 3.4 Activation of prodrugs by targeted enzymes. In part A, the same drug (doxorubicin) is released from two different prodrugs (doxorubicin glucuronide and cephalosporin doxorubicin) by two different enzymes (*β*-glucuronidase or *β*-lactamase). Part B illustrates that the same enzyme (*β*-lactamase) can catalyze the release of two different drugs (doxorubicin and phenylenediamine mustard) from two different prodrugs (cephalosporin doxorubicin and cephalosporin mustard).

specific for a single amino acid. The resulting conjugate is actually a mix of conjugate proteins consisting of antibody and enzyme linked at different sites and containing hybrid proteins with different molar ratios of enzyme to antibody.[62] The recombinant antibody–enzyme fusion proteins for prodrug activation produced to date are listed in Table 3.1 along with their corresponding prodrug/drug combinations.

3.2 IMMUNOENZYME STRUCTURE

Recombinant DNA technology affords the opportunity to design fusion proteins for specific applications by altering the features of the antibody or enzyme domains. For instance, the nature of the antibody used, i.e., $F(ab')_2$, Fab, or sFv, depends on the application intended. For many applications, only the antigen-binding domains of the antibody are required. Fv fragments are the smallest (25 kDa) antibody fragments that contain the complete antigen binding site.[63,64] Fv fragments are noncovalently associated V_H and V_L domains that tend to dissociate at low protein concentrations,[65] but in their single chain Fv (sFv) analogs, the two variable domains are coupled by peptide linkers[66,67] (reviewed in 68, 69). The linker needs to be of sufficient length to bridge the distance between the C-terminus of the first V domain and the N-terminus of the second V domain. Several studies have examined the characteristics of sFvs that are affected by the length and composition of the peptide linker.[70–73] Overall, the nature of the linker can affect affinity, dimerization, and aggregation of the sFv. Indeed, the length of the linker can be adjusted to prevent the V_H and V_L domains on the same chain from pairing with each other.[74] This can be exploited to create bispecific binding proteins. Linkers should not interfere with the association of the two domains. Glycine provides flexibility to the $(GGGGS)_3$ peptide linker originally used by Huston et al.[67] It is also devoid of charged and hydrophobic residues that might interact with the V domain surfaces and interfere either with the binding of these domains to each other or with the binding of the sFv to the antigen.[68] Although different linker peptides between V_H and V_L domains have been used without affecting the in vitro activities of sFv immunotoxin fusion proteins,[75] the nature of the linker was found to markedly affect biological properties of an RNase sFv[27] (see Sections 3.4 and 3.5). Recently the introduction of an interchain disulfide bond has been used to link V_H-V_L domains after genetically modifying each domain to introduce opposing cysteine residues.[45,65,76] The advantages of disulfide-linked single chain Fvs (dsFv) compared to sFvs include enhanced serum stability, decreased tendency to aggregate, increased production, similar or increased antigen binding, and improved antitumor activity in animals (reviewed in 77). Disadvantages include the requirement for additional protein engineering and for experiments to investigate possible effects of the introduced disulfide bond on the affinity of the Fv. A fusion protein for prodrug activation has recently been constructed with a dsFv.[45]

TABLE 3.1 Components of Fusion Protein Activated Prodrug Systems

Nomenclature	Antigen	Enzyme	Prodrug	Drug	Reference
L6-sFv-BCβL	Ganglioside antigen (breast, lung, colon, ovarian)	β-Lactamase II (*B. cereus*)	Cephalosporin mustard	Phenylenediamine mustard	108
L49-sFv-bL	p97 (melanotransferrin)	r2-1β-Lactamase (*E. cloacae*)	Cephalosporin nitrogen mustard	Phenylenediamine mustard	110
dsFv3-β-Lactamase	$p185^{HER2}$ (breast, ovarian)	β-Lactamase RTEM-1	Cephalosporin doxorubicin	Doxorubicin	45
dsFv3-β-Lactamase	$p185^{HER2}$	β-Lactamase RTEM-1	Cephalosporin taxol	Taxol	168
hu431β-gluc	CEA (carcinoembryonic antigen)	Human β-glucuronidase	Doxorubicin-glucuronide	Doxorubicin	43

Finally, in designing the Fv binding unit, the choice of the V region order may be important, i.e., V_H-linker-V_L or V_L-linker-V_H. Both the affinity and secretion level of the protein can be influenced by the V domain order (reviewed in 78).

Natural antibodies as well as $F(ab')_2$ fragments are bivalent and usually bind polyvalent cellular antigens with higher affinities than the monovalent Fab or sFv fragments.[79] Generally it is recognized that high affinity antibodies are preferable, yet valency and affinity have to be balanced against factors such as size, which affect tumor penetration and plasma clearance rate. Comparative studies show that small antibody fragments clear from the blood faster.[80–82] Rapid clearance of the fusion protein would be advantageous for prodrug administration since the prodrug could be administered sooner without activation by residual enzyme in the circulation. The disadvantage is that more fusion enzyme might have to be administered since rapid clearance reduces adequate tumor uptake. Yet due to the smaller size, sFvs exhibit better tumor penetration and are more evenly distributed throughout the tumor compared to intact IgGs.[83] Adding to the complexity of antibody pharmacokinetics is the enzyme moiety in each fusion protein. Ultimately, the ideal antibody fragment used with a particular enzyme may have to be determined for individual fusion proteins by in vivo testing.

One of the major limitations in the clinical use of the antibody–enzyme fusion proteins in humans is the generation of an immune response against the xenogeneic protein. Although allergic reactions can occur upon retreatment of patients who have developed antibodies against the enzyme conjugates, the major effect of the induced antibodies appears to be a shortening of the half-life of the conjugate, which impedes targeting and tumor uptake (reviewed in 84, 85). In some of the studies described in this chapter, chimeric or humanized antibodies are being used in the antibody domain in an attempt to reduce this problem. Chimeric antibodies are built by incorporating entire murine variable regions within human constant regions, while in humanized antibodies the only murine sequences are in the CDRs (complementarity determining regions) (reviewed in 86). Thus far it appears that the immune response is attenuated in patients that receive humanized antibodies (84, 87 and references therein). For these reasons, the fusion protein effector domain in some of the studies of each of the targeting strategies described herein is comprised of a human enzyme.

3.2.1 Architecture of Targeted Nucleases for Cancer Therapy.

Both chemical conjugates and recombinant fusion proteins consisting of antitransferrin antibodies linked to RNase proteins or fused to RNase genes have been made (88 and references therein). RNase fusion proteins have been constructed with several different architectures. In one variation the 5′ region of the ANG gene was fused to the 3′ region of the CH2 domain[29] of a chimeric antihuman-transferrin receptor antibody without a spacer peptide (E6)[89] (Fig. 3.5A). The same chimeric antibody had previously been fused to the gene for

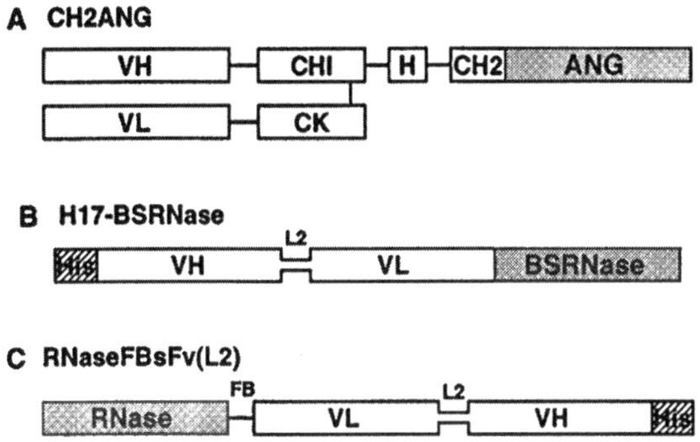

Figure 3.5 Configuration of RNase-based immunofusions. *A*. CH2ANG, the gene for ANG was fused to the 3′ end of the CH2 domain of a chimeric antihuman transferrin receptor antibody (E6).[89] This construct was expressed in myeloma cells that also secreted the light chain of the same antibody.[29] *B*. H17-BSRNase, the gene for BSRNase was fused to the 3′ end of the V_L domain of an sFv directed against hPLAP.[21,58] *C*. the genes for three different human RNases, EDN,[30] ANG,[27] or hpRNase,[30] were fused to the 5′ end of the V_L domain of an sFv directed against the human transferrin receptor (E6). L2, $(GGGGS)_3$ flexible linker originally described by Huston et al.;[67] His, histidyl residues; FB, 13 amino acid residue spacer.[94]

tumor necrosis factor[90] in a manner that left the hinge region unaffected and dimerization of the heavy chain possible. In this way $F(ab')_2$ antibody-ANG fusions could have been generated.

Another RNase fusion protein, H17-BSRNase,[21,58] also was designed to allow RNase dimerization since bovine seminal RNase(BSRNase) itself dimerizes by virtue of two intersubunit disulfide bonds and an exchanged amino terminus.[91] Variable regions cloned from MAb H17E2[92] that targets the human tumor-associated antigen placental alkaline phosphatase (hPLAP) were joined by a flexible linker, L2 $(GGGGS)_3$,[67] to produce a single-chain Fv fragment that was not separated from the RNase by a spacer peptide[21] (Fig. 3.5B).* The configuration of the Fv was V_H-V_L with the 3′ region of V_L fused to the 5′ region of the BSRNase gene. The monomeric form is shown. Histidyl residues were incorporated at the 5′ or the 3′ end of the V_H gene in the BSRNase construct (Fig. 3.5B) or the fusion proteins derived from monomeric RNases described in the next paragraph (Fig. 3.5C). The histidyl tag was incorporated to facilitate purification by metal chelate chromatography. Subsequently several other derivatives were constructed with additional peptides that were designed to improve folding and intracellular cytotoxicity.[22] These variations are discussed in the section pertaining to function (Section 3.5).

Single-chain Fv antibody fusions have also been constructed with three different human RNases, EDN,[30] hpRNase,[30] or ANG[27] with the architecture shown in Figure 3.5C. This was determined to be the optimum configuration

*For ease of comparison, the $(GGGGS)_3$ linker is referred to as L2 throughout this chapter.

for generating stable active protein for RNase sFvs expressed as insoluble protein in inclusion bodies.[27] The sFv was composed of antibody V_L and V_H domains[93] from the same chimeric antitransferrin receptor antibody used for the $F(ab')_2$ antibody-ANG fusion protein shown in Figure 3.5A. The nature of the linker connecting the Fv domains as well as the presence of a spacer peptide (FB, residues 48-60 of staphylococcal protein A,(AKKLNDAQA-PKSD)[94] between the V_L and RNase proved to be important for the function of the antibody and RNase portion of the molecule (see Sections 3.4 and 3.5).

3.2.2 Architecture of Targeted Plasminogen Activators for Thrombolysis.

To deliver plasminogen activators specifically to intravascular clots, both t-PA and scu-PA have been linked chemically as well as recombinantly to antibodies that are fibrin-specific and do not cross react with fibrinogen (the circulating precursor of fibrin), thus allowing the conversion of plasminogen to be limited to the clot surface and decreasing systemic fibrinogenolysis (reviewed in 32, 35, 36). In the studies described here, two antibodies have been used. The first antibody, 59D8, was raised by immunization with a synthetic heptapeptide based on a unique epitope in the amino-terminal sequence of the fibrin β chain.[95] The second antibody, MA-15C5, is specific for fragment D-dimer of cross-linked human fibrin.[96]

Three recombinant proteins based on the 59D8 antibody were designed (Fig. 3.6). Initially sequences coding for the β chain of t-PA were fused to the heavy chain gene of 59D8 replacing the CH2 and CH3 domains (not shown).[97] Although this fusion protein, r59D8-tPA, was capable of high affinity fibrin binding and plasminogen activation, it was subsequently found to be ineffective in lysing a plasma clot, discussed in (32). Therefore the amino acid sequences coding for the β-chain of t-PA were replaced with amino acid residues 144-411 that encode a low M_r form (32 kDa) of the catalytic domain of scu-PA (Box 3.2, reference 98). The resulting fusion protein r-scuPA(32kDa)-59D8[99] is shown in Figure 3.6A. The enzyme is joined to the V_H behind a CH2 domain. This configuration was modified by Yang et al.[100] who deleted the CH2 domain and fused the enzyme (amino acid residues 144-411) directly behind the hinge region (59D8-scuPA-T, Fig. 3.6B). 59D8-scuPA-T also contains alterations in the enzyme domain. Scu-PA can be cleaved by plasmin, which results in enhanced plasminogen activation in vitro.[101] Thrombin cleavage results in the generation of an incorrect amino terminus and inactive scu-PA.[102] The thrombin and plasmin cleavage sites are indicated in the schematic drawing of scu-PA and scu-PA-32kDa enzymatic domains (Box 3.2). However, Yang et al. (100 and references therein) postulated that thrombin rather than plasmin activation could potentially distinguish a newly formed thrombus from an established clot (see Sections 3.4 and 3.5). Therefore the gene for scu-PA was mutated such that thrombin cleavage would result in the generation of the correct amino terminus and an active enzyme (Fig. 3.6B).

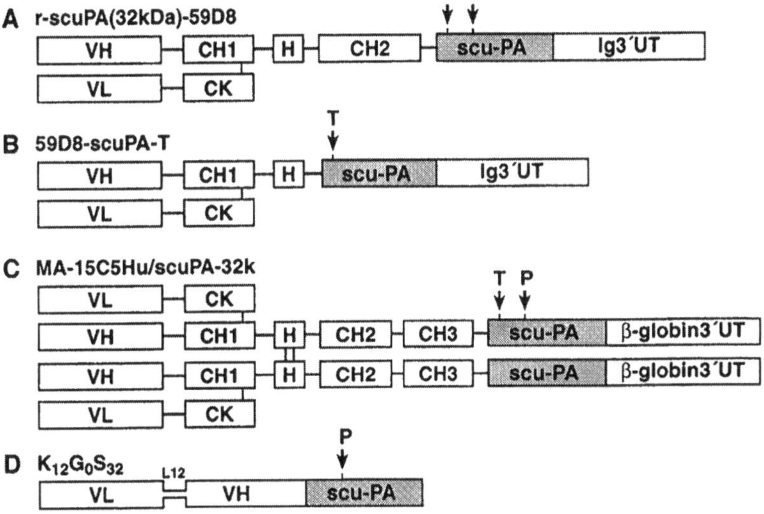

Figure 3.6 Configuration of plasminogen activator fusion proteins and enzymes. *A.* r-scuPA (32kDa)-59D8, the gene encoding amino acid residues 144-411 of scu-PA was fused to the 3′ end of the CH2 domain of an anti-fibrin antibody, 59D8. This gene was expressed in heavy chain loss variant hybridoma cells that expressed the light chain of the same antibody.[99] *B.* 59D8-scuPA-T, a thrombin activatable low molecular weight scu-PA obtained by deletion of F157 and K158. The gene for the mutated scu-PA-32kDa was fused to the hinge region of the same antibody used in A, 59D8.[100] *C.* MA-15C5Hu/scu-PA-32k, DNA encoding scu-PA (144-411) was fused to the CH3 domain of a humanized antibody specific for fragment D-dimer of cross-linked human fibrin, MA-15C5. This construct was coexpressed with DNA encoding the light chain of the same antibody in CHO cells.[103] *D.* $K_{12}G_0S_{32}$, DNA encoding amino acid residues 132-411 of scu-PA was fused to the V_H domain of the sFv derived from the murine antibody MA-15C5. The V_L and V_H chains were joined together by L12, a 7 amino acid synthetic linker (AGQGSSV).[105] T, thrombin cleavage site R156-F157; P, plasmin cleavage site K158-I159. Ig3′ UT and β-globin 3′ UT represent the 3′ untranslated (UT) region of the mouse immunoglobulin or β-globin genes.[99]

Fusion proteins based on the second humanized antifibrin antibody (MA-15C5Hu) were also designed and constructed. The humanized heavy chain consisted of the V_H and CH1, CH2, and CH3 domains of MA-15C5Hu fused to the same region of scu-PA (amino acid residues 144-411) used in the r-scuPA(32kDa)-59D8 fusion to create MA-15C5Hu/scu-PA-32k[103] (Fig. 3.6C). The valency of the fusion proteins shown in Figure 3.6 represents the molecular forms of these proteins after expression and purification. The presence of the hinge region allowed for the formation of two interchain disulfide bonds generating a bivalent molecule only in MA-15C5Hu/scu-PA-32k. Inclusion of 3′ untranslated regions (UT) in the design of these constructs is relevant to their expression and is discussed in Section 3.3.

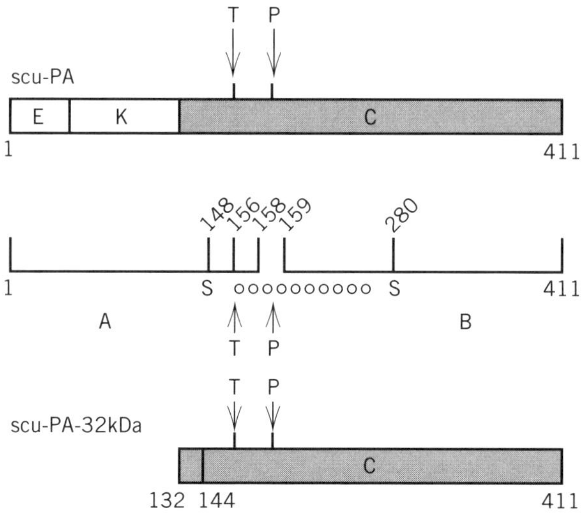

Box 3.2 Structure and Function of Urokinase-Type Plasminogen Activator. Single chain (sc) u-PA is a physiological plasminogen activator secreted into the circulation from endothelial and other cell types. The amino terminus of scu-PA contains EGF-like (E) and kringle (K) domains that are involved in fibrin selectivity while the catalytic (C) domain resides in the carboxyl terminus. During clot lysis plasminogen is converted to plasmin by scu-PA. Plasmin, in turn, cleaves the single chain form of scu-PA between amino acid residues 158 and 159 (P) to generate a two chain derivative (A&B) linked by disulfide bonds (SS) between Cys148 in the A chain and Cys280 in the B chain. Although both the single and two chain forms of the protein are active catalytically, they express different properties. In contrast to the two chain form, single chain scu-PA exhibits fibrin selectivity and is resistant to inhibition by plasminogen activator inhibitor 1. Since two chain scu-PA is irreversibly inhibited by plasminogen activator inhibitor 1, it is cleared from the circulation more rapidly. Mutant proteins engineered to be resistant to plasmin cleavage after amino acid residue 158 do not lyse plasma clots in vitro indicating that conversion to the two chain form may be needed for scu-PA to function optimally in fibrinolysis. However, the physiological importance of the single chain vs. two chain forms of scu-PA is not fully understood. Thrombin (T) cleavage of scu-PA between amino acid residues Arg156 and Phe157 results in an inactive protein. However, deletion of amino acid residues 157 and 158 destroys the plasmin cleavage site simultaneously causing thrombin cleavage to generate a viable two chain enzyme. Thrombin-activatable scu-PA might be more selective for blood clots that are newly forming rather than previously established. Lysis of established blood clots during thrombolytic therapy can cause excessive bleeding. A low M_r form of scu-PA (scu-PA-32 kDa) retains all of the properties of scu-PA but is more suitable for large-scale recombinant production and is the form of the molecule used in the construction of the plasminogen activator fusion enzymes described in this chapter.

Additionally, constructs utilizing the Fvs of the MA-15C5 antibody and scu-PA were also made. The sFv was constructed[104] from the Fvs of MA-15C5 by connecting the carboxyl-terminal end of the V_L chain to the amino-terminal end of the V_H chain via L12, a 7 amino acid synthetic linker (AGQGSSV) which had been designed by a computer-assisted method.[105] The sFv was fused without a spacer sequence to the 5′ end of nucleotides encoding amino acid residues 132-411 of the catalytic domain of scu-PA to form the single chain fusion protein $K_{12}G_0S_{32}$[106] (Fig. 3.6D).

3.2.3 Architecture of Fusion Enzymes to Activate Prodrugs

Several distinct architectures were employed to construct fusion proteins suited to activate anticancer prodrugs (Fig. 3.7). A fusion protein consisting of the humanized Fab fragment of the anti-CEA MAb BW 431 and human β-glucuronidase was isolated as two different M_r forms of the antibody–enzyme fusion.[43] Under denaturing conditions, a monovalent molecule of M_r

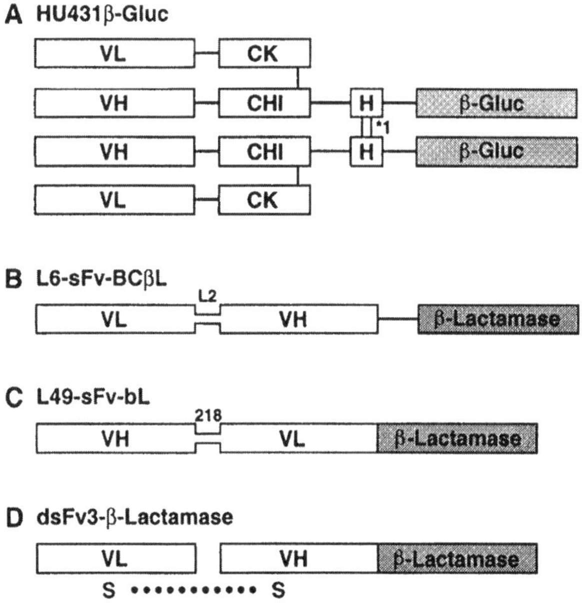

Figure 3.7 Configuration of fusion proteins designed to activate prodrugs. *A*. The gene encoding human β-glucuronidase was fused to the hinge region of a humanized anti-CEA antibody, BW 431. This construct was coexpressed with the vector containing the humanized light chain gene to the same antibody in BHK cells.[43] The two hinge region cysteines (*1) can form intrachain disulfide bonds resulting in a 125 kDa protein or as shown here two interchain disulfide bonds (250 kDa protein). *B–D*. The gene for β-lactamase was fused to the 3′ end of either V_H or V_L domains of sFvs derived from *B* L6[108], *C* L49[110] or *D* humanized anti-p185^{HER2}.[45] The V_L and V_H domains were either connected by a peptide linker L2[67] or 218[70] or by a disulfide bond.[45,168]

125 kDa contained the light chain of the humanized MAb consisting of the V_L and C_L domains covalently linked by an interchain disulfide bond to the humanized heavy chain. The humanized heavy chain consisted of the humanized V_H domain of MAb BW 431 and the CH1 and hinge region derived from human IgG_3 separated from human β-glucuronidase by a spacer peptide. The presence of the hinge region also allowed for the formation of two interchain disulfide bonds, generating a 250 kDa bivalent molecule (Fig. 3.7A). In the monovalent molecule, the two hinge region cysteines formed an intrachain disulfide bond.[43]

Other fusion proteins designed for prodrug activation utilized Fv antibody fragments. Variable regions cloned from L6, a MAb recognizing a tumor-associated antigen present on breast, colon, lung, and ovarian carcinoma cells,[107] were joined by a flexible linker (L2)[67] to produce a single-chain Fv fragment that was separated by a six amino acid spacer (GSGGSG) from β-lactamase (L6-sFv-BCβL) (Fig. 3.7B).[108] Subsequently the same laboratory modified this design by joining the Fv antibody fragments of L49, an antibody against the human p97 (melanotransferrin) tumor antigen,[109] with a different linker sequence (GSTSGSGKPGSGEGSTKG; known as the "218" linker)(70) between the V_H and V_L based on its ability to reduce sFv protein aggregation (L49-sFv-bL).[110] In the latter construct the orientation of the Fv fragments was reversed (V_H, V_L), and the enzyme was fused to the V_L without an intervening peptide spacer (Fig. 3.7C).

The development of a humanized disulfide-stabilized sFv (dsFv) anti-$p185^{HER2}$ Fv-β-lactamase fusion protein has also been accomplished.[45] Potential interchain disulfide bonding pairs were identified based on (1) distance found in natural disulfide bonds, (2) residues buried at the V_H-V_L interface (to minimize disulfide bond cleavage in serum), (3) favorable geometry to avoid strain, and (4) presence in framework regions. Three different dsFvs were constructed and characterized. Of these, only one (dsFv3) had similar binding for $p185^{HER2}$ compared to the wild type (wt) Fv and was chosen for the construction of the fusion protein in which β-lactamase RTEM-1 was joined to the carboxy terminus of V_H (dsFv3-β-lactamase) (Fig. 3.7D).

3.2.4 Summary.

Overall the configurations of the fusion proteins used for the three antibody–enzyme targeting strategies shown in this section were similar. They all contained examples of sFv enzyme fusions as well as Fab-enzyme fusions that exhibited the possibility of assuming bivalent forms.

The Fvs were joined by a variety of linkers including the classic $(GGGGS)_3$ originally designed by Huston et al.[67] and the "218" linker[70] that had been shown to reduce aggregation and proteolysis. Another fusion protein, dsFv3-β-lactamase, was built with Fvs that were joined by a disulfide bond instead of a linker peptide. The fusion proteins ranged in size from about 43 kDa of the RNase sFv fusions to 250 kDa of the β-glucuronidase truncated IgG fusion enzyme for prodrug activation. In one interesting design, an sFv fusion protein was induced to dimerize because the enzyme chosen (BSRNase) exists as a natural dimer. sFv fusion domains were oriented as V_H-V_L or the reverse. In the case of human RNase fusions, the choice was empirical and orientation was not addressed in the other studies. In one case, β-lactamase was fused to the V_H of L6 sFv via a spacer peptide but later to the V_L of the L49 or V_H of the anti-p185^{HER2} antibody without any intervening sequence. It is difficult to generalize the importance of separating the enzyme from the antibody domain. This appears to depend on the individual fusion proteins. For example, the CH2ANG expressed in myeloma cells is the most potent fusion protein to date and it is not separated from the antibody by a spacer peptide. The sFv form of this fusion protein expressed in bacteria requires a spacer peptide between the RNase and sFv for optimal activity. This may pertain to the orientation of ANG which is fused at the N-terminus to the CH2 region but to the C-terminus of the sFv. In that regard, the addition of a spacer between the C-terminus of the H17sFv and N-terminus of BSRNase did not affect expression, purification, or activity of that fusion protein, perhaps because the N-terminus of BSRNase is very flexible.[111]

3.3 PRODUCTION OF ANTIBODY-ENZYME FUSION PROTEINS

Several expression systems have been used for the expression of the antibody-enzyme fusion proteins described in this chapter: mammalian cells, bacteria, baculovirus-infected insect cells, and transgenic animals. Each system has its advantages and its limitations for the expression of recombinant antibodies (reviewed in 112). DNA-encoding antibody genes[2,3] and antibody–enzyme fusion genes[4] was stably transfected into myeloma cells, making it possible to express recombinant antibodies or fusion proteins in cells that may be best suited to assemble, posttranslationally modify (e.g., glycosylate), and secrete these molecules. Secretion through the appropriate intracellular compartments in mammalian cells also protects against proteolysis and results in a protein with the correct amino terminus.[113] Moreover, since the proteins are secreted into culture medium, purification is simplified because the product is not co-localized with intracellular proteins. One problem with expressing proteins in mammalian cells can be the yield, however, cost efficient production of antibody–enzyme fusion proteins became available with the development of hollow-fiber bioreactors.[114,115] This type of bioreactor was used to produce large quantities of a targeted plasminogen activator fusion protein (50–300 mg/L).[98] A possible drawback to expressing proteins for clinical use in mammalian cells is a requirement to remove contaminants such as viral DNA or products associated with serum containing medium. Mammalian cells can now be grown in serum-free production media, further simplifying the recovery process by removing potentially contaminating serum proteins (reviewed in 116). Furthermore, antibodies can be expressed in nonlymphoid cells such as

Chinese hamster ovary (CHO)[117] which can be grown under serum-free conditions.[118]

The first reports[63,119] using *E. coli* to produce antibody fragments ushered in a new era in recombinant antibody technology and has been recently reviewed.[120,121] The well-established procedures for genetic manipulation of DNA for expression in bacteria as well as the rapid growth of bacteria that generates a high yield of recombinant material compared to other expression systems have fueled studies of engineered antibodies and antibody fusion proteins.

Recombinant proteins can be expressed in the bacterial cytoplasm where they often are incorporated into inclusion bodies as protein aggregates, necessitating solubilization with denaturants and reducing agents followed by refolding. Yields, especially of structurally complex hybrid fusion proteins, are often low after refolding and the N-terminal is modified with a methionine. Furthermore, conditions such as protein concentration, salt concentration, and composition of the redox-buffer may have to be optimized for each recombinant protein, particularly antibodies fused to various effector molecules that affect the physiochemical characteristics of the entire protein. The formation and purification of recombinant proteins from inclusion bodies is thoroughly discussed in Ref. 122.

In another expression strategy, recombinant proteins fused to a leader sequence can be directed to the *E. coli* periplasmic space between the inner and outer bacterial membranes. In this case, several of the problems noted previously can be avoided. Recombinant proteins retain the native amino terminus after the leader is processed. Furthermore the periplasmic space is an oxidizing environment that favors the formation of disulfide bonds and correct protein folding. A significant disadvantage of expression in *E. coli* is the absence of posttranslational modification. Although deglycosylation of antibodies has been shown not to affect antigen binding, natural effector functions can be hampered (121 and references therein). This problem is minimized if the desired product is an antibody fragment or a fusion protein that does not require glycosylation.

Recombinant baculoviruses can also be used to express antibodies[123,124] and sFv fusion proteins [106,125] in cultured insect cells. Eukaryotic baculovirus-infected cells offer many of the advantages of mammalian cell expression to yield correctly processed and folded proteins. For this reason this system is often used when proteins cannot be expressed in bacteria, and when mammalian cell expression is inefficient. Despite its potential, the baculovirus system is still limited compared to mammalian expression systems because of the narrow range of available promoters, selectable markers, viral vectors as well as problems with the ease of manipulation and scale up.

Once a recombinant product is identified, commercialization may ultimately require transgenic organisms as bioreactors to cost-effectively generate large amounts of material. With regard to immunoglobulin expression, the possibility of producing full-length, functional antibodies by coexpressing the V_L and

V_H genes in transgenic plants has been demonstrated.[126,127] However, glycosylation with regard to the types of sugars attached is different in plant and mammalian cells (128 and references therein). For this reason, transgenic mammals may be more appropriate for the expression of antibodies and antibody fusion proteins. The production of other types of recombinant proteins in the milk of transgenic animals has already been accomplished (reviewed in 129, 130). The genomes of transgenic animals such as mice, rabbits, goats, and cows have been modified by genetic manipulation to contain foreign DNA encoding proteins of interest, and the integrated DNA can be passed to successive generations. Generally proteins expressed in transgenic animals have been directed to the mammary gland with promoters such as the goat β-casein gene[131,132] because, among other reasons, milk can be produced in large quantities and is easily harvested. As described in the following section, $F(ab')_2$ antibody–enzyme fusion proteins have also been produced in the milk of transgenic animals.

3.3.1 Targeted Nucleases Expressed in Cultured Cells, Bacteria, and the Milk of Transgenic Mice

The first recombinant RNase fusion protein (CH2ANG) was expressed in myeloma cells[29] that had been engineered to secrete an antitransferrin receptor chimeric light-chain E12B5.[89] The myeloma cells were transfected with the pSV2-derived vector containing the gene for the heavy chain of the antitransferrin receptor antibody fused at the CH2 domain to human ANG (Fig. 3.5A). After selecting for the *gpt* gene, culture supernatants of clones testing positive for human IgG were followed for reproducible human IgG activity. The highest producing clones were subcloned by limiting dilution. The amount of this $F(ab')_2$ RNase fusion protein secreted into culture medium ranged from 1–5 ng/ml, based on the reactivity to antihuman IgG antibodies; 1–2 ng/ml based on ANG immunoreactivity. This low level of production precluded purification to homogeneity of the antitransferrin receptor targeted RNase from these supernatants.

To increase the level of expression, an RNase-antibody construct was prepared for expression in bacteria.[28] The gene for the human RNase, EDN, was fused to the gene coding for an antitransferrin receptor sFv composed of the V_L and V_H domains of the same antitransferrin receptor antibody used for the $F(ab')_2$ RNase fusion protein. The EDN sFv gene was first fused to the *E. coli* omp A signal sequence to secrete soluble protein to the periplasm. Analysis of the periplasmic fraction, however, revealed the presence of very little protein. Decreasing the temperature during the induction period from 37°C to 18°C did not lead to an increase in the level of soluble protein. The EDN sFv and subsequently ANG[27] and hpRNase sFv[30] constructs were then modified for expression as insoluble proteins in inclusion bodies in *E. coli* BL21(DE3), using the pET-11d vector, a system designed for the expression of toxic gene products.[133] Details of the pitfalls and their solutions encountered in the expression and purification of RNase fusion proteins have recently been compiled.[134] Briefly, after induction with IPTG (800 μM), the inclusion bodies were vigorously washed, the proteins denatured with 6 M guanidine-HCl, and renatured by rapid dilution into a Tris/L-arginine buffer containing 4 mM oxidized glutathione as described by Brinkmann et al.[135] Renatured RNase fusion proteins were first chromatographed on a heparin-Sepharose column (EDN sFv) or CM Sephadex C-50 (ANG sFv and hpRNase sFv), followed by final purification by immobilized metal affinity chromatography on Ni^{2+}-NTA agarose. Yields of purified protein from shake flasks varied for each construct: ANG sFv, 750 μg/L; EDN sFv, 950 μg/L; hpRNase sFv, 3500 μg/L.

The single chain BSRNase fusion protein (H17-BSRNase) (Fig. 3.5B) was expressed in *E. coli* BL21(DE3) with the pel B leader sequence to secrete the protein directly to the periplasm.[21] As with the human RNase antitransferrin receptor chimeras, little protein was found to be soluble; rather, the fusion protein was found in insoluble inclusion bodies necessitating denaturation and refolding to generate active BSRNase sFv. Investigation of different refolding methods showed that the protocol described by Buchner et al.[136] with slight modifications[22] yielded the highest amount of active material, most likely because the method employs a redox system to help form correct disulfide pairings. This was particularly important for BSRNase sFv with 14 disulfide bonds per dimer.

To date all RNase fusion proteins reported, including those constructed with peptide targeting ligands such as EGF,[31] IL2 (K. Psarras, pers. communic), or FGF (M. Seno, pers. communic.) have been expressed as insoluble proteins. Typically not all of the refolded fusion protein is active due to the presence of incorrectly folded molecular species and aggregates. Recently transgenic plants and animals have been exploited as biorectors for antibodies and other recombinant proteins (reviewed in 128 and 129, respectively). We therefore sought to take this approach with CH2ANG. Each of the DNAs encoding the antibody heavy chain-ANG fusion or the light chain was cloned between exons 2 and 7 of the goat β-casein gene, as described for other proteins.[132] Mouse embryos were coinjected and reimplanted with the two transgenes that had been purified free of prokaryotic DNA. Transgenic mice were identified by Southern blot analysis, and founder mice were bred to produce lactating transgenic animals. Milk was collected, diluted with PBS and fractionated on a TSK-3000 HPLC (high performance liquid chromatography) column. A peak of approximately the correct size containing CH2ANG was identified. This material was toxic to human tumor cells that overexpressed the human transferrin receptor while a similar preparation from control, nontransgenic mouse milk exhibited no toxicity.[137]

3.3.2 Expression of Plasminogen Activator Fusions in Cell Culture or Insect Cells

The DNA encoding the r-scuPA(32 kDa)-59D8 heavy chain enzyme fusion (Fig. 3.6A) was electroporated into heavy chain loss variant hybridoma cell lines that expressed the light chain of the same antibody, and drug resistant cells were selected.[99] Recombinant protein was purified from the cell supernatant. Initially, levels of r-scuPA(32 kDa)-59D8[99] (as well as the first recombinant fusion protein, r59D8-tPA[97]) that were secreted into the cell culture supernatant were very low—only about 1% of the amount secreted by the original hybridoma cell line (0.008-0.06 μg/mL recombinant protein compared to 8–10 μg/mL parental antibody). Northern blot analysis showed low steady-state mRNA levels that were due to mRNA instability rather than a reduced transcription rate.[99,138] Replacing the 3′ UT region of t-PA with either 3′UT of human β-globin gene or mouse γ2b heavy chain gene resulted in a 68–100-fold increase in protein expression.[138] Production was enhanced further by growing cells to high density in the extrafiber space of CellMax (Type B) bioreactor containing cellulose acetate hollow fibers.[99] Using this system, multiple liters of supernatant could easily be processed to generate milligram quantities of protein (50–300 mg/L).

MA-15C5Hu/scu-PA-32k (Fig. 3.6C) was expressed in CHO cells by co-transfecting the DNA encoding the fusion protein with the DNA encoding the light chain of the same antibody.[103] Stable transfectants were selected for dihydrofolate reductase expression and G418 resistance and the best producing cell line was used for production in roller bottles. The final yield of purified protein was about 1 mg/L of conditioned media.

Purification was achieved by sequential two-step affinity chromatography essentially as described.[139] The first step selected for functional antibody. Columns for selecting functional antibody consisted of either the synthetic heptapeptide, GHRPLDK, used to generate the 59D8 antibody, or the fragment D-dimer of fibrin, used to generate the MA-15C5 antibody, linked to Sepharose. The second purification step used immobilized plasminogen activator inhibitors such as benzamidine[99] or a metal resin such as Zn^{2+}-chelate Sepharose[103] to select for the enzyme portion of the molecule. Fusion protein preparations that contained both plasmin and thrombin cleavage sites in the enzymatic domain (see Fig. 3.6) contained some fusion protein with an intact enzymatic domain (desired product) and some fusion protein containing plasmin- or thrombin-cleaved enzyme. The cleaved enzyme forms were generated by proteolysis from contaminating enzymes during expression and purification. Incorporation of the plasmin inhibitor, aprotinin, did not prevent the spurious cleavage. Consequently, chromatography using reagents such as benzamidine[99] or antibody affinity columns that recognized only plasmin cleaved enzyme[103] were used to remove that molecular species.

The sFv-plasminogen activator fusion protein ($K_{12}G_0S_{32}$) (Fig. 3.6D) was secreted by baculovirus-infected *Spodoptera frugiperda* (Sf9) insect cells.[106] The recombinant protein was purified by ion-exchange chromatography on SP-Sephadex followed by gel filtration on Sephadex G100. Final yield was about 1.2 mg/L of conditioned media. Analysis of this recombinant protein revealed that the fusion protein was a single 57 kDa species containing the desired uncleaved plasminogen activator domain. Isolation of a homogeneous species was due to the removal of the thrombin cleavage site. When this site was restored for direct comparison of biological activities of the two fusion proteins, only 40% of the final purified fusion protein constituted uncleaved enzyme.[125]

3.3.3 Expression of Fusion Enzymes for Prodrug Activation in Cultured Cells and Bacteria

The anti-CEA human β-glucuronidase was expressed in BHK cultured cells.[43] The fusion gene consisting of the humanized V_H gene connected to the enzyme was cotransfected into BHK cells with a second vector containing the humanized light chain gene and positive selection was accomplished with methotrexate. Supernatants containing the fusion protein were purified by anti-idiotype affinity chromatography. Sufficient highly purified material was obtained for extensive in vitro and in vivo characterization.

Other fusion enzymes for activation of prodrugs have been expressed in bacteria as soluble proteins. L6-sFv-BCβL (Fig. 3.7B) was secreted directly into *E. coli* XL1-Blue-A3 culture supernatants (40%) and periplasmic space (60%) after IPTG induction (200 μM).[108] Both the supernatant and cell sonicates were subjected to immunoaffinity purification using the anti-idiotypic antibody, 13B. L6-sFv-BCβL was determined to be approximately 95% pure and the yield of purified protein was 2.5 mg/L of starting culture volume. In a variation on this protocol, L49-sFv-bL was expressed in *E. coli* strain BL21 (DE3).[110] Notably in this system, significant levels of toxicity were observed when the IPTG concentration exceeded 90 μM, resulting in inhibition of bacterial growth and eventual outgrowth of cell populations that did not express fusion protein. Optimal conditions were determined to be 50 μM IPTG and a temperature of 23°C. In shake flask cultures, 80% of the active fusion protein was found in the periplasm of bacterial cells with the remainder present in the culture supernatant. Under these conditions, the maximal yield of fusion protein was obtained by detergent mediated release with Nonidet-P-40.

The fusion protein was purified using a two-stage affinity chromatography method. The periplasmic preparations were first applied to an affinity column composed of a soluble form of the antigen followed by binding the material to immobilized phenylboronic acid, a resin that binds to the active sites of β-lactamase.

The dsFv3-β-lactamase fusion protein targeting the p185^{HER2} product of the HER2 proto-oncogene was expressed in a soluble form in *E. coli* using a dicistronic vector that contained the heat-stable enterotoxin II (*stII*) signal sequence to direct secretion to the periplasmic space.[45] Following fermentation, the fusion protein was extracted from the paste and purified using two affinity chromatography steps: immobilized phenylboronic acid to select for the active enzyme followed by protein A.

3.3.4 Summary

Immunoenzymes have been successfully expressed in a variety of systems. With the exception of RNase fusion proteins, all of the fusion proteins can be expressed and secreted from bacteria and mammalian or insect cells without refolding. In two independent studies[21,28] attempts to secrete soluble RNase sFv fusion proteins in bacteria resulted in very little protein in the periplasmic fraction or media; consequently they have been expressed as insoluble proteins in inclusion bodies and refolded before purification. Although the first recombinant RNase fusion (CH2ANG) was expressed as soluble protein in myeloma cells, levels were too low to enable purification. Interestingly, very low level expression of plasminogen activator fusion proteins was found to correspond to the nature of the 3′ UT. Since the CH2ANG constructs did not contain any 3′ sequences, it is possible that inclusion of such sequences might significantly increase expression levels.

3.4 CHARACTERIZATION OF BINDING AND ENZYMATIC ACTIVITY

This section addresses the in vitro activity of the targeting and effector domains of the hybrid fusion proteins described in this chapter. The targeting domain in most of these fusions comprises various forms of antibody fragments that recognize antigens best suited for the action of the particular enzyme to which they are attached. Properties of the ideal antigens vary for each application. For instance, for targeting nucleases the antigens should be located on the cell surface and should efficiently and rapidly internalize, while for prodrug therapy the cell surface antigens should not be internalized (Fig. 3.1). For both of these anticancer strategies the ideal antigen is specific only to the tumor cell. Thus far, the best antigens discovered for tumor targeting are not ideal—they are merely amplified on tumor cells relative to normal tissue, and success depends on achieving a large therapeutic window between targeting normal and tumor tissue. Two different avenues to widen the therapeutic window for cancer therapy are presented in this chapter: (1) the use of enzymes that express their toxicity mainly inside the cell to which they are targeted by the ligand, and (2) saturating the surface of tumor cells with enzymes to activate prodrugs only at tumor sites.

3.4.1 Antigens and Enzymes for Targeted Nucleases

Human RNase sFvs were targeted to the transferrin receptor with a chimeric sFv derived from the E6 parental antibody.[89] The transferrin receptor is an integral membrane glycoprotein that binds and internalizes transferrin–iron complexes into cells. It is widely expressed on both tumor and normal cells, but the number of transferrin receptors is coupled to growth rate.[140] Most tumor cells proliferate more rapidly than normal cells and express more transferrin receptors, thus this antigen was shown to have potential for tumor targeting.[141,142] Effector proteins linked to transferrin are rapidly internalized by receptor-mediated endocytosis, an important criteria for generation of potent selective cell killing agents.[143] A complication of targeting the transferrin receptor is that antitransferrin antibody drug conjugates cross the blood-brain barrier[144] and transferrin receptors are present on the luminal side of capillary endothelial cells.[145] Thus systemic administration of an antibody linked to a toxin could target brain endothelium unless the endothelial cells were not sensitive to killing by the effector protein. Even though systemic administration is precluded, interstitial infusion of transferrin conjugated to a modified diphtheria toxin (CRM107) in patients with brain tumors was shown to be possible without serious toxicities, presumably because the conjugate did not interact with transferrin receptors inside capillaries.[146] This indicates that transferrin receptor-based therapies may be feasible in certain compartments such as the brain depending on the mode of administration. (Exploitation of the transferrin receptor to deliver drugs to the brain is described in Chapter 2). Other human RNase fusion proteins have been constructed with growth factors or cytokines such as EGF, FGF or IL2 as the targeting agent. These fusion proteins are fully humanized since both domains are built from human proteins. Possible drawbacks to using these types of specific cell binding ligands, in addition to receptor expression on normal tissues, are competitive inhibition by endogenous circulating ligand and the expression of natural biological activity of the ligand at concentrations too low to kill the cell but suitable for receptor binding and activation. Bovine seminal RNase was targeted against the tumor associated human hPLAP that is an isoform of placental alkaline phosphatase. The sFv[92] was constructed from the H17E2 antibody[147] that is present on solid carcinomas such as ovarian and testicular as well as on some bladder and head and neck cancers.[148] It was found to localize to human xenografts in a murine model of human cancer more rapidly than the IgG form.[22]

The choice of enzymes (discussed in this chapter) for direct tumor targeting has been centered on members of the pancreatic ribonuclease family for the reasons expressed in the overview (see Section 3.1). The preferred RNases are human enzymes because, like humanized antibodies, they would most likely be less immunogenic than enzymes from other species. Yet, as previously discussed, there are RNases from nonmammalian species that express inherent cytotoxicity, albeit less virulently than plant and bacterial toxins. One of these, the frog protein Onconase, can be administered to humans as a single agent.[12] Because Onconase shares significant homology with the human enzymes (Fig. 3.2), it may be possible to humanize it by substituting various domains with those from human enzymes, particularly as the antigenic determinents on Onconase are identified. Studies have already shown the feasibility of this approach.[149,150] Also, the use of other nucleases or other human enzymes that express cytotoxicity intracellularly can be envisioned. For example, work is in progress to construct an sFv fusion with genes encoding the antibody directed against hPLAP and bovine pancreatic DNase I.[151] This enzyme cleaves double-stranded DNA and is present in cells known to be particularly sensitive to apoptosis thereby causing speculation that it may induce this pathway when targeted to tumor cells.[151] Targeting is expected to require that nuclear targeting signals be incorporated into the fusion protein. Whether the use of an intracellular enzyme would generate antiself antibodies is not known.

Three different human RNases have been fused to a single chain antibody derived from a chimeric antibody against the human transferrin receptor. The nature of the peptide linkers and spacers between fusion protein domains markedly affected the activity of both the antibody and enzyme domains. Three ANG single-chain fusion proteins were constructed with variations in the type of linker connecting the V_L and V_H chain [EGKSSGSGSESKEF, L1 or $(GGGGS)_3$,L2] as well as with or without a spacer (FB) connecting the ANG and sFv[27] (see Fig. 3.5C and Section 3.2). The ANGFB fusion protein with the L2 linker was 2.3 times more effective than the L1 containing ANGFB fusion protein in competing with the labeled monoclonal IgG1 antibody for binding to the transferrin receptor (Table 3.2). Similar results were observed with the L1[28] and L2[30] EDN sFv fusion proteins. Additionally, the ANG fusion protein without the FB spacer exhibited a 13-fold decrease in binding to the transferrin receptor as well as a decrease in RNase activity relative to ANGFBsFv(L2)

TABLE 3.2 Comparison of Fusion Protein Domains to Independent Antibody or RNases

	Binding[a] EC_{50} (nM)	Protein Synthesis[b] Inhibition IC_{50}(nM)	tRNA[c] Degradation	Reference
ANGFBsFv(L2)	30	45	0.01	27
ANGFBsFv(L1)	70	60	0.01	27
ANGsFV(L2)	400	300	<0.002	27
EDNFBsFv(L1)	70	N.D.	2.2	28
EDNFBsFv(L2)	4	8	110	30
hpRNaseFBsFv(L2)	N.D.	4	210	30
E6IgG	1	N/A	N/A	27
ANG	N/A	9	0.02	27
EDN	N/A	0.2	900	30
hpanc	N/A	8	270	30

[a]Competition between E6IgG and the fusion protein for binding to the human transferrin receptor.
[b]Determined from inhibition of protein synthesis in rabbit reticulocyte lysates.
[c]Specific activity is defined as ΔA_{260nm}/min/nmol protein with tRNA as the substrate.
N.D. = not determined.
N/A = not applicable.

(Table 3.2). For this reason, the FB spacer was used to construct fusion proteins with human EDN and pancreatic RNase (hpRNase[30]). In all of the RNase sFvs constructed to date, binding of the sFv fusion was less than that of the parental IgG. This was not surprising since the valency of the two differed (univalent vs. bivalent). However, in some of the fusions the activity of the RNase was also impaired. Protein synthesis inhibition in the rabbit reticulocyte lysate and tRNA degradation activities of the best ANG fusion protein were five and twofold less than free enzyme (Table 3.2). Moreover, EDNFBsFv(L2) exhibited 40- and 9-fold less enzymatic activity in the two assays relative to the free enzyme, but those activities were superior to the EDN fusion bearing the L1 linker in which the ability to degrade tRNA decreased about 400 times. Interestingly, only the hpRNase sFv retained most of its specific activity and was actually twofold more potent in the reticulocyte lysate assay than the unfused enzyme.

Antigen binding of monomeric H17-BSRNase approached that of the free sFv (less than twofold impaired).[22] Since this molecule can form dimers due to the dimeric nature of BSRNase, antigen binding of a dimeric fraction was compared to that of the parental H17E2 IgG. Affinity of the dimeric RNase fusion protein was decreased 25-fold compared to the native antibody due to the presence of aggregated material. The specific RNA-degrading activity of the fusion proteins was only 10–20% that of the native enzyme. Taken together, these results indicate that refolding is markedly influenced by RNase structure in addition to the overall architecture of the fusion protein.

3.4.2 Antigens and Enzymes for Targeted Thrombolysis

The antibodies used to target plasminogen activators for thrombolysis were selected for specificity against components of a blood clot in order not to react with soluble serum proteins or antigens on endothelial cells. Also, similar to the problem of antigen loss due to shedding or heterogeneity of expression that is problematic in tumor targeting, the antigen selected for thrombolysis must persist while the clot dissolves or at least decrease at the same rate as clot dissolution. In fact, this was shown for the antigen recognized by the 59D8 antibody.[152] Both antibodies used to target plasminogen activators were developed against the insoluble fibrin component of the thrombus and thus do not cross react with circulating fibrinogen (Fig. 3.3).[95,96]

Plasminogen activators are serine proteases comprised of both catalytic and functional domains, e.g., growth factor and kringle domains, that can act independently of each other (32, 33 and references therein). t-PA (70 kDa) used in the first recombinant plasminogen activator fusion protein[97] contains a fibrin-binding site in the amino terminus that increases the catalytic efficiency of conversion of plasminogen to plasmin most likely by increasing the local concentration of t-PA at the clot. Although scu-PA (54 kDa) exhibits fibrin selectivity, the molecular mechanisms differ since scu-PA does not contain a fibrin-binding site. Box 3.2 illustrates the shortened form (32 kDa) of scu-PA lacking the N-terminus[98] that was used in most of the fusion proteins for thrombolysis. The reasoning was that substitution of the amino terminal regions in either enzyme with fibrin-specific antibodies would enhance that natural function. The other domains constitute an EGF-like growth factor domain and "kringle" domains. In t-PA the amino terminal region has also been shown to modulate clearance rates by binding to both hepatocytes and endothelial cells.[153] The plasma half-life of t-PA is decreased both by hepatic clearance and circulating protease inhibitors which bind to and inactivate the catalytic domain, whereas scu-PA is not inhibited by plasma protease inhibitors—a decided advantage.

Scu-PA is a single chain enzyme (schematized in Box 3.2) and can be converted to a two-chain form by cleavage with plasmin or thrombin at the sites shown. The two-chain forms of these enzymes are connected by a disulfide bond. Detailed discussion of the properties of the two-chain vs. single chain enzymes is beyond the scope of this chapter (for further discussion see 32, 33). However, the cleavage sites are of particular relevance to the design, expression, and purification of recombinant plasminogen activator fusion proteins. As discussed in the section on expression and purification, the proteolytic cleavage sites complicate purification because of spurious proteolysis that occurs at the thrombin and plasmin cleavage sites. Scu-PA contains a thrombin-sensitive site at R156-F157 in addition to the K158-I159 plasmin cleavage site. Thrombin cleavage of unmodified scu-PA inactivates the catalytic domain. However, as hypothesized by Yang et al.[100] and references therein, there may be a physiological advantage to thrombin-mediated cleavage of a scu-PA fusion protein because it is thought that there is more thrombin than plasmin in newly forming blood clots. Newly formed blood clots are the desired targets of thrombolytic therapy in patients with acute myocardial infarction because lysis of established clots may contribute to excessive bleeding and strokes. Therefore, scu-PA was genetically modified by deleting F157 and K158 so that thrombin would not inactivate the catalytic domain. The modified enzyme, now sensitive to productive thrombin cleavage, was used to construct the 59D8-scu-PA-T recombinant fusion protein (Fig. 3.6B).[100]

The activities of both domains of the recombinant fusion proteins were tested for fibrin binding and catalytic activities. The binding activities of r-scuPA(32kDa)-59D8 (Fig. 3.6A)[99] as well as MA-15C5Hu/scu-PA-32k (Fig. 3.6C)[103] were comparable to those of the native antibodies. Not surprisingly, due to its univalency, the sFv hybrid protein ($K_{12}G_0S_{32}$) (Fig. 3.6D) displayed reduced binding activity compared to the native antibody, MA-15C5 (Ka, $5.5 \times 10^9\,M^{-1}$ vs. $2 \times 10^{10}\,M^{-1}$, respectively[106]).

To determine catalytic activity of the enzymatic domain, the antibody plasminogen activator fusion proteins and native enzymes were incubated with plasminogen. The amount of plasmin generated was determined by monitoring the liberation of paranitroaniline from S-2251.[106] The Km's for activation of plasminogen by r-scuPA(32 kDa)-59D8 (Km 16.6 μM) or MA-15C5Hu/scu-PA-32k (Km 6 μM) vs. unfused enzyme (Km 9–12 μM) did not differ significantly from each other.[99,103] The fusion protein that was productively cleaved by thrombin (59D8-scuPA-T) (Fig. 3.6B) exhibited a higher Km (66 μM) than the fusion protein that retained the plasmin cleavage site (59D8-scuPA, 22 μM) (not shown, 100). The authors speculated that selective cleavage by thrombin might have sterically interfered with the catalytic site in the modified scu-PA.[100] In contrast to the results previously described, the catalytic domain of the low M_r form of scu-PA in $K_{12}G_0S_{32}$ was more efficient compared to the unfused enzyme (2.9 vs. 12 μM).[106]

3.4.3 Antigens and Enzymes for Targeted Prodrug Therapy.

Carcinoembryonic antigen (CEA) is not a tumor-specific antigen but its concentration has been shown to be increased in some gastrointestinal, gynecological, colon, and solid carcinomas. In addition to being widely expressed on normal tissues, the antigen is shed into the serum, albeit at low antigen to tumor ratios. Despite these limitations, anti-CEA antibodies can localize to tumors and have been used for tumor imaging in humans (reviewed in 154).

Three fusion proteins constructed with β-lactamase for prodrug activation also targeted antigens amplified on solid tumors relative to normal tissue. Originally a monoclonal mouse antibody (L6) was raised against a human lung carcinoma,[155] and this MAb was subsequently found to be highly reactive with lung, breast, colon, and ovarian neoplasias. It recognizes a ganglioside antigen and can mediate antibody-dependent cellular cytotoxicity (ADCC), eliciting antitumor effects in mice[156] and some responses in humans when combined with IL-2.[157] Another fusion protein containing the L49 monoclonal antibody[109] was targeted to the p97 melanoma-associated antigen that is structurally and functionally related to transferrin.[158] The third β-lactamase fusion protein described herein localizes to tumors that overexpress a tyrosine kinase encoded by the HER2 protooncogene ($p185^{HER2}$). This cell surface antigen is overexpressed in about 30% of human breast and ovarian cancers and correlates with a poor prognosis.[159,160] Since this subpopulation of patients may be relatively nonresponsive to conventional chemotherapy, targeting the cells that overexpress $p185^{HER2}$ with antibodies carrying toxic enzymes[161,162] or enzymes to activate prodrugs[45,50] may selectively decrease tumor cells overexpressing this antigen. Interestingly, in the aforementioned studies, the same antigen ($p185^{HER2}$) is targeted with two strategies that require opposing properties, i.e., the ideal antigen for direct targeting with immunotoxins is rapidly internalized; conversely prodrug activation requires stable cell surface expression. That $p185^{HER2}$ can be targeted with both toxins and prodrugs may mean that different epitopes of the same antigen and/or antibody isotype may confer properties suitable for varying targeting strategies.

Certain enzyme characteristics are required for prodrug activation. (1) It is crucial that the prodrug only be activated at the tumor. This excludes circulating or tissue accessible human enzymes as well as enzymes from other species that express similar catalytic activities; (2) yet the enzyme should express optimal catalytic activity under human physiological conditions without interference from endogenous inhibitors. (3) The enzymes should be relatively nontoxic and exhibit favorable kinetics with respect to plasma residence. (4) The enzymes should be able to activate a variety of prodrugs to generate active drugs with different mechanisms. (5) The enzymes should be nonimmunogenic since treatment of humans with bacterial or plant-derived enzymes elicit a strong immune response that interferes with subsequent readministration (reviewed in 85).

Human β-glucuronidase fulfills many of the criteria just listed. Fused to a humanized antibody, it is likely to be minimally immunogenic as a recombinant fusion protein for prodrug activation. Furthermore, since β-glucuronidase is a sequestered lysosomal enzyme, the endogenous enzyme may not be accessible to a prodrug that can not enter cells. Intracellular localization of the enzyme is likely to confer the required specificity, although some association of the prodrug with normal tissue would not be unexpected. In this regard, β-glucuronidase was found to be present at high concentrations in necrotic tissue such as exists in tumors. For that reason, the prodrug alone has significant localized antitumor effects and could conceivably be used without the fusion protein.[163] Still, that strategy depends on areas of tumor necrosis that would be present in more advanced cancer. The treatment of minimal or metastatic disease would still require activation by a targeted enzyme (K. Bosslet, pers. communic.).

The other fusion enzymes for prodrug activation presented here employ members of the bacterial β-lactamase family that are essentially host defense proteins in bacteria functioning to inactivate antibiotics that contain a β-lactam ring. As described herein, they are able to activate a panel of prodrugs to generate different cytotoxic agents. They are active and stable under physiological conditions and are catalytically sufficiently different from human enzymes that nonspecific activation of their prodrugs would not be expected to occur. Three enzymes differing in sequence and structure within this family have been used. L6 fused to the mature form of β-lactamase II (BCβL)[164] from *Bacillus cereus* was used to activate a cephalosporin mustard prodrug.[108] The recombinant L49 fusion protein[110] contained a genetically modified form of β-lactamase (r2-1 β-lactamase[165]) encoded by the P99 bL gene from *Enterobacter cloacae.* In that study, mutants generated from a phage display system were selected to determine the effects of kinetics on efficacy. Similar studies[166,167] endeavoring to broaden substrate specificities have also been carried out on the RTEM-1 β-lactamase used to generate doxorubicin from a cephalosporin doxorubicin prodrug[45] or taxol from a cephalosporin taxol prodrug.[168]

Should any of the β-lactamase fusion proteins enter clinical trials, immunogenicity may be a problem. This is surmised from the results of clinical testing of a first generation chemical conjugate comprised of an anti-CEA antibody conjugated to a bacterial enzyme, carboxypeptidase G2.[52,169] However, immunosuppression with cyclosporin delayed the immune response in human patients to the antibody-enzyme conjugate[170] (see Section 3.5 for further discussion).

Independent functional testing of each domain of the fusion protein consisting of the humanized Fab fragment of the anti CEA MAb BW 431 and human β-glucuronidase revealed that the specificity and avidity of the antibody as well as the enzymatic activity (Km, 1.3 mM), pH sensitivity, and stability of the enzyme were similar to the individual component proteins.[43] Thus fusion did not diminish binding or catalytic function. Similar results were noted for the individual domains of the two sFv fusion proteins, L6-sFv-BCβL[108] and L49-sFv-bL,[110] as well as for a humanized disulfide-stabilized anti-p185^{HER2} Fv fusion.[45] The latter three hybrid proteins were designed with β-lactamases as the enzymes for activating prodrugs (Kms, 0.02 mM) and targeted tumor-associated antigens on a variety of solid carcinomas (see Table 3.1). Of these, only the anti-CEA β-glucuronidase fusion protein bound to target antigen as well as parental whole antibody, most likely due to its ability to bind in a bivalent manner. Although the sFv and dsFv fusions bound less well than their bivalent parental antibodies, they retained completely the univalent antigen-binding activity of their respective unfused binding domains. Thus fusion did not diminish the univalent binding of the antibody domain.

3.4.4 Summary

The overall design of RNase sFvs was found to markedly affect both binding and enzymatic activities and this was reflected by the potency of the fusion protein in functional assays in vitro. The best activity was obtained when the V_H and V_L domains were joined by a flexible rather than rigid linker and when the RNase was separated by a spacer peptide from the sFv. No such constraints were reported in other studies that used a variety of designs with regard to spacer peptides and linkers. Moreover, the RNase enzymatic domain did not retain full enzymatic activity compared to the unfused enzyme, unlike the results reported in the other studies. This may reflect suboptimal refolding of the RNase as a fusion protein vs. fusion proteins secreted as soluble refolded proteins and/or the purification methods used. Purification strategies for plasminogen activator chemical conjugates and recombinant fusion proteins were based on functional criteria.[139] Two sequential affinity methods were employed, one using a specific enzyme inhibitor that only bound to catalytic regions, the other selecting for functional antibody. One or two stage affinity chromatographic methods were also employed for purification of fusion proteins to activate prodrugs. Taken together these results suggest that purification of hybrid proteins by methods only recognizing functional domains will preselect for fully active fusion proteins.

Strategies varied with regard to humanization of the fusion proteins. In choosing the enzyme for prodrug activation, the issues of immunogenicity vs. specificity had to be considered. The activation of the prodrug must take place only at the tumor for this type of therapy to be successful. The use of an endogenous human enzyme poses the risk of activation in normal tissues, therefore, bacterial and other nonmammalian sources were used. β-glucuronidase is an intracellular enzyme in normal tissues but appears in the extracellular fluid in tumors due to necrosis, thereby raising the possibility of a therapeutic window for this enzyme. Since the humanized Fab fragment of the anti-CEA MAb BW 431 was used as the antibody domain, this particular hybrid is composed of human and humanized domains. Similarly targeted plasminogen activators and RNases were built from chimeric, humanized, and human building blocks.

3.5 PROGRESS IN ANTIBODY-ENZYME FUSION PROTEIN APPLICATION

The immunoenzyme fusion proteins described in this chapter have been functionally characterized in vitro and in animal models and those results are presented in this section. Overall, the results concur with the hypothesis that targeting enzymes to directly kill cancer cells, or activate plasminogen or prodrugs yields superior therapeutic results with fewer toxic side effects compared to nontargeted enzymes. The translation of these encouraging results to clinical utility awaits future evaluation.

Some insight into possible problems or success of at least one of these immunoenzyme systems may be inferred from the results of two clinical trials of the prodrug strategy using a chemical antibody–enzyme conjugate. The antibody-directed enzyme prodrug therapy (ADEPT) components were administered to patients with advanced colorectal cancer in three steps: (1) a murine monoclonal antibody against CEA conjugated via a thio-ether linkage to carboxypeptidase G2 (CPG2) was administered i.v. followed by (2) a slow i.v. infusion of an antibody against CPG2 to decrease the levels of conjugate not localized in the tumor (clearing antibody). This was followed by (3) the prodrug (4-[(2-chloroethyl) (2-mesyloxyethyl) amino] benzoyl-L-glutamic acid (CMDA).[52] In the first stage of the trial, patients received only the prodrug in escalating doses since it had not been administered to humans previously. The dose-limiting toxicity associated with the prodrug (nausea) appeared to be associated with the vehicle in which it had to be administered, dimethyl sulfoxide (DMSO). Some of the patients proceeded to the second phase in which they received all of the components of ADEPT with escalating doses of the prodrug. The highest prodrug doses were best tolerated when it was infused 2–3 days postantibody–enzyme administration at a constant rate in the continued presence of the clearing antibody. When the immunosuppressant cyclosporin was added to the protocol, some patients were able to receive up to three courses of treatment. Partial or mixed responses to the therapy were reported in some of the patients. This trial demonstrated the feasibility of the prodrug approach in that the relatively nontoxic prodrug could be well tolerated under certain routes of administration in the constant presence of the antibody that inactivated CPG2 suggesting conversion of the drug took place mainly in the tumor and not in the circulation. The major limitations centered around the immunogenicity of the conjugate and the chemistry and pharmacokinetics of the activated drug. It was not very reactive and the biological half-life was too long (about 20 minutes). The immune response limited repetitive treatment and possibly interfered with activation of the prodrug since anti-CPG2 antibodies in the serum of one patient inhibited CPG2 activity in vitro.[54]

Cyclosporin delayed but did not prevent the appearance of the immune response to the murine antibody-bacterial enzyme conjugate while increasing the nausea associated with DMSO.[170] The pharmacokinetics and plasma levels of the same prodrug/drug system in a second Phase I clinical trial have recently been reported.[55] Improved methods for extracting these agents from plasma and for measuring their activity were used to obtain the data. For that reason, levels of the conjugate could be measured in biopsy samples and localization to tumor firmly established since conjugate was not found in normal tissues. Although the active drug was generated only in the tumor, active drug was present in the blood due to leakage from tumor sites. Clinical results are to be reported separately.

3.5.1 In vitro and in vivo Functional Assays: Targeted RNases.

In vitro cytotoxicity assays were performed by plating cells into 96 well microtiter plates in complete growth medium.[23] Sample or control additions were added and the plates incubated at 37°C for varying times. Protein synthesis was measured by adding [^{14}C]leucine for 1 to 4 h followed by harvesting the cells onto glass fiber filters using a PHD cell harvester. Cytotoxicity was also assessed using the MTT (3-[4,5-dimethylthiazol-2-yl]2,5-diphenyltetrazolium bromide; thiazolyl blue) assay to measure cell viability.[171] Results of these assays demonstrated that the CH2ANG and ANG sFv fusions were 4 and 2 logs, respectively, more cytotoxic to tumor cells overexpressing the human transferrin receptor than a disulfide linked chemical conjugate comprised of transferrin or an antibody against the transferrin receptor and bovine RNase A (reviewed in 88). Fusion proteins are homogeneous molecules since individual domains are linked at a unique site and the activity of the antibody or the enzyme is not reduced by the chemical coupling procedure. That most likely explains why human RNase sFv fusion proteins are more potent in vitro than their corresponding chemical conjugates.[88] Although RNase chemical conjugates exhibited low potency in vitro (0.2 μM vs. 0.0002 μM of the same antibody conjugated to ricin A-chain), they may be useful because comparable doses achieved in vivo have been shown to extend survival in animals.[24]

RNase fusion proteins known to date are listed in Tables 3.3 and 3.4. As described in Section 3.4 and shown in Table 3.2, the nature of the RNase and sFv architecture markedly affected the function of the fusion protein domains.

TABLE 3.3 Toxicity of RNase sFvs to Human Carcinoma Cell Lines in vitro

RNasesFv	EDNFBsFv(L2)	ANGFBsFv(L2)	hpRNaseFBsFv(L2)	ANGFBsFv(L1)	ANGsFv(L2)
Cell Line			IC_{50}(nM)		
MDA-MB-231^{mdr1} (multidrug resistant breast carcinoma	3	10	10	30	40
HT-29^{mdr1} (multidrug resistant colon carcinoma)	5	7	N.D.	>100	>100
HS578T (breast carcinoma)	16	>100	90	>100	>100
ACHN (renal carcinoma)	1	4	5	20	>100
SF539 (CNS carcinoma)	17	17	N.D.	>100	>100
Malme (melanoma)	8	>100	8	>100	>100

Source: Refs. 27 and 30.

N.D. = not determined.

> = highest concentration tested.

TABLE 3.4 In vitro Cytotoxicity of RNase Fusion Proteins on Various Cell Lines

Fusion Protein	Cell Line	IC_{50}(nM)	Reference
H17-BSRNase	H.Ep-2	35	22
	KB (epidermoid carcinoma)	4	
CH2ANG	K562 (erythroleukemia)	0.05	29
hpRNase-hIL2[a]	MJ (T-cell)	30	
hpRNasel-hEGF	A431 (squamous carcinoma)	250	31
bFGF-hRNase1[b]	(mesangial)	5000	
bFGF-hRNase1[b]	B16/BL6 (murine melanoma)	>5000	
bFGF-des.1-7hRNase1[b,c]	B16/BL6	3000	

[a]K. Psarras, pers. communic.
[b]M. Seno, pers. communic.
[c]bFGF-des.1-7hRNase1, human FGF fused to human pancreatic RNase from which the first seven amino acids were deleted.

The hpRNase and EDN fusions constructed with the flexible L2 peptide linking V_H and V_L and an FB spacer peptide separating the antibody and enzyme maintained the best functional activity of each separate domain. As expected, these fusion derivatives also displayed the best in vitro specific cytotoxicity (Table 3.3). Cytotoxicity also varied when the same sFv fusion architecture was used but the human RNase changed. EDNFBsFv(L2) was more potent than ANGFBsFv(L2) in 5/6 human tumor cell lines (IC_{50}s, 1–16 nM, Table 3.3) and equivalent to the ANG fusion on SF539 CNS carcinoma cells (IC_{50}, 17 nM). EDNFBsFv(L2) was more potent than hpRNaseFBsFv(L2) on 3/4 tumor cell lines and equivalent to that fusion protein on Malme melanoma cells (IC_{50}, 8 nM).

The most potent RNase fusion protein CH2ANG,[29] killed human leukemia cells at an IC_{50} of 1–5 ng/ml (0.05 nM) as estimated by ELISA using antibodies against ANG and IgG[29] (Table 3.4). ANG was specifically targeted to the tumor cells because cytotoxicity to K562 cells was completely blocked by an excess of the E6 parental antibody that competed for binding to the human transferrin receptor. Also, a nonhuman cell line lacking the human transferrin receptor was not affected. The potency of this fusion most likely reflects the bivalent structure of the protein, which could have increased the avidity of binding. Additionally, ANG may have dimerized, which could have enhanced translocation. This reasoning is based on studies performed in the 1960s and 1970s that demonstrated that dimeric bovine pancreatic RNase A was cytotoxic to cells, a property not shared by the monomeric enzyme (reviewed in 172). Furthermore, the natural dimer of bovine seminal RNase is cytotoxic to cells while monomeric forms are not (reviewed in 91). In that regard, the fusion protein containing that enzyme, H17-BSRNase, killed tumor cells that expressed the hPLAP antigen with IC_{50}s in the nM range[22] (Table 3.4). Although designed to be a dimeric fusion protein, most of the correctly folded material was monomeric, most likely due to the complexity of refolding and reconfiguring the dimeric fusion. Specificity was demonstrated since the IC_{50} of the fusion on antigen negative cells was $>10^5$ nM. Variations in design of this fusion protein were made to potentiate the cytotoxicity of the molecule.[22] A diphtheria toxin (DT) disulfide loop was inserted between the sFv and enzyme to increase translocation[173] with and without a KDEL sequence known to improve the cytotoxicity of a ricin based immunotoxin.[174] The addition of KDEL to the C-terminus of BSRNase improved the cytotoxicity fourfold while that of a fusion protein containing the DT disulfide loop and KDEL was improved about 10-fold.

Human pancreatic RNase fused to human EGF[31] or IL2 (K. Psarras, pers. communic.) have also been constructed (Table 3.4). Interestingly, the cytotoxicity of the IL2 fusion occurs over the same molar concentration range as human RNase sFv while that of the EGF fusion requires 10-fold higher amounts of fusion protein. The lower activity of the EGF fusion was thought to be related in part to a substantial amount of incorrectly folded, inactive material in the preparation.[31] An FGF-RNase fusion protein (M. Seno, pers. communic.) manifested the least cytotoxicity. Indeed, B16/BL6 murine melanoma cells were growth inhibited (cytostatic) and not killed (cytotoxic). In spite of the high IC_{50}s, FGF fusion proteins suppressed the proliferation of mesangial cells in vivo since growth inhibitory concentrations were easily achieved due to the lack of toxic effects to normal issue.

Since the concentration of EDNFBsFv(L2) (1–10 nM) required to achieve 50% protein synthesis inhibition in cell culture assays compared favorably to concentrations published for antitransferrin receptor antibody conjugates made with plant toxins such as gelonin (2–5 nM)[175] or modified diphtheria toxin (CRM107, IC_{50} 1 nM),[93] the ability of that RNase fusion protein to prevent or delay formation of tumors in a clinically relevant model for brain cancer was tested. The model used assesses the efficacy of direct intratumoral therapy with targeted protein toxins for solid human gliomas and was developed by Laske et al.[176] U251 log-phase human glioma cells were injected subcutaneously into the right flank of athymic mice with reconstituted basement membrane (matrigel), which markedly enhanced the growth of the cells.[176] Mice that had been innoculated with U251 human glioma cells were randomly assigned to treatment groups. Treatment began 24 h after injection of tumor cells. Three injections of EDNFBsFv(L2) (14 μg/dose) were administered and compared to a control group of animals treated with buffer alone.[177] By day 27, a difference in the rate of growth of the control and EDN fusion protein-treated tumors became significant as the control tumors began to increase steadily in size while the sFv-treated tumors remained essentially the same size (0.5–1 cm). This pattern of growth continued until day 45, the last day of observation and 40 days after the last treatment, at which time the control tumors were four times the size of the human RNase sFv-treated tumors.

Human RNase-based immunoenzymes are expected to be less toxic and less immunogenic in humans. One advantage to using related members of the RNase A family is their homology across species (reviewed in 178). Therefore the action of isologous fusion proteins can be tested in syngeneic models. A fusion protein was constructed with DNA encoding murine ANG (mANG, gift from R. Shapiro, Harvard Medical School) and the E6sFv. Following a protocol that previously demonstrated immunogenicity of a diphtheria toxin-based immunotoxin,[179] mANG sFv was injected into male or female outbred mice, one or four times (10 μg/dose). No significant immune response was generated against the murine ANG portion of the fusion protein.[177]

Although no signs of toxicity in mice were observed after injection of the EDN sFv, the sFv against the human transferrin receptor does not recognize the murine transferrin receptor. Therefore, nonspecific targeting of normal receptor-bearing mouse tissues does not occur, which can lead to an under-estimation of toxic side effects. To assess nonspecific toxicity using a targeting agent that does recognize a mouse receptor, murine ANG was fused to the gene for murine IL2. No toxic side effects were observed with a bolus dose of 500 μg of mIL2-mANG.[177] Thus mice tolerate at least 10 times the dose of IL2 fused to a bacterial enzyme.[180]

Taken together, these preliminary preclinical experiments support our hypothesis that RNase-based immunotoxins can elicit antitumor effects in mice and that they could be less toxic and less immunogenic in humans. Furthermore, they indicate that effective and selective RNase-based fusion proteins can be developed using sFvs as the targeting moieties.

3.5.2 In vitro and in vivo Functional Assays: Targeted Plasminogen Activators

The in vitro activity of the total fusion protein was determined by examining the ability of the fusion protein to lyse ^{125}I-fibrin-labeled human plasma clots immersed in human citrated plasma. With one exception, all the fusion proteins made with scu-PA were more active than the unfused enzyme to lyse blood clots in vitro, reflecting targeting by the antibody. The specificity of targeting was demonstrated by competition with free antigen.[103,106] The fold increase of r-scuPA(32 kDa)-59D8, MA15C5Hu/scu-PA-32K and $K_{12}G_0S_{32}$ over the unfused enzyme was 6, 12, and 13, respectively. Two variations in the design of $K_{12}G_0S_{32}$ were made.[125] One fusion protein contained the kringle domain (Box 3.2) and another was designed with intact scu-PA. Both of these had a similar fibrinolytic activity as $K_{12}G_0S_{32}$ that contained only scu-PA 32 kDa. Similar results were noted for scuPA (32kDa)-59D8 with regard to spacer sequences between the antibody and enzyme.[36] Based on these results it was concluded that the functional activity of the sFv fusion protein was not impaired by spatial constraints due to lack of a spacer between the sFv and catalytic domain.

Only the fusion protein designed with the thrombin cleavage site (59D8-scuPA-T) was less active than the unfused scu-PA or the fusion protein with a plasmin cleavage site (59D8-scuPA) in this human plasma clot assay. After 5 h, 230 nM 59D8-scuPA-T generated approximately 25% clot lysis whereas 34 nM 59D8-scuPA generated 25% clot lysis in less than 1 h.[100] This was most likely due to the combination of decreased catalytic activity of the enzymatic domain (see Section 3.4) as well as the nature of the assay, which was configured to assess the lysis of blood clots that contained more plasmin than thrombin. When assayed on clots that contained more thrombin, 59D8-scuPA-T but not 59D8-scuPA or scuPA was effective.

Pharmacokinetic and thrombolytic properties of chimeric plasminogen activator fusion proteins were determined in three different animal species: hamsters, rabbits, and baboons (Table 3.5). Results from these studies were expressed as the fold increase in half-life or decrease in clearance rate vs. the unfused enzyme. Only the sFv fusion protein, $K_{12}G_0S_{32}$, with a molecular weight similar to intact scu-PA (57 versus 54 kDa) exhibited a similar $t_{1/2}$ (α) as well as a similar plasma clearance rate to the unfused enzyme in hamsters.[181] Deglycosylation of the sFv ($K_{12}G_2S_{32}$) increased the $t_{1/2}(\alpha)$ and reduced plasma clearance rates.[181] Not surprisingly, Ma-15C5Hu/scu-PA-32k (215 kDa) was cleared more slowly in rabbit, hamster, and baboon[182] and both the $F(ab')_2$ fusion protein and r-scuPA(32 kDa)-59D8 (103 kDa) had a $t_{1/2}(\alpha)$

TABLE 3.5 Pharmacokinetic and Thrombolytic Properties of Targeted Plasminogen Activators Relative to Unfused Enzyme

Fusion Proteins (animal model)	M_r (kDa)	$t_{1/2}(\alpha)$	CL_p	Thrombolytic Potency	Reference
r-scuPA (32kDa)-59D8 (rabbit)	103	5	N.D.	20	99
MA-15C5HU/scu-PA-32K	215				
(Rabbit)		5	30	11	182
(Hamster)		6	10	23	182
(Baboon)		4	10	5	182
$K_{12}G_0S_{32}$ (Hamster)	57	1	1	6	181
$K_{12}G_2S_{32}$ (Hamster)		2	4	11	181

Notes: Fold increase in $t_{1/2}(\alpha)$, decrease in CLp (clearance from plasma) or increase in the ability to lyse a ^{125}I-fibrin labeled clot relative to scu-PA.

N.D. = not determined.

4–6-fold longer than scu-PA.[99,182] Thrombolytic potency was determined in the same three animal models by measuring lysis of [^{125}I]fibrin-labeled clots. In the hamster pulmonary embolism model,[183] a [^{125}I]fibrin-labeled human plasmin clot was produced in vitro and injected into the jugular vein of heparinized outbred hamsters.[182] Thrombolytic agents were infused intravenously over 60 min and lysis was measured 30 min after the termination of the infusion by measuring the difference in radioactivity initially incorporated into the clot and the residual radioactivity found in the lungs and heart. In the rabbit jugular vein thrombosis model[184] and the baboon femoral vein thrombosis model,[185] the [^{125}I]fibrin-labeled clot was produced in the vein using $CaCl_2$, thrombin, and either human plasma (rabbit model) or baboon blood (baboon model) containing traces of [^{125}I]-labeled fibrinogen.[182] The clots were aged for 30 min before the i.v. administration of the thrombolytic agents. As shown in Table 3.5, all fusion proteins were more effective thrombolytic agents than scu-PA. With the exception of $K_{12}G_0S_{32}$, the increase in potency also correlated with the increase in the $t_{1/2}(\alpha)$ and decrease in the plasma clearance rate. The influences of antibody targeting as well as prolongation of plasma levels on thrombolytic potency is exemplified by the glycosylated sFv fusion protein and its deglycosylated analog $K_{12}G_2S_{32}$. Since there is no difference in half-life or plasma clearance rates between the sFv fusion with the low M_r enzyme or scu-PA, the sixfold increase in thrombolytic potency must be caused by targeting the fusion protein to the clot. The same binding domain with a deglycosylated enzyme persists for longer times in the circulation and exhibits an 11-fold increase in thrombolytic activity. Thus both targeting and clearance rates that are decreased by size or other factors contribute to thrombolytic activity in these models. Another baboon model developed to more accurately reflect thrombosis, thrombolysis, and hemostasis in humans was developed.[186] In this model, a labeled clot was preformed in vivo on a dacron graft incorporated into a chronic AV shunt and maintained with infusions of heparin to prevent occlusion. The fusion protein used, AFA-scuPA (91 kDa), was similar to r-scuPA(32 kDa)-59D8 except that the entire Fc portion was deleted. It was directly compared to scuPA or t-PA. AFA-scuPA lysed clots faster and more completely than either of the unfused plasminogen activators without causing bleeding, whereas comparable doses of either scuPA or t-PA caused unacceptable bleeding. Therefore, antibody targeting of scu-PA to fibrin results in increased thrombolytic potencies with less impairment of hemostasis compared to current therapeutic agents.

3.5.3 In vitro and in vivo Functional Assays: Targeted Enzymes for Prodrug Activation.

The in vitro cytotoxicity of prodrugs activated by enzyme fusions was determined using direct cytotoxicity assays essentially as described for targeted RNases. Tumor cells were plated into 96 well tissue culture plates and treated with enzyme or enzyme fusion for 0.5–2 h, washed, and treated with test medium containing the drug or prodrug for a similar length of time. After washing the wells, fresh growth medium was returned to the cells and the plates were incubated overnight[108,110] or for 3 days[45] before terminating the experiment and measuring DNA synthesis[108,110] or cell number.[45] In the studies where it was reported, the prodrug was less cytotoxic than the free drug on cultured cells. Cytotoxicity was completely restored by prior treatment of the cells with the enzyme fusion protein (Table 3.6). The studies were also designed to assess various parameters of specificity. β-lactamase fusion proteins that targeted lung[108] or melanoma cells[110] activated the prodrug to the same degree as the corresponding chemically produced Fab′ conjugates. The activation was immunologically specific since the L49-sFv-bL fusion protein did not activate the prodrug on cells that were saturated with unconjugated antibody before being exposed to the fusion protein.[110] In another study, the prodrug was more toxic when activated by dsFv3-β-lactamase on SK-BR-3 breast cancer cells that expressed 130 times the receptor protein of MCF7 breast cancer cells, demonstrating that the prodrug was specifically targeted to cells overexpressing $p185^{HER2}$ (Table 3.6). Interestingly, cells treated with enzyme alone also activated the prodrug, most likely reflecting a small degree of nonspecific protein binding exacerbated by the high catalytic efficiency of the enzymes.[45] The same enzyme fusion protein was used to generate taxol from PROTAX (Table 3.6, Ref. 168). Although taxol generated from the prodrug exhibited toxicity that approached the parent drug, a prolonged incubation was required.[168]

Pharmacokinetic parameters and therapeutic efficacy have been determined for two of the fusion protein–prodrug systems presented in this chapter. In the presence of the anti-CEA-β-glucuronidase fusion protein,[43] a doxorubicin prodrug elicited impressive therapeutic effects in nude mice bearing CEA expressing LoVo colon carcinoma xenografts.[44] Treatment was initiated by injecting the prodrug (250 mg/kg, $<16\%$ of the maximum tolerated dose, MTD) 7 days after administration of the fusion protein (20 mg/kg) when the blood:tumor ratio of the fusion protein was greater than 100. Therapeutic efficacy was assessed by monitoring tumor growth. The results were reported as %T/C (percentage of treated tumor size versus control on day 24 of the experiment). Significant antitumor effects with one course of treatment were noted only in animals treated with prodrug and fusion protein. Prodrug alone (68% T/C) or doxorubicin (61% T/C) had no significant antitumoral effects against this tumor compared to the group receiving prodrug and fusion protein (26% T/C).

The therapeutic effects of L49-sFv-bL combined with the prodrug cephalosporin nitrogen mustard (CCM) in two courses of treatment are shown in Table 3.7.[110] Variations in experimental design included two different doses of the fusion protein (1 and 4 mg/kg), different times of addition of the prodrug after administration of the fusion protein (0.5–2 days) and different prodrug doses (75–275 mg/kg, 25–92% MTD). These different schedules are shown in Table 3.7. Significant numbers of cures were obtained (100% in some groups). This

TABLE 3.6 Activation of Prodrug by Antibody-Enzyme Fusions in vitro

Fusion Protein	Cells	Prodrug	Drug	Prodrug and Fusion Protein	Prodrug and Enzyme	Reference
				IC_{50}(nM)		
L6-sFv-BCβL	H2981 (lung carcinoma)	>10,000	N.D.	5	100	108
L6-Fab′-BCβL	H2981	>10,000	N.D.	2.5	100	108
L49-sFv-bL	3677 (melanoma)	16,000	300	300	N.D.	110
L49-Fab′-bL	3677	>16,000	300	300	N.D.	110
dsFv3-β-Lactamase	SK-BR-3 (breast carcinoma)	>10,000	1,000[a]	1,000	6,000	45
dsFv3-β-Lactamase	MCF7 (breast carcinoma)	10,000	1,000[a]	10,000	10,000	45
dsFv3-β-Lactamase	SK-BR-3	1,000	200[b]	200	N.D.	168

Note: N.D. = not determined.

[a]Drug, doxorubicin.

[b]Drug, taxol.

TABLE 3.7 Therapeutic Effects of L49-sFv-bL in Nude Mice with Melanoma Xenografts

Fusion Protein Dose (mg/kg)	1					4			
Prodrug Administration (D)[a]	.5			1		1		2	
Tumor Blood Ratio	66			105		141		150	
Prodrug Dose (mg/kg)	75	125	175	150	225	75	110	150	275
(%) Mean Tolerated Dose	25	42	56	50	75	25	37	50	92
Therapeutic Efficacy (% cures)	66	100	100	50	100	50	80	0	50

Source: Ref. 110.

[a]D, days after fusion protein administration.

experiment illustrates some of the complexities of the prodrug approach in vivo. In general, the more fusion protein administered, the longer the clearance time and the time to obtain optimal tumor:blood ratios. Since the amount of active fusion protein in the tumor was found to decrease with time,[110] prodrug conversion would be expected to be less. This is illustrated by comparing the effect of prodrug administration 12 h after the addition of 1 mg/kg of fusion protein (66% cures) to that of administering the same dose of prodrug (75 mg/kg) 24 h after the addition of 4 mg/kg fusion protein (50% cures). Thus more cures were obtained when the interval between fusion protein and prodrug was short even though the blood:tumor ratio was more favorable with the administration of fourfold more fusion enzyme (141 vs. 66, respectively). For that reason, even a very high tumor to blood ratio[150] achieved two days after administration of 4 mg/kg of fusion protein yielded 0 cures compared to 50% cures when a lower dose of fusion protein (1 mg/kg) was used and the same prodrug dose was added 1 day after the sFv-fusion.

3.5.4 Summary

The ultimate goal of devising targeted nucleases is to develop more potent and specific drugs for the treatment of cancer or other diseases where the elimination of pathogenic cells is desirable. The eventual feasibility of this approach appears favorable since there is precedence for the clinical use of nontargeted RNase (see Section 3.1). This antitumor RNase, Onconase, can sensitize colon carcinoma cells to vincristine in mice even in the presence of P-glycoprotein-mediated multidrug resistance.[187] Moreover, the specific cytotoxicity of Onconase can be enhanced 2000-fold by conjugating it to an internalizing antibody present on B-cell lymphomas and that increase in activity is manifested by potent antitumor effects against disseminated lymphoma in SCID mice without causing toxic side effects.[188] Taken together, these results demonstrate that RNase can be as potent as a plant or bacterial toxin when combined with an appropriate antibody and this bodes well for the eventual clinical utility of targeted RNase fusions. Since problems in expressing and purifying sufficient amounts of material for animal studies have been solved, preclinical evaluation of RNase fusions is underway.

Targeted plasminogen activators have been studied in an extensive series of functional experiments in vitro and in several animal models. These chimeric proteins are composed of a high affinity antibody that recognizes fibrin to confer enhanced selectivity while the increased size of the fusion protein relative to the native plasminogen activators beneficially prolongs the circulating half-life. Significantly, in a baboon model in which preformed blood clots were treated with the fusion protein or unfused plasminogen activators, the fusion protein was found to be a superior thrombolytic agent. Although potency was increased, hemostatic function was less impaired in baboons receiving the fusion protein than in those receiving comparable doses of nontargeted plasminogen activators. Thus improved efficacy without unacceptable bleeding in a relevant primate model provides a sound rationale for future clinical investigation.

Recombinant fusion proteins combined with prodrugs were shown to exhibit antitumor activity in animal models that exceeded that obtained with the parent drug. Results such as these stimulated clinical evaluation of a chemical antibody–enzyme conjugate that activated a benzoic acid mustard prodrug. Thus far 15 patients have received that prodrug therapy in two separate Phase I clinical trials. Although the reported results from the first clinical trial were encouraging, they also identified facets needing improvement to obtain better therapeutic effects. Some of these are addressed by the enzyme/prodrug combinations elucidated in this chapter.

The active drug generated by the anti-CEA-CPG2 fusion protein was not very reactive; generation of chemotherapeutic drugs such as taxol, which is approved for treatment of breast cancer at the site of the tumor, might be a significant advancement. The small size of sFv chimeric proteins should improve tumor penetration resulting in enhanced conversion of prodrug to drug at the tumor. Additionally, sFv fusion proteins clear from blood and tissues rapidly while still achieving relatively high tumor to blood ratios. In this regard an sFv fusion protein displayed improved pharmacokinetic properties when directly compared to its Fab′ homolog.[110] Another significant advantage is that recombinantly produced fusion proteins are homogeneous compared to chemical conjugates that are not linked at a specific site of the antibody or enzyme, thus displaying more reproducible and uniform functional characteristics.

Finally, clinically related problems of immunogenicity in humans associated with administration of murine and bacterial proteins was clearly illustrated in the first ADEPT trial. A humanized antibody against the p185^{HER2} antigen was incorporated into one of the prodrug fusions to address the immunogenicity of the antibody. However another of the fusion proteins was designed with a humanized anti-CEA antibody and a human enzyme to address the immunogenicity of both domains.

In conclusion, engineering antibodies and antibody fusions is a new field that is exploring and exploiting new and ever newer technologies. As problems with engineered antibody fragments including their expression and purification have been elucidated and solved, the possibility of adding novel effector domains such as those described herein has emerged and their application is being rigorously pursued.

REFERENCES

1. Ochi, A., R. G. Hawley, T. Hawley, M. J. Schulman, A. Traunecker, G. Kohler, N. Hozumi. 1983. Functional immunoglobulin M production after transfection of cloned immunoglobulin heavy and light chain genes into lymphoid cells. *Proc. Natl. Acad. Sci. USA.* 80: 6351–6355.
2. Oi, V. T., S. L. Morrison, L. A. Herzenberg, P. Berg. 1983. Immunoglobulin gene expression in transformed lymphoid cells. *Proc. Natl. Acad. Sci. USA* 80: 825–829.
3. Neuberger, M. S. 1983. Expression and regulation of immunoglobulin heavy chain gene transfected into lymphoid cells. *EMBO J.* 2: 1373–1378.
4. Neuberger, M. S., G. T. Williams, R. O. Fox. 1984. Recombinant antibodies possessing novel effector funcions. *Nature* 312: 604–608.
5. Williams, G. T., M. S. Neuberger. 1986. Production of antibody-tagged-enzymes by myeloma cells: application to DNA polymerase I Klenow fragment. *Gene* 43: 319–324.
6. Ledoux, L. 1955. Action of ribonuclease on two solid tumors in vivo. *Nature* 176: 36–37.
7. Ledoux, L. 1955. Action of ribonuclease on certain ascites tumours. *Nature* 175: 258–259.
8. Aleksandrowicz, J. 1958. Intracutaneous ribonuclease in chronic myelocytic leukemia. *Lancet* 2: 420.
9. Glukhov, B. N., A. P. Jerusalimsky, V. M. Canter, R. I. Salganik. 1976. Ribonuclease treatment of tick-borne encephalitis. *Arch. Neurol.* 33: 598–603.
10. Darzynkiewicz, Z., S. P. Carter, S. M. Mikulski, W. J. Ardelt, K. Shogen. 1988. Cytostatic and cytotoxic effects of Pannon (P-30 protein) a novel anti-cancer agent. *Cell Tissue Kinet.* 21: 169–182.
11. Mikulski, S. M., W. Ardelt, K. Shogen, E. H. Bernstein, H. Menduke. 1990. Striking increase of survival of mice bearing M109 Madison Carcinoma treated with a novel protein from amphibian embryos. *J. Natl. Cancer Inst.* 82: 151–153.
12. Mikulski, S. M., A. M. Grossman, P. W. Carter, K. Shogen, J. J. Costanzi. 1993. Phase 1 human clinical trial of ONCONASE (P-30 protein) administered intravenously on a weekly schedule in cancer patients with solid tumors. *Int. J. Oncol.* 3: 57–64.
13. Youle, R. J., D. L. Newton, Y. N. Wu, M. Gadina, S. M. Rybak. 1993. Cytotoxic ribonucleases and chimeras in cancer therapy. *Crit. Rev. Therapeut. Drug Carrier Systems* 10: 1–28.

14. Conde, F. P., R. Orlandi, S. Canevari, D. Mezzanzanica, M. Ripamonti, S. M. Munoz, P. Jorge, M. I. Colnaghi. 1989. The Aspergillus toxin restriction is a suitable cytotoxic agent for generation of immunoconjugates with monoclonal antibodies directed against human carcinoma cells. *Eur. J. Biochem.* 178: 795–802.

15. Prior, T. I., D. J. Fitzgerald, I. Pastan. 1991. Barnase toxin: a new chimeric toxin composed of *Pseudomonas* exotoxin A and Barnase. *Cell* 64: 1017–1023.

16. Fett, J. W., D. J. Strydom, R. R. Lobb, E. M. Alderman, J. L. Bethune, J. F. Riordan, B. L. Vallee. 1985. Isolation and characterization of angiogenin, an angiogenic protein from human carcinoma cells. *Biochemistry* 24: 5480–5486.

17. Gleich, G. J., D. A. Loegering, M. P. Bell, J. L. Checkel, S. J. Ackerman, D. J. McKean. 1986. Biochemical and functional similarities between human eosinophil-derived neurotoxin and eosinophil cationic protein: Homology with ribonuclease. *Proc. Natl. Acad. Sci. USA* 83: 3146–3150.

18. Rybak, S. M., D. L. Newton, S. M. Mikulski, A. Viera, R. J. Youle. 1993. Cytotoxic Onconase and ribonuclease A chimeras: comparison and in vitro characterization. *Drug Delivery* 1: 3–10.

19. D'Alessio, G. 1993. New and cryptic biological messages from RNases. *Trends in Cell Biol.* 3: 106–109.

20. Laccetti, P., D. Spalletti-Cernia, G. Portella, P. DeCorato, G. D'Alessio, G. Vecchio. 1994. Seminal RNase inhibits tumor growth and reduces the metastatic potential of Lewis lung carcinoma. *Cancer Res.* 54: 4253–4256.

21. Deonarain, M. P., A. A. Epenetos. 1995. Construction, refolding and cytotoxicity of a single chain Fv-seminal ribonuclease fusion protein expressed in *Escherichia coli. Tumor Targeting* 1: 177–182.

22. Deonarain, M. P., A. A. Epenetos. 1998. Design, characterization and antitumor cytotoxicity of a panel of recombinant, mammalian ribonuclease-based immunotoxins. *Br. J. Cancer. 77:* 537–546.

23. Rybak, S. M., S. K. Saxena, E. J. Ackerman, R. J. Youle. 1991. Cytotoxic potential of ribonuclease and ribonuclease hybrid proteins. *J. Biol. Chem.* 266: 21202–21207.

24. Newton, D. L., O. Ilercil, D. W. Laske, E. Oldfield, S. M. Rybak, R. J. Youle. 1992. Cytotoxic ribonuclease chimeras: Targeted tumoricidal activity in vitro and in vivo. *J. Biol. Chem.* 267: 19572–19578.

25. Jinno, H., M. Ueda, S. Ozawa, K. Kikuchi, T. Ikeda, K. Enomoto, M. Kitajima. 1996. Epidermal growth factor receptor-dependent cytotoxic effect by an EGF-ribonuclease conjugate on human cancer cell lines: a trial for less immunogenic chimeric toxin. *Can. Chemother. Pharmacol.* 38: 303–308.

26. Jinno, H., M. Ueda, S. Ozawa, T. Ikeda, K. Enomoto, K. Psarras, K. Kitajima, H. Yamada, M. Seno. 1996. Epidermal growth factor receptor-dependent cytotoxicity for human squamous carcinoma cell lines of a conjugate composed of human EGF and RNase 1. *Life Sci.* 58: 1901–1908.

27. Newton, D. L., Y. Xue, K. A. Olsen, J. W. Fett, S. M. Rybak. 1996. Angiogenin single-chain immunofusions: Influence of peptide linkers and spacers between fusion protein domains. *Biochemistry* 35: 545–553.

28. Newton, D. L., P. J. Nicholls, S. M. Rybak, R. J. Youle. 1994. Expression and characterization of recombinant human eosinophil-derived neurotoxin and

eosinophil-derived neurotoxin anti-transferrin receptor sFv. *J. Biol. Chem.* 269: 26739–26745.

29. Rybak, S. M., H. R. Hoogenboom, H. M. Meade, J. C. Raus, D. Schwartz, R. J. Youle. 1992. Humanization of immuntoxins. *Proc. Natl. Acad. Sci. USA* 89: 3165–3169.
30. Zewe, M., S. M. Rybak, S. Dubel, J. F. Coy, M. Welschof, D. L. Newton, M. Little. 1997. Cloning and cytotoxicity of a human pancreatic RNase immunofusion. *Immunotechnology.* 3: 127–136.
31. Psarras, K., M. Ueda, T. Yamamura, H. Jinno, S. Ozawa, M. Kitajima, S. Komatsu, M. Seno. Human pancreatic RNase1-human epidermal growth factor fusion becomes cytotoxic against squamous cell carcinomas, submitted.
32. Haber, E., T. Quertermous, G. R. Matsueda, M. R. Runge. 1989. Innovative approaches to plasminogen activator therapy. *Science* 243: 51–56.
33. Lijnen, H. R., D. Collen. 1995. Molecular and cellular basis of fibrinolysis. In *Hematology: Basic Principles and Practice*, ed. R. Hoffmann et al., New York, Edinburgh, London, Madrid, Melbourne, Milan, Tokyo: Churchill Livingstone, pp. 1588–1596
34. Bode, C., G. R. Matsueda, K. Y. Hui, E. Haber. 1985. Antibody-directed urokinase: a specific fibrinolytic agent. *Science* 229: 765–767.
35. Collen, D. 1993. Towards improved thrombolytic therapy. Lancet 342: 34–36.
36. Haber, E. 1994. Antibody-plasminogen activator conjugates and recombinant proteins. *Methods in Mol. Genet.* 5: 111–127.
37. Collen, D. 1997. Thrombolytic therapy. *Thrombosis and Haemostasis* 78: 742–746.
38. Philpott, G. W., W. T. Shearer, R. W. Bower, C. W. Parker. 1973. Selective cytotoxicity of hapten-substituted cells with an antibody-enzyme conjugate. *J. Immunol.* 111: 921–929.
39. Bagshawe, K. D., C. J. Springer, F. Searle, P. Antoniw, S. K. Sharma, R. G. Melton, R. F. Sherwood. 1988. A cytoxic agent can be generated selectively at cancer sites. *Br. J. Cancer* 58: 700–703.
40. Senter, P. D., M. G. Saulnier, G. J. Schreiber, D. L. Hirschberg, J. P. Brown, I. Hellstrom, K. E. Hellstrom. 1988. Anti-tumor effects of antibody-alkaline phosphatase conjugates in combination with etoposide phosphate. *Proc. Natl. Acad. Sci. USA* 85: 4842–4846.
41. Senter, P. D., G. J. Schreiber, D. L. Hirschberg, S. A. Ashe, K. E. Hellstrom, I. Hellstrom. 1989. Enhancement of the in vitro and in vivo antitumor activities of phosphorylated mitomycin C and etoposide derivatives by monoclonal antibody-alkaline phosphatase conjugates. *Cancer Res.* 49: 5789–5792.
42. Senter, P. D. 1990. Activation of prodrugs by antibody-enzyme conjugates: A new approach to cancer therapy. *FASEB J.* 4: 188–193.
43. Bosslet, K., J. Czech, P. Lorenz, H. H. Sedlacek, M. Schuermann, G. Seemann. 1992. Molecular and functional characterization of a fusion protein suited for tumor specific prodrug activation. *Br. J. Cancer* 65: 234–238.
44. Bosslet, K., J. Czech, D. Hoffmann. 1994. Tumor-selective prodrug activation by fusion protein-mediated catalysis. *Cancer Res.* 54: 2151–2159.
45. Rodrigues, M. L., L. G. Presta, C. E. Kotts, C. Wirth, J. Mordenti, G. Osaka, W. L. T. Wong, A. Nuijens, B. Blackburn, P. Carter. 1995. Development of a

humanized disulfide-stabilized anti-p185 HER2 Fv-β-Lactamase fusion protein for activation of a cephalosporin doxorubicin prodrug. *Cancer Res.* 55: 63–70.

46. Wallace, P. M., J. F. MacMaster, V. F. Smith, D. E. Kerr, P. D. Senter, W. L. Cosand. 1994. Intratumoral generation of 5-flourouracil mediated by an antibody-cytosine deaminase conjugate in combination with 5-fluorocytosine. *Cancer Res.* 54: 2719–2723.

47. Svensson, H. P., V. M. Vrudhula, J. E. Emswiler, J. F. MacMaster, W. L. Cosand, P. D. Senter, P. M. Wallace. 1995. In vitro and in vivo activities of a doxorubicin prodrug in combination with monoclonal antibody β-lactamase conjugates. *Cancer Res.* 55: 2357–2365.

48. Springer, C. J., K. D. Bagshawe, S. K. Sharma, F. Searle, J. A. Boden, P. Antoniw, P. J. Burke, G. T. Rogers, R. F. Sherwood, R. G. Melton. 1991. Ablation of human choriocarcinoma xenografts in nude mice by antibody-directed enzyme prodrug therapy (ADEPT) with three novel compounds. *Eur. J. Cancer* 27: 1361–1366.

49. Meyer, D. L., L. N. Jungheim, K. L. Law, S. D. Mikolajczyk, T. A. Shepherd, D. G. Mackensen, S. L. Briggs, J. J. Starling. 1993. Site-specific prodrug activation by antibody-β-lactamase conjugates: regression and long-term growth inhibition of human colon carcinoma xenograft models. *Cancer Res.* 53: 3956–3963.

50. Eccles, S. A., W. J. Court, G. A. Box, C. J. Dean, R. G. Melton, C. J. Springer. 1994. Regression of established breast carcioma xenografts with antibody-directed enzyme prodrug therapy against c-erbB2 p185. *Cancer Res.* 54: 5171–5177.

51. Kerr, D. E., G. J. Schreiber, V. M. Vrudhula, H. P. Svensson, I. Hellstrom, K. E. Hellstrom, P. D. Senter. 1995. Regressions and cures of melanoma xenografts following treatment with monoclonal antibody β-lactamase conjugates in combination with anticancer prodrugs. *Cancer Res.* 55: 3558–3563.

52. Bagshawe, K. D., S. K. Sharma, C. J. Springer, P. Antoniw, J. A. Boden, G. T. Rogers, P. J. Burke, R. G. Melton, R. F. Sherwood. 1991. Antibody directed enzyme prodrug therapy (ADEPT) clinical report. *Dis. Markers.* 9: 233–238.

53. Springer, C. J., G. K. Poon, S. K. Sharma, K. D. Bagshawe. 1993. Identification of prodrug, active drug, and metabolites in an ADEPT clinical study. *Cell Biophys.* 22: 9–26.

54. Sharma, S. K., K. D. Bagshawe, R. G. Melton, R. F. Sherwood. 1992. Human immune response to monoclonal antibody-enzyme conjugates in ADEPT pilot clinical trial. *Cell Biophys.* 21: 109–120.

55. Martin, J., S. M. Stribbling, G. K. Poon, R. H. Begent, M. Napier, S. K. Sharma, C. J. Springer. 1997. Antibody-directed enzyme prodrug therapy: pharmacokinetics and plasma levels of prodrug and drug in a phase I clinical trial. *Cancer Chemother. Pharmacol.* 40: 189–201.

56. Bagshawe, K. D. 1994. Antibody-directed enzyme prodrug therapy. *Clin. Pharmacokinetic Concepts.* 27: 368–376.

57. Bagshawe, K. D., S. K. Sharma, C. J. Springer, G. T. Rogers. 1994. Antibody directed enzyme prodrug therapy (ADEPT). A review of some theoretical, experimental and clinical aspects. *Ann. Oncol.* 5: 879–891.

58. Deonarain, M. P., A. A. Epenetos. 1994. Targeting enzymes for cancer therapy:old enzymes in new roles. *Br. J. Cancer* 70: 786–794.

59. Senter, P. D. 1990. Activation of prodrugs by antibody-enzyme conjugates: A new approach to cancer therapy. *FASEB J.* 4: 188–193.

60. Senter, P. D., P. M. Wallace, H. P. Svensson, V. M. Vrudhula, D. E. Kerr, I. Hellstrom, K. E. Hellstrom. 1993. Generation of cytotoxic agents by targeted enzymes. *Bioconjugate Chem.* 4: 3–9.

61. Hellstrom, K. E., P. D. Senter. 1991. Activation of prodrugs by targeted enzymes. *Eur. J. Cancer* 27: 1342–1343.

62. Carlsson, J., H. Drevin, R. Axen. 1978. Protein thiolation and reversible protein-protein conjugation. *Biochem. J.* 173: 723–737.

63. Skerra, A., A. Pluckthun. 1988. Assembly of a functional immunoglobulin Fv fragment in Escherichia coli. *Science* 240: 1038–1041.

64. Field, H., G. T. Yarranton, A. R. Rees. 1990. Expression of mouse immunoglobulin light and heavy chain variable regions in *Escherichia coli* and reconstitution of antigen-binding activity. *Protein Eng.* 3: 641–647.

65. Glockshuber, R., M. Malia, I. Pfitzinger, A. Pluckthun. 1990. A comparison of strategies to stabilize immunoglobulin Fv-fragments. *Biochemistry* 29: 1362–1367.

66. Bird, R. E., K. D. Hardman, J. W. Jacobson, S. Johnson, B. M. Kaufman, S. M. Lee, T. Lee, S. H. Pope, G. S. Riordan, M. Whitlow. 1988. Single-chain antigen-binding proteins. *Science* 242: 423–426.

67. Huston, J. S., D. Levinson, M. Mudgett-Hunter, M. S. Tai, J. Novotny, M. N. Margolies, R. J. Ridge, R. E. Bruccoleri, E. Haber, R. Crea, H. Oppermann. 1988. Protein engineering of antibody binding sites: recovery of specific activity in an anti-digoxin single-chain Fv analogue produced in Escherichia coli. *Proc. Natl. Acad. Sci. USA* 85: 5879–5883.

68. Huston, J. S., M. Mudgett-Hunter, M. S. Tai, J. McCartney, F. Warren, E. Haber, H. Oppermann. 1991. Protein engineering of single-chain Fv analogs and fusion proteins. *Methods Enzymol.* 203: 46–88.

69. Huston, J. S., M. N. Margolies, E. Haber. 1996. Antibody binding sites. *Adv. Prot. Chem.* 49: 329–450.

70. Whitlow, M., B. A. Bell, S. L. Feng, D. Filpula, K. D. Hardman, S. L. Hubert, M. L. Rollence, J. F. Wood, M. E. Schott, D. E. Milenic, T. Yokota, J. Schlom. 1993. An improved linker for single-chain Fv with reduced aggregation and enhanced proteolytic stability. *Protein Eng.* 6: 989–995.

71. Desplancq, D., D. J. King, A. D. G. Lawson, A. Mountain. 1994. Multimerization behaviour of single chain Fv variants for the tumour-binding antibody B72.3. *Protein Eng.* 7:H 1027–1033.

72. Kortt, A. A., M. Lah, G. W. Oddie, C. L. Gruen, J. E. Burns, L. A. Pearce, J. L. Atwell, A. J. McCoy, G. J. Howlett, D. W. Metzger, R. G. Webster, P. J. Hudson. 1997. Single-chain Fv fragments of anti-neuraminidase antibody NC10 containing five- and ten-residue linkers form dimers and with zero-residue linker a trimer. *Protein Eng.* 10: 423–433.

73. Alfthan, K., K. Takkinen, D. Sizmann, H. Soderlund, T. T. Teeri. 1995. Properties of a single-chain antibody containing different linker peptides. *Protein Eng.* 8: 725–731.

74. Holliger, P., T. Prospero, G. Winter. 1993. "Diabodies": Small bivalent and bispecific antibody fragments. *Proc. Natl. Acad. Sci. USA* 90: 6444–6448.

75. Batra, J. K., D. FitzGerald, M. Gately, V. K. Chaudhary, I. Pastan. 1990. Anti-Tac(Fv)-PE40, a single chain antibody *Pseudomonas* fusion protein directed at interleukin 2 receptor bearing cells. *J. Biol. Chem.* 265: 15198–15202.

76. Brinkmann, U., Y. Reiter, S. H. Jung, B. Lee, I. Pastan. 1993. A recombinant immunotoxin containing a disulfide-stabilized Fv fragment. *Proc. Natl. Acad. Sci. USA* 90: 7538–7542.

77. Reiter, Y., I. Pastan. 1996. Antibody engineering of recombinant Fv immunotoxins for improved targeting of cancer: Disulfide-stabilized Fv immunotoxins. *Clin. Cancer Res.* 2: 245–252.

78. Huston, J. S., J. McCartney, M. S. Tai, C. Mottola-Hartshorn, D. Jin, F. Warren, P. Keck, H. Oppermann. 1993. Medical applications of single-chain antibodies. *Intern. Rev. Immunol.* 10: 195–217.

79. Crothers, D. M., H. Metzger. 1972. The influence of polyvalency on the binding properties of antibodies. *Immunochem.* 9: 341–357.

80. Colcher, D., R. Bird, M. Roselli, K. D. Hardman, S. Johnson, S. Pope, S. W. Dodd, M. W. Pantoliano, D. E. Milenic, J. Schlom. 1990. In vivo tumor targeting of a recombinant single-chain antigen-binding protein. *J. Natl. Cancer Inst.* 82: 1191–1197.

81. Milenic, D. E., T. Yokota, D. R. Filpula, M. A. J. Finkelman, S. W. Dodd, J. F. Wood, M. Whitlow, P. Snoy, J. Schlom. 1991. Construction, binding properties, metabolism, and tumor targeting of a single-chain Fv derived from pancarcinoma monoclonal antibody CC49. *Cancer Res.* 51: 6363–6371.

82. King, D. J., A. Turner, A. P. H. Farnsworth, J. R. Adair, R. J. Owens, R. B. Pedley, D. Baldock, K. A. Proudfoot, A. D. G. Lawson, N. R. A. Beeley, K. Millar, T. A. Millican, B. A. Boyce, P. Antoniw, A. Mountain, R. H. J. Begent, D. Shochat, G. T. Yarranton. 1994. Improved tumor targeting with chemically cross-linked recombinant antibody fragments. *Cancer Res.* 54: 6176–6185.

83. Yokota, T., D. E. Milenic, M. Whitlow, J. Schlom. 1992. Rapid tumor penetration of a single-chain Fv and comparison with other immunoglobulin forms. *Cancer Res.* 52: 3402–3408.

84. Khazaeli, M. B., R. M. Conry, A. F. LoBuglio. 1994. Human immune response to monoclonal antibodies. *J. Immunotherapy* 15: 42–52.

85. Rybak, S. M., R. J. Youle. 1991. Clinical use of immunotoxins: Monoclonal antibodies conjugated to protein toxins. *Immunol. and Allergy Clinics of North America* 11: 359–380.

86. Jolliffe, L. K. 1993. Humanized antibodies: Enhancing therapeutic utility through antibody engineering. *Intern. Rev. Immunol.* 10: 241–250.

87. Stephens, S., S. Emtage, O. Vetterlein, L. Chaplin, C. Bebbington, A. Nesbitt, M. Sopwith, D. Athwal, C. Novak, M. Bodmer. 1995. Comprehensive pharmacokinetics of a humanized antibody and analysis of residual anti-idiotypic responses. *Immunology* 85: 668–674.

88. Rybak, S. M., D. L. Newton, Y. Xue. 1995. RNase and RNase immunofusions for cancer therapy. *Tumor Targeting* 1: 141–147.

89. Hoogenboom, H. R., J. C. M. Raus, G. Volckaert. 1990. Cloning and expression of a chimeric antibody directed against the human transferrin receptor. *J. Immunol.* 144: 3211–3217.

90. Hoogenboom, H. R., G. Volckaert, J. C. M. Raus. 1991. Construction and expression of antibody-tumor necrosis factor fusion proteins. *Mol. Immunol.* 28: 1027–1037.

91. D'Alessio, G., A. Di Donato, A. Parente, R. Piccoli. 1991. Seminal RNase: a unique member of the ribonuclease superfamily. *Trends Biochem. Sci.* 16: 104–106.

92. Savage, P., G. Rowlinson-Busza, M. Verhoeyen, R. A. Spooner, A. So, J. Windust, P. J. Davis, A. A. Epenetos. 1993. Construction, characterisation and kinetics of a single chain antibody recognising the tumour-associated antigen placental alkaline phosphatase. *Br. J. Cancer.* 68: 738–742.

93. Nicholls, P. J., V. G. Johnson, S. M. Andrew, H. R. Hoogenboom, J. C. Raus, R. J. Youle. 1993. Characterization of single-chain antibody (sFv)-toxin fusion proteins produced in vitro in rabbit reticulocyte lysate. *J. Biol. Chem.* 268: 5302–5308.

94. Tai, M. S., M. Mudgett-Hunter, D. Levinson, G.-M. Wu, E. Haber, H. Oppermann, J. S. Huston. 1990. A bifunctional fusion protein containing Fc-binding fragment B of staphylococcal protein A amino terminal to antidigoxin single-chain Fv. *Biochemistry* 29: 8024–8030.

95. Hui, K. Y., E. Haber, G. R. Matsueda. 1983. Monoclonal antibodies to a synthetic fibrin-like peptide bind to human fibrin but not fibrinogen. *Science* 222: 1129–1132.

96. Holvoet, P., J. M. Stassen, Y. Hashimoto, D. Spriggs, P. Devos, D. Collen. 1989. Binding properties of monoclonal antibodies against human fragment D-dimer of cross-linked fibrin to human plasma clots in an in vivo model in rabbits. *Thromb. Haemostasis.* 61: 307–313.

97. Schnee, J. M., M. S. Runge, G. R. Matsueda, N. W. Hudson, J. G. Seidman, E. Haber, T. Quertermous. 1987. Construction and expression of a recombinant antibody-targeted plasminogen activator. *Proc. Natl. Acad. Sci. USA* 84: 6904–6908.

98. Stump, D. C., H. R. Lijnen, D. Collen. 1986. Purification and characterization of a novel low molecular weight form of single-chain urokinase-type plasminogen activator. *J. Biol. Chem.* 261: 17120–17126.

99. Runge, M. S., T. Quertermous, P. J. Zavodny, T. W. Love, C. Bode, M. Freitag, S.-Y. Shaw, P. L. Huang, C.-C. Chou, D. Mullins, J. M. Schnee, C. E. Savard, M. E. Rothenberg, J. B. Newell, G. R. Matsueda, E. Haber. 1991. A recombinant chimeric plasminogen activator with high affinity for fibrin has increased thrombolytic potency in vitro and in vivo. *Proc. Natl. Acad. Sci. USA* 88: 10337–10341.

100. Yang, W. P., J. Goldstein, R. Procyk, G. R. Matsueda, S. Y. Shaw. 1994. Design and evaluation of a thrombin-activable plasminogen activator. *Biochemistry* 33: 2306–2312.

101. Lijnen, H. R., B. VanHoef, F. DeCock, D. Collen. 1989. The mechanism of plasminogen activation and fibrin dissolution by single chain urokinase-type plasminogen activator in a plasma milieu in vitro. *Blood* 73: 1864–1872.

102. Ichinose, A., K. Fujikawa, T. Suyama. 1986. The activation of pro-urokinase by plasma kallikrein and its inactivation by thrombin. *J. Biol. Chem.* 261: 3486–3489.

103. Vandamme, A. M., M. Dewerchin, H. R. Lijnen, H. Bernar, F. Bulens, L. Nelles, D. Collen. 1992. Characterization of a recombinant chimeric plasminogen ac-

tivator composed of a fibrin fragment-D-dimer-specific humanized monoclonal antibody and a truncated single-chain urokinase. *Eur. J. Biochem.* 205: 139–146.

104. Laroche, Y., M. Demaeyer, J. M. Stassen, Y. Gansemans, E. Demarsin, G. Matthyssens, D. Collen, P. Holvoet. 1991. Characterization of a recombinant single-chain molecule comprising the variable domains of a monoclonal antibody specific for human fibrin fragment D-dimer. *J. Biol. Chem.* 266: 16343–16349.

105. Claessens, M., E. VanCutsem, I. Lasters, S. Wodak. 1989. Modelling the polypeptide backbone with 'spare parts' from known protein structures. *Protein Eng.* 2: 335–345.

106. Holvoet, P., Y. Laroche, H. R. Lijnen, R. V. Cauwenberge, E. Demarsin, E. Brouwers, G. Matthyssens, D. Collen. 1991. Characterization of a chimeric plasminogen activator consisting of a single-chain Fv fragment derived from a fibrin fragment D-dimer-specific antibody and a truncated single-chain urokinase. *J. Biol. Chem.* 266: 19717–19724.

107. Hellstrom, I., D. Horn, P. Linsley, J. P. Brown, V. Brankovan, K. E. Hellstrom. 1986. Monoclonal antibodies raised against human lung carcinoma. *Cancer Res.* 46: 3917–3923.

108. Goshorn, S. C., H. P. Svensson, D. E. Kerr, J. E. Somerville, P. D. Senter, H. P. Fell. 1993. Genetic construction, expression and characterization of a single chain anti-carcinoma antibody fused to β-lactamase. *Cancer Res.* 53: 2123–2127.

109. Brown, J. P., K. Nishiyama, I. Hellstrom, K. E. Hellstrom. 1981. Structural characterization of human melanoma-associated antigen p97 with monoclonal antibodies. *J. Immunol.* 127: 539–546.

110. Siemers, N. O., D. E. Kerr, S. Yarnold, M. R. Stebbins, V. M. Vrudhula, I. Hellstrom, K. E. Hellstrom, P. D. Senter. 1997. Construction, expression and activities of L49–sFv-β Lactamase, a single-chain antibody fusion protein for anticancer prodrug activation. *Bioconjugate Chem.* 8: 510–519.

111. Cafaro, V., C. DeLorenzo, R. Piccoli, A. Bracale, M. R. Mastronicola, A. DiDonato, G. D'Alessio. 1995. The anti-tumor action of seminal ribonuclease and its quaternary conformations. *FEBS Letts.* 359: 31–34.

112. Wright, A., S.-U. Shin, S. L. Morrison. 1992. Genetically engineered antibodies: Progress and Prospects. *Crit. Rev. Immunol.* 12: 125–168.

113. Gething, M. J., J. Sambrook. 1992. Protein folding in the cell. *Nature* 355: 33–45.

114. Knazek, R. A., P. M. Gullino, P. O. Kohler, R. L. Dedrick. 1972. Cell culture on artificial capillaries: an approach to tissue growth in vitro. *Science* 178: 65–66.

115. Arathoon, W. R., J. R. Birch. 1986. Large-scale cell culture in biotechnology. *Science* 232: 1390–1395.

116. Mather, J. P., A. Moore, R. Shawley. 1997. Optimization of growth, viability, and specific productivity for expression of recombinant proteins in mammalian cells. *Meth. Molecular Biol.* 62: 369–382.

117. Cattaneo, A., M. S. Neuberger. 1987. Polymeric immunoglobulin M is secreted by transfectants of non-lymphoid cells in the absence of immunoglobulin J chain. *EMBO J.* 6: 2753–2758.

118. Etcheverry, T. 1996. Expression of engineered proteins in mammalian cell culture. In *Protein Engineering: Principles and Practice*, ed. J. L. Cleland, C. S. Craik, New York: Wiley-Liss, pp. 163–181.

119. Better, M., C. P. Chang, R. R. Robinson, A. H. Horwitz. 1988. *Escherichia coli* secretion of an active chimeric antibody fragment. *Science* 240: 1041–1043.

120. Skerra, A. 1993. Bacterial expression of immunoglobulin fragments. *Curr. Opin. Immunol.* 5: 256–262.

121. Pluckthun, A. 1991. Antibody engineering: Advances from the use of *Escherichia coli* expression systems. *Biotechnology* 9: 545–551.

122. Rudolph, R. 1996. Successful protein folding on an industrial scale. In *Protein Engineering: Principles and Practice*, ed. J. L. Cleland, C. S. Craik, New York: Wiley-Liss, pp. 283–298.

123. Hasemann, C. A., J. D. Capra. 1990. High-level production of a functional immunoglobulin heterodimer in a baculovirus expression system. *Proc. Natl. Acad. Sci. USA* 87: 3942–3946.

124. zu Putliz, J., W. L. Kubasek, M. Duchene, M. Marget, B. U. von Specht, H. Domdey. 1990. Antibody production in baculovirus-infected insect cells. *Bio/Technology* 8: 651–654.

125. Holvoet, P., Y. Laroche, H. R. Lijnen, B. Van Hoef, E. Brouwers, F. DeCock, M. Lauwereys, Y. Gansemans, D. Collen. 1992. Biochemical characterization of single-chain chimeric plasminogen activators consisting of a single-chain Fv fragment of a fibrin-specific antibody and single-chain urokinase. *Eur. J. Biochem.* 210: 945–952.

126. Hiatt, A., R. Cafferkey, K. Bowdish. 1989. Production of antibodies in transgenic plants. *Nature* 342: 76–78.

127. During, K., S. Hippe, F. Kreuzaler, J. Schell. 1990. Synthesis and self-assembly of a functional monoclonal antibody in transgenic *Nicotiana tabacum. Plant Mol. Biol.* 15: 281–293.

128. Ma, J. K. C., M. B. Hein. 1995. Immunotherapeutic potential of antibodies produced in plants. *TIBTECH.* 13: 522–527.

129. Colman, A. 1996. Production of proteins in the milk of transgenic livestock: problems, solutions and successes. *Am. J. Clin. Nutr.* 63: 639S–645S.

130. Logan, J. S. 1993. Transgenic animals: beyond "funny milk." *Cur. Opin. Biotechnol.* 4: 591–595.

131. Ebert, K. M., P. DiTullio, C. A. Barry, J. E. Schindler, S. L. Ayres, T. E. Smith, L. J. Pellerin, H. M. Meade, J. Denman, B. Roberts. 1994. Induction of human tissue plasminogen activator in the mammary gland of transgenic goats. *Bio/Technology.* 12: 699–702.

132. DiTullio, P., S. H. Cheng, J. Marshall, R. J. Gregory, K. M. Ebert, H. M. Meade, A. E. Smith. 1992. Production of cystic fibrosis transmembrane conductance regulator in the milk of transgenic mice. *Bio/Technology* 10: 74–77.

133. Studier, F. W., A. H. Rosenberg, J. J. Dunn, J. W. Dubendorff. 1990. Use of T7 RNA polymerase to direct expression of cloned genes. *Methods Enzymol.* 185: 60–89.

134. Newton, D. L., S. M. Rybak. Construction of ribonuclease chimeras for selective cytotoxicity; preparation of RNase-sFv fusion proteins. *Methods in Molec. Targeting: Drug Targeting*, in press.

135. Brinkmann, U., J. Buchner, I. Pastan. 1992. Independent domain folding of *Pseudomonas* exotoxin and single-chain immunotoxins: Influence of interdomain

connections. *Proc. Natl. Acad. Sci. USA* 89: 3075–3079.

136. Buchner, J., I. Pastan, U. Brinkmann. 1992. A method for increasing the yield of properly folded recombinant fusion proteins: Single-chain immunotoxins from renaturation of bacterial inclusion bodies. *Anal. Biochem.* 205: 263–270.

137. Newton, D. L., S. M. Rybak. 1996. Novel therapeutics for cancer and AIDS based on human and humanized ribonuclease-antibody chimeric proteins. In: *10th International Biotechnology Symposium,* Sydney, Australia.

138. Love, T. W., T. Quertermous, P. J. Zavodny, M. S. Runge, C. C. Chou, D. Mullins, P. L. Huang, J. M. Schnee, A. S. Kestin, C. E. Savard, K. D. Michelson, G. R. Matsueda, E. Haber. 1993. High-level expression of antibody-plasminogen activator fusion proteins in hybridoma cells. *Thromb. Res.* 69: 221–229.

139. Bode, C., M. S. Runge, E. Haber. 1992. Purifying antibody-plasminogen activator conjugates. *Bioconj. Chem.* 3: 69–272.

140. Trowbridge, I. S., M. B. Omary. 1981. Human cell surface glycoprotein related to cell proliferation is the receptor for transferrin. *Proc. Natl. Acad. Sci. USA* 78: 3039–3043.

141. Trowbridge, I. S., D. L. Domingo. 1981. Anti-transferrin receptor monoclonal antibody and toxin-antibody conjugates affect growth of human tumor cells. *Nature* 294: 171–173.

142. Trowbridge, I. S. 1988. Transferrin receptor as a potential therapeutic agent. *Prog. Allergy.* 45: 121–146.

143. Taetle, R., J. Castagnola, J. Mendelsohn. 1986. Mechanisms of growth inhibition by anti-transferrin receptor monoclonal antibodies. *Cancer Res.* 46: 1759–1763.

144. Friden, P. M., L. R. Walus, G. F. Musso, M. A. Taylor, B. Malfroy, R. M. Starzyk. 1991. Anti-transferrin receptor antibody and antibody-drug conjugates cross the blood-brain barrier. *Proc. Natl. Acad. Sci. USA* 88: 4771–4775.

145. Jefferies, W. A., M. R. Brandon, S. V. Hunt, A. F. Williams, K. C. Gatter, D. Y. Mason. 1984. Transferrin receptor on endothelium of brain capillaries. *Nature* 312: 162–163.

146. Laske, D. W., R. J. Youle, E. H. Oldfield. 1997. Tumor regression with regional distribution of the targeted toxin TF-CRM107 in patients with malignant brain tumors. *Nat. Med.* 3: 1362–1368.

147. Travers, P., W. Bodmer. 1984. Preparation and characterization of monoclonal antibodies against placental alkaline phosphatase and other human trophoblast-associated determinants. *Int. J. Cancer.* 33: 633–641.

148. Epenetos, A. A., P. Travers, K. C. Gatter, R. D. Oliver, D. Y. Mason, W. F. Bodmer. 1984. An immunohistological study of testicular germ cell tumours using two different monoclonal antibodies against placental alkaline phosphatase. *Br. J. Cancer* 49: 11–15.

149. Newton, D. L., Y. Xue, L. Boque, A. Wlodawer, H. F. Kung, S. M. Rybak. 1997. Expression and characterization of a cytotoxic human-frog chimeric ribonuclease: Potential for cancer therapy. *Protein Eng.* 10: 463–470.

150. Boix, E., Y. Wu, V. M. Vasandani, S. K. Saxena, W. Ardelt, J. Ladner, R. J. Youle. 1996. Role of the N terminus in RNase A homologues: Differences in catalytic activity, ribonuclease inhibitor interaction and cytotoxicity. *J. Mol. Biol.* 257: 992–1007.

151. Linardou, H., M. P. Deonarain, R. A. Spooner, A. A. Epenetos. 1994. Deoxyribonuclease I (DNaseI): A novel approach for targeted cancer therapy. *Cell Biophysics* 24/25: 243–248.

152. Chen, F., E. Haber, G. R. Matsueda. 1992. Availability of the β-beta (15–21) epitope on cross-linked human fibrin and its plasmic degradation products. *Thromb. Haemostasis.* 67: 335–340.

153. Kuiper, J., M. Otter, D. C. Rijken, T. J. van Berkel. 1988. Characterization of the interaction in vivo of tissue-type plasminogen acivator with liver cells. *J. Biol. Chem.* 263: 18220–18224.

154. Goldenberg, D. M. 1992. Cancer imaging with CEA antibodies: historical and current perspectives. *Int. J. Biol. Markers.* 7: 183–188.

155. Hellstrom, I., D. Horn, P. Linsley, J. P. Brown, V. Brankovan, K. E. Hellstrom. 1986. Monoclonal mouse antibodies raised against human lung carcinoma. *Cancer Res.* 46: 3917–3923.

156. Hellstrom, I., P. L. Beaumier, K. E. Hellstrom. 1986. Antitumor effects of L6, an IgG2a antibody that reacts with most human carcinomas. *Proc. Natl. Acad. Sci. USA* 83: 7059–7063.

157. Ziegler, L. D., P. Palazzolo, J. Cunningham, M. Janus, K. Itoh, K. Hayakawa, I. Hellstrom, K. E. Hellstrom, C. Nicaise, R. Dennin, J. L. Murray. 1992. Phase I trial of murine monoclonal antibody L6 in combination with subcutaneous interleukin-2 in patients with advanced carcinoma of the breast, colorectum, and lung. *J. Clin. Oncol.* 10: 1470–1478.

158. Brown, J. P., R. M. Hewick, I. Hellstrom, K. E. Hellstrom, R. F. Doolittle, W. J. Dreyer. 1982. Human melanoma-associated antigen p97 is structurally and functionally related to transferrin. *Nature* 296: 171–173.

159. Slamon, D. J., G. M. Clark, S. G. Wong, W. J. Levin, A. Ullrich, W. L. McGuire. 1987. Human breast cancer: correlation of relapse and survival with amplification of the HER-2/neu oncogene. *Science* 235: 177–182.

160. Berchuck, A., A. Kamel, R. Whitaker, B. Kerns, G. Olt, R. Kinney, J. T, Soper, R. Dodge, D. L. Clarke-Pearson, P. Marks, M. S. McKenzie, S. Yin, R. C. Bast Jr. 1990. Overexpression of *HER-2/neu* is associated with poor survival in advanced epithelial ovarian cancer. *Cancer Res.* 50: 4087–4091.

161. Batra, J. K., P. G. Kasprzyk, R. E. Bird, I. Pastan, C. R. King. 1992. Recombinant anti-erbB2 immunotoxins containing *Pseudomonas* exotoxin. *Proc. Natl. Acad. Sci. USA* 89: 5867–5871.

162. Wels, W., I. M. Harwerth, M. Mueller, B. Groner, N. E. Hynes. 1992. Selective inhibition of tumor cell growth by a recombinant single-chain antibody-toxin specific for the erbB-2 receptor. *Cancer Res.* 52: 6310–6317.

163. Bosslet, K., J. Czech, D. Hoffman. 1995. A novel one-step tumor-selective prodrug activation system. *Tumor Targeting* 1: 45–50.

164. Hussain, M., A. Carlino, M. J. Madonna, J. O. Lampen. 1985. Cloning and sequencing of the metallothioprotein beta-lactamase II gene of *Bacillus cereus* 569/H in *Escherichia coli*. *J. Bacteriol.* 164: 223–229.

165. Siemers, N. O., D. E. Yelton, J. Bajorath, P. D. Senter. 1996. Modifying the specificity and activity of the Enterobacter cloacae P99 β-lactamase by mutagenesis within an M13 phage vector. *Biochemistry.* 35: 2104–2111.

166. Adachi, H., T. Ohta, H. Matsuzawa. 1991. Site-directed mutants, at position 166, of RTEM-1 beta-lactamase that form a stable acyl-enzyme intermediate with penicillin. *J. Biol. Chem.* 266: 3186–3191.

167. Healey, W. J., M. R. Labgold, J. H. Richards. 1989. Substrate specificities in class A beta-lactamases: preference for penams vs. cephems. The role of residue 237. *Proteins* 6: 275–283.

168. Rodrigues, M. L., P. Carter, C. Wirth, S. Mullins, A. Lee, B. K. Blackburn. 1995. Synthesis and β-lactamase-mediated activation of a cephalosporin-taxol prodrug. *Chemistry and Biol.* 2: 223–227.

169. Bagshawe, K. D., S. K. Sharma, C. J. Springer, P. Antoniw. 1995. Antibody directed enzyme prodrug therapy: a pilot-scale clinical trial. *Tumor Targeting* 1: 17–29.

170. Bagshawe, K. D., S. K. Sharma. 1996. Cyclosporine delays the host immune response to antibody enzyme conjugate in ADEPT. *Transplan. Proc.* 28: 3156–3158.

171. Mosmann, T. 1983. Rapid colorimetric assay for cellular growth and survival: application to proliferation and cytotoxicity assays. *J. Immunol. Methods.* 65: 55–63.

172. Roth, J. S. 1963. Ribonuclease activity and cancer: A review. *Cancer Res.* 23: 657–666.

173. O'Hare, M., A. N. Brown, K. Hussain, A. Gebhart, G. Watson, L. M. Roberts, E. S. Vitetta, P. E. Thorpe, J. M. Lord. 1990. Cytotoxicity of a recombinant ricin-A-chain fusion protein containing a proteolytically-cleavable spacer sequence. *FEBS Lett.* 273: 200–204.

174. Wales, R., L. M. Roberts, J. M. Lord. 1993. Addition of an endoplasmic reticulum retrieval sequence to ricin A chain significantly increases its cytotoxicity to mammalian cells. *J. Biol. Chem.* 268: 23986–23990.

175. Yazdi, P. T., L. A. Wenning, R. M. Murphey. 1995. Influence of cellular trafficking on protein synthesis inhibition of immunotoxins directed against the transferrin receptor. *Cancer Res.* 55: 3763–3771.

176. Laske, D. W., O. Ilercil, A. Akbasak, R. J. Youle, E. H. Oldfield. 1994. Efficacy of direct intratumoral therapy with targeted protein toxins for solid human gliomas in nude mice. *J. Neurosurg.* 80: 520–526.

177. Newton, D. L., S. M. Rybak. 1996. Single-chain immunofusions engineered with human RNases (Abstract). In *Exploring and Exploiting Antibody and Ig Superfamily Combining Sites.* Taos, NM, Keystone Symposia p. 20.

178. Beintema, J. J., C. Schuller, M. Irie, A. Carsana. 1988. Molecular evolution of the ribonuclease superfamily. *Prog. Biophys. Mol. Biol.* 51: 165–192.

179. Jin, F. S., R. J. Youle, V. G. Johnson, J. Shiloach, R. Fass, D. L. Longo, S. H. Bridges. 1991. Suppression of the immune response to immunotoxins with anti-CD4 monoclonal antibodies. *J. Immunol.* 146: 1806–1811.

180. Lorberboum-Galski, H., L. Y. Barrett, R. L. Kirkman, M. Ogata, M. C. Willingham, D. J. FitzGerald, I. Pastan. 1989. Cardiac allograft survival in mice treated with IL-2–PE40. *Proc. Natl. Acad. Sci. USA* 86: 1008–1012.

181. Holvoet, P., Y. Laroche, J. M. Stassen, H. R. Lijnen, B. van Hoef, F. DeCock, A. van Houtven, Y. Gansemans, G. Matthyssens, D. Collen. 1993. Pharmacokinetic

and thrombolytic properties of chimeric plasminogen activators consisting of a single-chain Fv fragment of a fibrin-specific antibody fused to single-chain urokinase. *Blood* 81: 696–703.

182. Dewerchin, M., A. M. Vandamme, P. Holvoet, F. DeCock, G. Lemmens, H. R. Lijnen, J. M. Stassen, D. Collen. 1992. Thrombolytic and pharmacokinetic properties of a recombinant chimeric plasminogen activator consisting of a fibrin fragment D-dimer specific humanized monoclonal antibody and a truncated single-chain urokinase. *Thrombosis and Haemostasis* 68: 170–179.

183. Stassen, J. M., I. Vanlinthout, H. R. Lijnen, D. Collen. 1990. A hamster pulmonary embolism model for the evaluation of the thrombolytic and pharmacokinetic properties of thrombolytic agents. *Fibrinolysis* 4: 15–21.

184. Collen, D., J. M. Stassen, M. Verstraete. 1983. Thrombolysis with human extrinsic (tissue-type) plasminogen activator in rabbits with experimental jugular vein thrombosis. Effect of molecular form and dose of activator, age of the thrombus and route of administration. *J. Clin. Invest.* 71: 368–376.

185. Collen, D., M. Dewerchin, H. J. Rapold, H. R. Lijnen, J. M. Stassen. 1990. Thrombolytic and pharmacokinetic properties of a conjugate of recombinant single-chain urokinase-type plasminogen activator with a monoclonal antibody specific for cross-linked fibrin in a baboon venous thrombosis model. *Circulation* 82: 1744–1753.

186. Runge, M. S., L. A. Harker, C. Bode, J. Ruef, A. B. Kelly, U. M. Marzec, E. Allen, R. Caban, S. Y. Shaw, E. Haber, S. R. Hanson. 1996. Enhanced thrombolytic and antithrombotic potency of a fibrin-targeted plasminogen activator in baboons. *Circulation* 94: 1412–1422.

187. Rybak, S. M., J. W. Pearson, W. F. Fogler, K. Volker, S. E. Spence, D. L. Newton, S. M. Mikulski, W. Ardelt, C. W. Riggs, H. F. Kung, D. L. Longo. 1996. Enhancement of vincristine cytotoxicity in drug-resistant cells by simultaneous treatment with Onconase, an anti-tumor ribonuclease. *J. Natl. Cancer Inst.* 88: 747–753.

188. Newton, D. L., H. J. Hansen, S. M. Mikulski, D. M. Goldenberg, S. M. Rybak. 1998. In vitro and in vivo characterization of LL2-RNase conjugates against the CD22 antigen on human B-cell lymphoma. *Proc. Am. Assoc. Cancer Res.* 39: 435, New Orleans, LA.

189. Strydom, D. J., J. W. Fett, R. R. Lobb, E. M. Alderman, J. L. Bethune, J. F. Riordan, B. L. Vallee. 1985. Amino acid sequence of human tumor derived angiogenin. *Biochemistry* 24: 5486–5494.

190. Rosenberg, H. F., D. G. Tenen, S. J. Ackerman. 1989. Molecular cloning of the human eosinophil-derived neurotoxin: A member of the ribonuclease gene family. *Proc. Natl. Acad. Sci. USA* 86: 4460–4464.

191. Rosenberg, H. F., S. J. Ackerman, D. G. Tenen. 1989. Human eosinophil catinonic protein. Molecular cloning of a cytotoxin and helminthotoxin with ribonuclease activity. *J. Exp. Med.* 170: 163–176.

192. Ardelt, W., S. M. Mikulski, K. Shogen. 1991. Amino acid sequence of an anti-tumor protein from *Rana pipiens* oocytes and early embryos. Homology to pancreatic ribonucleases. *J. Biol. Chem.* 266: 245–251.

4

RECOMBINANT IMMUNOTOXINS

DAVID FITZGERALD
National Cancer Institute
Bethesda, MD 20892

4.1 INTRODUCTION AND BACKGROUND

The product of joining an antibody molecule with a toxin molecule is termed an *immunotoxin.* Antibodies are large multifunctional proteins that protect mammals from pathogens. The IgG type of antibody, which binds antigen at two sites per molecule, participates in the memory response of the immune system. This often results in the production of antibodies with high binding affinity. Structure–function studies of IgG molecules have identified the sequences that contribute to antigen binding, and a combination of approaches has led to the identification of the Fv fragment (see Fig 4.1) as the smallest unit of antibody structure that mediates antigen binding.[1–3] The sequence localization of other antibody functions such as complement fixation and Fc receptor binding has also been reported.

Gene cloning techniques have allowed for the expression and characterization of isolated antibody fragments. While the Fv fragment retains binding activity, it contains none of the effector functions that can render an antibody cytotoxic. By using gene fusion techniques, it is possible to join the Fv fragment with functional domains from enzymes, toxins, and cytokines and produce novel therapeutic molecules. Toxins, the subject of this chapter, are cytotoxic proteins made by bacteria and plants, and the most potent among them can destroy mammalian cells with delivery of only a few molecules per cell.[4]

Antibody Fusion Proteins, Edited by Steven M. Chamow and Avi Ashkenazi
ISBN 0471-18358-X

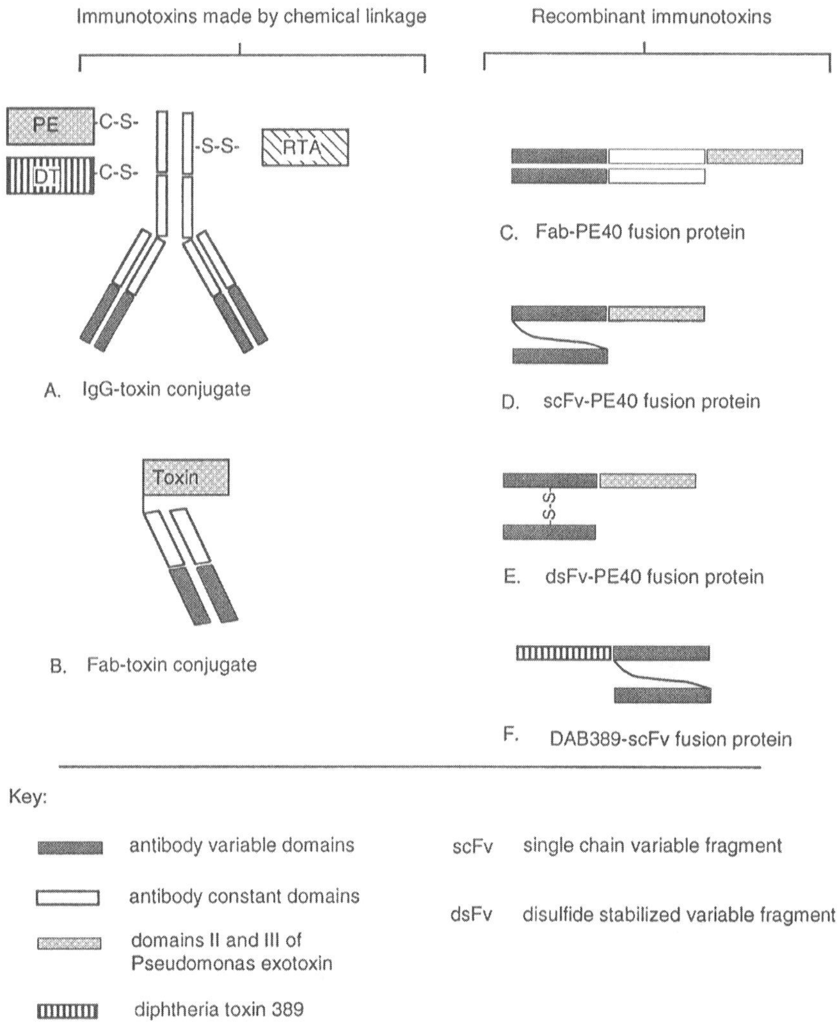

Figure 4.1 Immunotoxin design. Immunotoxins made by chemical linkage are shown on the left, those made recombinantly are shown on the right. *A*. This is the first generation of immunotoxin. The antibody is purified, the toxin is purified, and the two are joined together using bifunctional cross-linking agents. Shown is a disulfide-linked ricin A chain (RTA) immunotoxin, a thioether-linked DT immunotoxin, and a thioether-linked PE immunotoxin. *B*. Shown is the Fab fragment of an antibody linked to a toxin by a disulfide bond. *C*. This is a Fab fragment constructed by fusing the cDNA encoding the Fd (the variable and CH1 domains of a heavy chain) with DNA encoding PE40 (a truncated form of Pseudomonas exotoxin composed of domains II and III—the translocating and ADP-ribosylating domains, respectively). The cDNA for the light chain is expressed from a separate plasmid (Fig. 4.3). The final construct is assembled by refolding the component parts in a redox shuffling buffer.[59] Recently, Chen et al. expressed a recombinant Fab toxin construct in mammalian cells where

Typically, toxins are enzymes that disrupt a vital cellular function. Because of their potency, toxins are attractive agents for destroying cancer cells. At the most basic level, immunotoxins are designed to allow the antibody fragment to bind to a cell surface antigen or receptor, while the toxin portion mediates cell death (see also Chapter 3).

Immunotoxins, made using mouse monoclonal antibodies, were first described about 20 years ago.[5] Initially, these bifunctional molecules were constructed by purifying the full-sized antibody from hybridoma supernatants, the toxin from bacterial or plant extracts, and then combining the two using chemical cross-linking reagents.[6–10] The goal was to produce a one-to-one antibody–toxin conjugate. Routinely, this ratio produced active hybrids that retained most of the antibody's original binding characteristics and the enzymatic activity associated with the lethal part of the toxin. Since an IgG antibody molecule has a molecular weight (MW) of 150 kDa and with toxins in the range of 30–60 kDa, the combined size of early immunotoxins was approximately 200 kDa (Fig. 4.1A). These conjugates were first constructed using thiol chemistries that produced disulfide bonds. Later, depending on the nature of the toxin, it was possible to make more stable thioether bonds (Fig. 4.1A).

Because many cancers are not cured with existing therapies, immunotoxins have become prominent in the field of experimental therapeutics. Clinical trials were initiated about 10 years ago and continue today (see following and Ref. 11). When immunotoxins are used as anticancer agents, the antibody is selected based on its ability to bind to a surface antigen that is expressed preferentially on cancer tissue. Antibodies that cross react with normal tissues are discarded, lest they produce unanticipated side effects. Also, the antibody must mediate transport of the toxin to the inside of the target cell. Thus antireceptor antibodies are often chosen. The toxin is selected from a small collection of known bacterial or plant toxins. Particular attention is made in choosing an agent that is both potent and broadly active on different cells and tissues. Since cancer arises from many different cell types, it is important to have a toxin that can be used universally. Toxins with enzymatic domains that inflict biochemi-

---

Figure 4.1 *continued*
assembly occurred in the endoplasmic reticulum.[62] *D*. A single chain Fv (scFv) portion of an antibody[2,3] is fused directly with PE40.[51] This construct has the advantage of being expressed from one plasmid. However, not all scFv contructs encode antibody fragments that retain full and stable antigen-binding activity. Sometimes a more stable construct is necessary. *E*. The introduction of cysteine residues into the framework regions of the variable portions of the heavy and light chains has allowed for the assembly of very stable antibody fragments. This construct, like the recombinant Fab immunotoxin, must be expressed from two plasmids and then refolded to the final product. *F*. Toxin-derived gene fusions began with DT.[12] Here the DT is truncated at residue 389. Placed at the C-terminus, in place of the toxins binding domain, is the scFv.

cal lesions are preferred to those that create pores or other kinds of damage. Three of these enzymatic toxins, diphtheria toxin (DT), pseudomonas exotoxin (PE), and ricin, have emerged as standards against which all new toxins are likely to be judged. For reviews of immunotoxins made with these toxins see Refs 12–14. These particular toxins kill cells by shutting down protein synthesis, causing a biochemical lesion that ultimately leads to apoptotic cell death. Typically, immunotoxins made from one of these toxins with a cancer-binding monoclonal antibody have IC_{50} values in the range of 10^{-10}-10^{-12} molar. IC_{50} values are generated from dose-dependent killing curves, where increasing concentrations of immuntoxin are added to cell lines that express the target antigen. In this type of assay, an antigen-negative cell line is used as a background control. An antigen-negative line will usually be at least 1000-fold less sensitive to the same immuntoxin.

4.2 TOXIN STRUCTURE AND FUNCTION

Toxins such as DT, PE, and ricin are multidomain proteins composed of functionally similar subunits. To kill cells, these toxins require minimally three domains: cell surface binding, translocation to the cell cytosol, and enzymatic inactivation of protein synthesis. Because crystal structures have been reported for each toxin,[15–18] it has been possible to analyze structure–function relationships. This has been aided considerably by the expression of recombinant forms of each protein, including single site mutations and truncations.[13,19–22] Despite a lack of sequence homology, these toxins share common themes of domain construction.

Each toxin has a binding domain that interacts with surface receptors on mammalian cells and tissues. DT binds to the heparin-binding epidermal growth factor-like (HB-EGF) precursor that is present on the surface of most mammalian cells and tissues.[23] Rodents such as mice and rats do not express a form of this protein that can bind DT and are therefore toxin resistant. Transfection of rodent cells with HB-EGF precursor restores toxin sensitivity.[23] PE binds to LRP (also known as the alpha 2 macroglobulin receptor).[24,25] Again, this is a receptor expressed on most cells and tissues. Mutant cells selected as toxin resistant and subsequently shown to be LRP-negative were shown to be 100-fold resistant compared to wild-type cells.[25] Transfection with chicken LRP can restore toxin sensitivity, confirming the role of the receptor in toxin uptake[74]. Ricin is a lectin and binds cells that display galactosylated proteins and lipids on their cell surface.[26] Because many different cell-surface proteins including receptors are glycoproteins that contain galactose residues, ricin is thought to enter cells via a variety of different receptors.[26]

Following binding to cell-surface proteins, toxin uptake occurs via the endocytic pathway. Delivery to the cytosol requires translocation. Because only a portion of each toxin is translocated (Fig. 4.2), a processing step (or

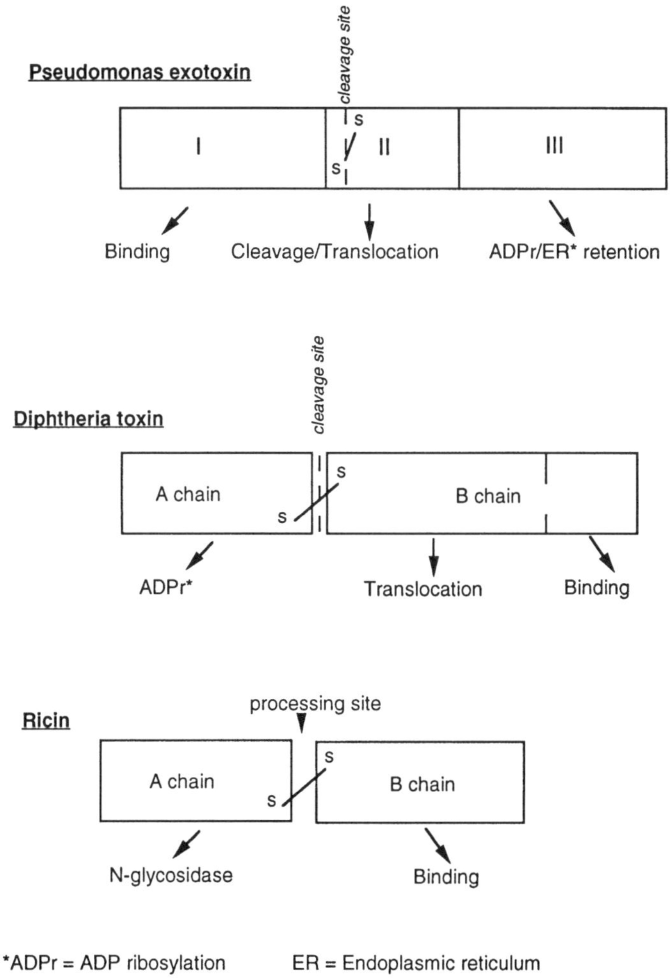

Figure 4.2 Organization of toxin functional domains. Shown is the location of the binding, translocation, processing and enzymatically active domains of DT, PE, and ricin.

steps) to produce the relevant fragment is required. So before considering translocation, we must examine the postendocytic processing events in the toxin pathway that produce active toxin fragments.

The processing of DT occurs in two steps: protease cleavage[27,28] followed by reduction of a key disulfide bond.[29] Cleavage can be accomplished in either of two locations. When secreted from *C. diphtheriae*, cleavage can occur in the medium by unidentified proteases or by cell-associated proteases such as furin.[30] Cleavage of DT occurs within an arginine-rich loop that contains a disulfide bond linking cysteines 182 and 197.[27,28] The subsequent reduction of this disulfide bond produces free A chain, the toxin subunit with enzymatic activity that is lethal to the cell. Transport of the A chain requires sequences in specific helices of the B chain.[22]

In contrast to DT, processing of PE happens entirely within mammalian cells.[31] Cleavage is by the cellular protease, furin, and appears to require an unfolding step that is mediated by the low pH environment of the endosome.[32,33] Like DT, PE is cleaved in an arginine-rich sequence (arginines at 274, 276, and 279) that contains a disulfide bond. Furin-mediated cleavage occurs between arginine 279 and glycine 280.[34] While processing of DT releases an NH-terminal fragment, the processing of PE produces a C-terminal 37 kDa fragment (Fig. 4.3). After cleavage by furin, reduction of the disulfide bond joining 265-287 releases this fragment.

The proteolytic processing of ricin is different from either of the two bacterial toxins. Ricin is produced from proricin by removal of a 12 amino acid linker peptide.[35] Processing happens within the germinating castor bean seeds

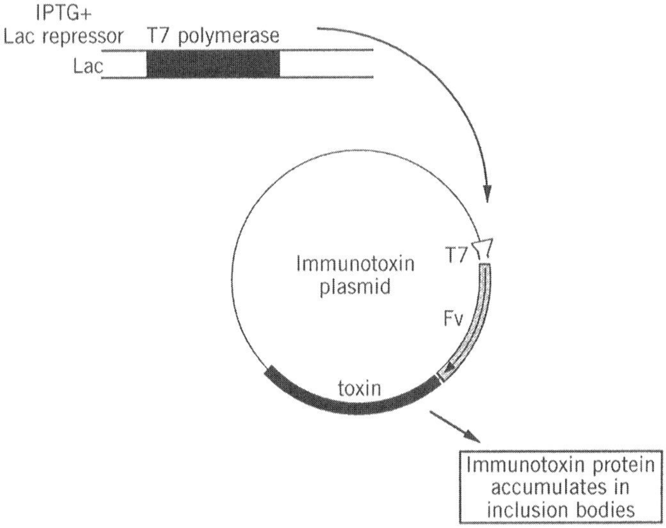

Figure 4.3 Expression of recombinant immunotoxins in *E. coli*. The T7 expression system, developed by Studier,[58] is composed of a relatively few elements. The T7 RNA polymerase is expressed from an integrated lambda phage under the control of a lac promoter. The addition of IPTG to the bacterial culture releases the negative control of the lac promoter and allows expression of this polymerase. An *E. coli* host (e.g., BL21 (λDE3) is transfected with a plasmid encoding the immunotoxin, preceeded by the T7 promoter. When IPTG is added to the culture, T7 polymerase transcribes many copies of toxin RNA. When this is translated into protein, it accumulates as inclusion bodies. In the case of Fab or dsFv constructs, separate cultures need to be made to express both heavy and light chain proteins.

leaving an A/B disulfide-linked heterodimer. Reduction of this disulfide bond produces free A chain for translocation.

Translocation follows toxin processing. Optimal translocation of DT requires an acidic pH environment.[22,36,37] Translocation of PE involves five of the six helices of B chain domain II and a C-terminal KDEL sequence.[38–40] For PE37, the need for a KDEL sequence has implicated the endoplasmic reticulum as the site of translocation. The secretory compartment is also thought to play an important role in the translocation of ricin to the cytosol. Monoclonal antiricin antibodies were used to block the toxicity of ricin and were thought to act in the secretory pathway because hybridoma cells synthesizing these antibodies became resistant to ricin.[41] While it does not have a conventional KDEL sequence, ricin A chain is rendered more toxic by artificially adding a KDEL sequence.[42,43] This result has bolstered the idea that ricin A chain also translocates from the endoplasmic reticulum to the cytosol. The role of ricin B chain in the translocation of the A chain is unclear. Immunotoxins made using whole ricin are usually more active than those with ricin A chain alone.[44,45] Clearly, the B chain does more than bind to a surface receptor: It may aid in the translocation process itself, it may act as an unfolding chaperone, or it may direct intracellular routing. Frankel et al. have presented data to show that a mutated B chain with reduced sugar binding can still enhance delivery of ricin A chain to the cell cytosol.[46]

These toxins cause enzymatic inhibition of protein synthesis. DT A chain and PE 37 enzymatically modify EF-2[47–49] by ADP-ribosylation so that it can no longer participate in the synthesis of new proteins. RTA is toxic because of its enzymatic attack on eukaryotic ribosomes.[50]

4.3 IMMUNOTOXIN DESIGN

Why choose toxins to kill cancer cells and other disease-causing cells? Because toxins are potent enzymes with high turnover rates. This means that only a few molecules need to be delivered to the cytosol to kill a target cell. Most endogenous enzyme systems have inhibitors and built-in regulation mechanisms that prevent them from uncontrolled and potentially destructive activity. This system of checks and balances is not operative when toxins are delivered to the cytosol. Apparently, there are no cellular inhibitors of toxin activity; moreover toxins are very stable to protease destruction. Thus, a strong case is made for exploiting toxins as the method of choice for killing diseased cells.

What is the role of the antibody? Antibodies must directly or indirectly deliver toxins to their final destination. The most useful antibodies react selectively with proteins on the surface of the target cell and mediate transport of the toxin to the appropriate intracellular destination (see Box 4.1 for intracellular immunotoxin pathway).

Recombinant immunotoxins follow one of three basic designs (see Fig. 4.1D–F). In the first example (Fig. 4.1D), an scFv-toxin is expressed from a

Box 4.1 Recombinant Immunotoxins: Pathways to the Cell Cytosol.

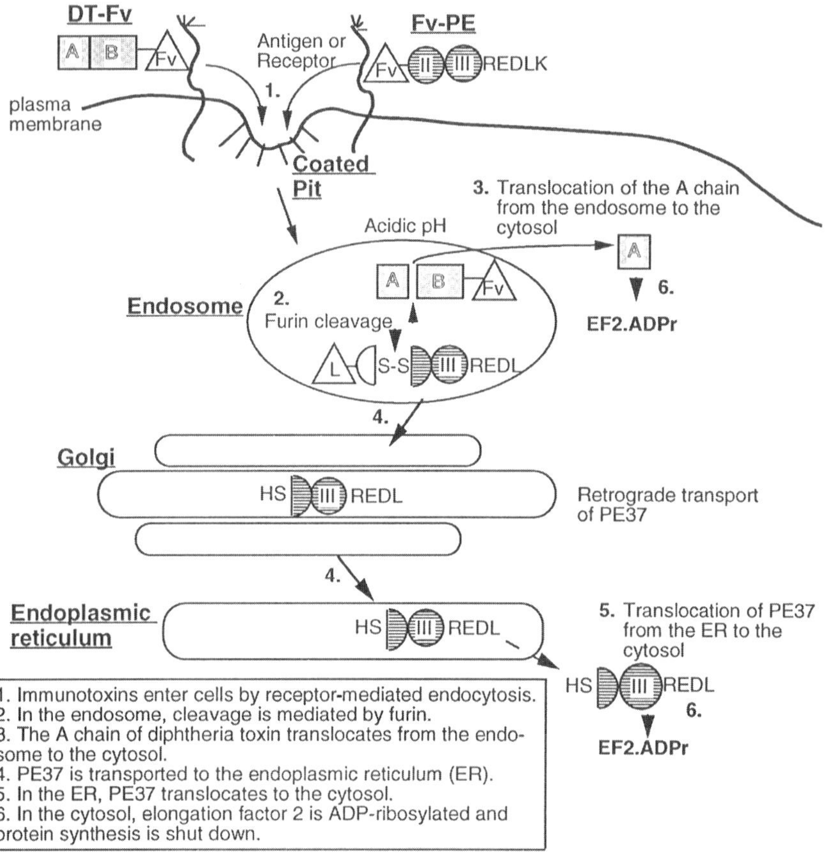
DT-Fv
A
B
Fv
Antigen or Receptor
Fv-PE
Fv
II
III
REDLK
plasma membrane
1.
Coated Pit
3. Translocation of the A chain from the endosome to the cytosol
Acidic pH
A
6.
EF2.ADPr
Endosome
2.
Furin cleavage
A
B
Fv
L
S-S
III
REDL
4.
Golgi
HS
III
REDL
Retrograde transport of PE37
4.
Endoplasmic reticulum
HS
III
REDL
5. Translocation of PE37 from the ER to the cytosol
HS
III
REDL
6.
EF2.ADPr
1. Immunotoxins enter cells by receptor-mediated endocytosis.
2. In the endosome, cleavage is mediated by furin.
3. The A chain of diphtheria toxin translocates from the endosome to the cytosol.
4. PE37 is transported to the endoplasmic reticulum (ER).
5. In the ER, PE37 translocates to the cytosol.
6. In the cytosol, elongation factor 2 is ADP-ribosylated and protein synthesis is shut down.

single transcript that encodes the variable portions of the heavy and light chains joined by a peptide linker.[51] The C-terminal end of one of the antibody chains is fused to sequences that encode the toxin. In the second example, dsFv-toxin (Fig. 4.1E) is produced from two different transcripts.[52–54] The variable portions of the heavy and light chains are each mutated to introduce a novel cysteine residue. Residues to be mutated are chosen based on their proximity to each other in antibody structures. Amino acids that are spaced about 6 Å apart are chosen most often. Each cDNA is mutated. The variable portion of either antibody chain is fused with the toxin to produce one of the coding plasmids; the other plasmid expresses only the varible portion of the antibody. The final product is realized after refolding under conditions to promote the formation of the novel disulfide bond. In the third example, recombinant Fab-toxin constructs (Fig. 4.1C) are similar to the dsFvs except the entire light chain and the first two domains of the heavy chain make up the antibody contribution. Because Fabs have "natural cysteines" that form disulfide bonds, there is no need to mutate sequences to produce covalently linked proteins. Again the toxin can be fused to either chain.

Until quite recently, recombinant immunotoxins were made using bacterial toxins only. Ricin was one of the first plant toxins to be evaluated as a fusion partner with another protein. However, when ricin A chain was fused to a cell binding ligand, it lost its ability to translocate to the cell cytosol. It has therefore been an important goal to find alternatives to this toxin from plant sources. The cloning, expression, and characterization of recombinant bryodin has introduced a new toxin for consideration.[55,56] The work of Siegall et al. have confirmed that plant toxins (from the ribosome inactivating family of toxins) can be used in the recombinant production of active immunotoxins both in *E. coli*[57] and in tobacco leaves (see next section).

4.4 IMMUNOTOXIN EXPRESSION

4.4.1 Immunotoxin Expression and Purification from *E. coli*

Immunotoxin sequences are cloned in an expression plasmid downstream of the T7 promoter. Plasmids are transformed into an expression host, typically *E. coli* strain BL21(λDE3), developed by Studier.[58] The strain produces T7 polymerase in response to induction of the lactose operon. BL21(λDE3), transformed with the appropriate plasmid, is grown in Luria-Bertani broth (LB) with ampicillin (100 μg/ml). At an absorbance (600 nm) of 0.5-2.0, the culture is induced with isopropyl-β-D-thiogalactopyranoside (IPTG) to bring about high level expression (Fig. 4.3). Recombinant immunotoxins are usually produced within the *E. coli* cell and are later recovered from inclusion bodies by denaturation and renaturation. The renatured material is then typically purified by Q Sepharose, Mono Q, and size exclusion high-pressure liquid chromatography.[59]

Immunotoxin Expression from Mammalian Cells

Because toxins are active against mammalian protein synthesis machinery, it was thought originally that expression in mammalian systems would result in "suicide" and death of the producing cells. In fact some gene therapy strategies deliberately use toxin genes in conjunction with specific promoters to selectively eliminate cells based on the initiation of promoter activity.[60,61] It was therefore quite surprising when Chen et al. announced that immunotoxins could be produced in mammalian cells.[62] They used a mammalian secretion signal sequence and cells that were immunotoxin receptor-negative to achieve their result. Their goal is to target T cells to tumors with local expression of immunotoxins, thus providing a way to reduce systemic exposure to these very toxic proteins.

Expression from Tobacco Leaves

The expression of plant toxins in most yeast, plant, and mammalian cells causes death of the host. Tagge et al. were the first to report the use of transgenic tobacco plants for expression of ricin-based toxins.[63] While expression levels were not very high, the ability to harvest thousands of tobacco leaves allows for considerable production. Siegall et al. have used this kind of system to produce their bryodin-derived immunotoxins.[64]

4.5 PRECLINICAL TESTING OF IMMUNOTOXINS

Purified immunotoxins are tested for cytotoxicity by measuring their ability to inhibit protein synthesis. Typically, immunotoxins in the range of 0.1–100 ng/ml are added to both target and nontarget cells. After an overnight incubation at 37°C, protein synthesis is measured by incubating cells with a radioactive amino acid. Immunotoxins that exhibit IC_{50} values below 10 ng/ml for target cells and above 1000 ng/ml for nontarget cells are considered potent and selective enough for testing in tumor models.

Usually, immunotoxin activity against growing tumors is assessed in a nude mouse or a SCID mouse model of the relevant cancer. Mice are first injected with the appropriate number of human tumor cells. After several days to allow the tumor to get established, immunotoxin treatments are given (immunotoxin injections on days 5, 7, and 9 are typical). Antitumor activity and toxicity in mice are assessed. Immunotoxins have an impressive record of being able to eliminate established tumors in such animal models. This capacity has been seen with treating both adenocarcinomas and hematologic tumors.[65–71] For animals in which complete tumor regression is observed, a therapeutic window is established by dividing the efficacious dose by the LD_{50} dose.

Immunotoxins with a good therapeutic window are then assessed in monkeys for toxicity to normal primate tissue. On completion of these preclinical evaluations, compounds with an acceptable therapeutic profile go foward to phase I trials in patients harboring cancers that have failed to respond adequately to conventional treatment.

4.6 CLINICAL USE OF IMMUNOTOXINS

The history of immunotoxin use in humans has been described in several recent articles.[11,72,73] These accounts tell of early clinical trials and progress made up to 1995. For the most part, phase I and II trials were conducted with immunotoxins constructed with full-sized antibodies covalently linked with one of the three toxins previously mentioned (or one of the plant toxins closely related to ricin, e.g., saporin). In almost every trial, tumors were seen to shrink and there was often a sense of immunotoxin efficacy. However, no trial has yet reported the dramatic "cures" that were seen when the same agents killed off rapidly growing tumors in nude mice. In fact, the most consistent observation has been one of unanticipated toxicities. These were often dose limiting at doses below those shown to be effective in model systems. Because the majority of the antibodies used in these trials were raised against human tissue (or proteins) injected into Balb/c mice, the antibodies were usually nonreactive with mouse tissue and reactive only with human (and sometimes monkey) tissue. Because of this disparity, it may be time to consider the production of antibodies by means other than those currently employed. For instance, the use of unimmunized phage display libraries may produce antibodies that react equally well with the mouse and human variants of the same receptor or antigen (see Chapter 6). This will allow for the generation of more meaningful preclinical toxicity data.

The future of immunotoxins as therapeutics is tied to our understanding of current clinical data and what they tell us about why immunotoxins have exhibited such promise in model systems, relatively modest toxicities in monkeys, and trial-ending toxicities in humans.

4.7 SUMMARY

The attraction of toxins as the killing component of an immunotoxin lies in the potency of these protein enzymes. When delivery of large molecules such as proteins to a tumor mass is limiting, the argument is made that we should employ the most potent toxic agent. In this way, only a few molecules need reach the intended target to gain therapeutic benefit. The expression of recombinant immunotoxins allows for the production of a homogeneous material that can be obtained in high yields. The obstacles remaining appear to be centered on dose-limiting damage to normal tissue, with its attendant side effects, and the immunogenicity of the foreign sequences in the toxin portion of the molecule. When these issues have been solved, current limitations may be overcome.

REFERENCES

1. Skerra, A., A. Plückthun. 1988. Assembly of a functional immunoglobulin Fv fragment in Escherichia coli. *Science* 240: 1038–1041.
2. Bird, R. E., K. D. Hardman, J. W. Jacobson, S. Johnson, B. M. Kaufman, S. M. Lee, T. Lee, S. H. Pope, G. S. Riordan, M. Whitlow. 1988. Single-chain antigen-binding proteins. *Science* 242: 423–426. [Published erratum appears in *Science* 1989 244(4903):409].

3. Huston, J. S., D. Levinson, H. M. Mudgett, M. S. Tai, J. Novotny, M. N. Margolies, R. J. Ridge, R. E. Bruccoleri, E. Haber, R. Crea, et al. 1988. Protein engineering of antibody binding sites: recovery of specific activity in an anti-digoxin single-chain Fv analogue produced in Escherichia coli. *Proc. Natl. Acad. Sci. USA* 85: 5879–5883.
4. Yamaizumi, M., E. Makada, T. Uchida, Y. Okada. 1978. One molecule of diphtheria toxin fragment A introduced into a cell can kill the cell. *Cell* 15: 245–250.
5. Blythman, H. E., P. Casellas, O. Gros, P. Gros, F. K. Jansen, F. Paolucci, B. Pau, H. Vidal. 1981. Immunotoxins: hybrid molecules of monoclonal antibodies and a toxin subunit specifically kill tumour cells. *Nature* 290: 145–146.
6. Cumber, A. J., J. A. Forrester, B. M. Foxwell, W. C. Ross, P. E. Thorpe. 1985. Preparation of antibody-toxin conjugates. *Methods Enzymol* 112: 207–225.
7. Ghetie, V., P. Thorpe, M. A. Ghetie, P. Knowles, J. W. Uhr, E. S. Vitetta. 1991. The GLP large scale preparation of immunotoxins containing deglycosylated ricin A chain and a hindered disulfide bond. *J. Immunol. Methods* 142: 223–230.
8. FitzGerald, D. J. 1987. Construction of immunotoxins using Pseudomonas exotoxin A. *Methods Enzymol.* 151: 139–145.
9. Kondo, T., D. FitzGerald, V. K. Chaudhary, S. Adhya, I. Pastan. 1988. Activity of immunotoxins constructed with modified Pseudomonas exotoxin A lacking the cell recognition domain. *J. Biol. Chem.* 263: 9470–9475.
10. Marsh, J. W., K. Srinivasachar, D. J. Neville. 1988. Antibody-toxin conjugation. *Cancer Treat. Res.* 37: 213–237.
11. Frankel, A. E., E. P. Tagge, M. C. Willingham. 1995. Clinical trials of targeted toxins. *Semin. Cancer Biol.* 6: 307–317.
12. Murphy, J. R., J. C. vanderSpek. 1995. Targeting diphtheria toxin to growth factor receptors. *Semin. Cancer Biol.* 6: 259–267.
13. Pastan, I. H., L. H. Pai, U. Brinkmann, D. J. FitzGerald. 1995. Recombinant toxins: new therapeutic agents for cancer. *Ann. N. Y. Acad. Sci.* 758: 345–354.
14. Vitetta, E. S., P. E. Thorpe. 1991. Immunotoxins containing ricin or its A chain. *Semin. Cell. Biol.* 2: 47–58.
15. Allured, V. S., R. J. Collier, S. F. Carroll, D. B. McKay. 1986. Structure of exotoxin A of Pseudomonas aeruginosa at 3.0 Å. *Proc. Natl. Acad. Sci.* 83: 1320–1324.
16. Choe, S., M. J. Bennett, G. Fujii, K. A. Kantardjieff, R. J. Collier, D. Eisenberg. 1992. The crystal structure of diphtheria toxin. *Nature* 357: 216–222.
17. Katzin, B. J., E. J. Collins, J. D. Robertus. 1991. Structure of ricin A-chain at 2.5 Å resolution. *Proteins:Structure, Function, and Genetics* 10: 251–259.
18. Rutenber, E., J. D. Robertus. 1991. Structure of ricin B-chain at 2.5 Å resolution. *Proteins:Structure Function and Genetics* 10: 260–269.
19. Frankel, A. E., C. Burbage, T. Fu, E. Tagge, J. Chandler, M. Willingham. 1996. Characterization of a ricin fusion toxin targeted to the interleukin-2 receptor. *Protein Eng.* 9: 913–919.
20. Frankel, A., P. Welsh, J. Richardson, J. D. Robertus. 1990. Role of arginine 180 and glutamic acid 177 of ricin toxin A chain in enzymatic inactivation of ribosomes. *Mol. Cell Biol.* 10: 6257–6263.
21. Shen, W. H., S. Choe, D. Eisenberg, R. J. Collier. 1994. Participation of lysine 516

and phenylalanine 530 of diphtheria toxin in receptor recognition. *J. Biol. Chem.* 269: 29077–29084.

22. Kaul, P., J. Silverman, W. H. Shen, S. R. Blanke, P. D. Huynh, A. Finkelstein, R. J. Collier. 1996. Roles of Glu 349 and Asp 352 in membrane insertion and translocation by diphtheria toxin. *Protein Sci.* 5: 687–692.
23. Naglich, J. G., J. E. Metherall, D. W. Russell, L. Eidels. 1992. Expression cloning of a diphtheria toxin receptor: identity with a heparin-binding EGF-like growth factor precursor. *Cell* 69: 1051–1061.
24. Kounnas, M. Z., R. E. Morris, M. R. Thompson, D. J. FitzGerald, D. K. Strickland, C. B. Saelinger. 1992. The alpha 2-macroglobulin receptor/low density lipoprotein receptor-related protein binds and internalizes pseudomonas exotoxin A. *J. Biol. Chem.* 267: 12420–12423.
25. FitzGerald, D. J., C. M. Fryling, A. Zdanovsky, C. B. Saelinger, M. Kounnas, J. A. Winkles, D. Strickland, S. Leppla. 1995. Pseudomonas exotoxin-mediated selection yields cells with altered expression of low-density lipoprotein receptor-related protein *J. Cell. Biol.* 129: 1533–1541. [published erratum appears in *J. Cell. Biol.* 1995 130(4):1015].
26. Newton, D. L., R. Wales, P. T. Richardson, S. Walbridge, S. K. Saxena, E. J. Ackerman, L. M. Roberts, J. M. Lord, R. J. Youle. 1992. Cell surface and intracellular functions for ricin galactose binding. *J. Biol. Chem.* 267: 11917–11922.
27. DeLange, R. J., R. E. Drazin, R. J. Collier. 1976. Amino-acid sequence of fragment A, an enzymically active fragment from diphtheria toxin. *Proc. Natl. Acad. Sci. USA* 73: 69–72.
28. Moskaug, J. O., K. Sletten, K. Sandvig, S. Olsnes. 1989. Translocation of diphtheria toxin A-fragment to the cytosol. Role of the site of interfragment cleavage. *J. Biol. Chem.* 264: 15709–15713.
29. Ryser, H. J., R. Mandel, F. Ghani. 1991. Cell surface sulfhydryls are required for the cytotoxicity of diphtheria toxin but not of ricin in Chinese hamster ovary cells. *J. Biol. Chem.* 266: 18439–18442.
30. Tsuneoka, M., K. Nakayama, K. Hatsuzawa, M. Komada, N. Kitamura, E. Mekada. 1993. Evidence for involvement of furin in cleavage and activation of diphtheria toxin. *J. Biol. Chem.* 268: 26461–26465.
31. Ogata, M., V. K. Chaudhary, I. Pastan, D. J. FitzGerald. 1990. Processing of Pseudomonas exotoxin by a cellular protease results in the generation of a 37,000-Da toxin fragment that is translocated to the cytosol. *J. Biol. Chem.* 265: 20678–20685.
32. Moehring, J. M., N. M. Inocencio, B. J. Robertson, T. J. Moehring. 1993. Expression of mouse furin in a Chinese hamster cell resistant to Pseudomonas exotoxin A and viruses complements the genetic lesion. *J. Biol. Chem.* 268: 2590–2594.
33. Chiron, M. F., C. M. Fryling, D. J. FitzGerald. 1994. Cleavage of Pseudomonas exotoxin and diphtheria toxin by a furin-like protease prepared from beef liver. *J. Biol. Chem.* 269: 18167–18176.
34. Ogata, M., C. M. Fryling, I. Pastan, D. J. FitzGerald. 1992. Cell-mediated cleavage of Pseudomonas exotoxin between Arg279 and Gly280 generates the enzymatically active fragment which translocates to the cytosol. *J. Biol. Chem.* 267: 25396–25401.

35. Harley, S. M., J. M. Lord. 1985. In vitro endoproteolytic cleavage of castor bean lectin precursors. *Plant Science* 41: 111–116.

36. Draper, R. K., M. I. Simon. 1980. The entry of diphtheria toxin into the mammalian cell cytoplasm: evidence for lyosomal involvement. *J. Cell. Biol.* 87: 849–854.

37. Sandvig, K., S. Olsnes. 1980. Diphtheria toxin entry into cells is facilitated by low pH. *J. Cell. Biol.* 87: 828–832.

38. Siegall, C. B., V. K. Chaudhary, D. J. FitzGerald, I. Pastan. 1989. Functional analysis of domains II, Ib, and III of Pseudomonas exotoxin. *J. Biol. Chem.* 264: 14256–14261.

39. Siegall, C. B., M. Ogata, I. Pastan, D. J. FitzGerald. 1991. Analysis of sequences in domain II of Pseudomonas exotoxin A which mediate translocation. *Biochem.* 30: 7154–7159.

40. Chaudhary, V. K., Y. Jinno, D. FitzGerald, I. Pastan. 1990. Pseudomonas exotoxin contains a specific sequence at the carboxyl terminus that is required for cytotoxicity. *Proc. Natl. Acad. Sci. USA* 87: 308–312.

41. Youle, R. J., M. Colombatti. 1987. Hybridoma cells containing intracellular anti-ricin antibodies show ricin meets secretory antibody before entering the cytosol. *J. Biol. Chem.* 262: 4676–4682.

42. Wales, R., J. A. Chaddock, L. M. Roberts, J. M. Lord. 1992. Addition of an ER retention signal to the ricin A chain increases the cytotoxicity of the holotoxin. *Exp. Cell. Res.* 203: 1–4.

43. Wales, R., L. M. Roberts, J. M. Lord. 1993. Addition of an endoplasmic reticulum retrieval sequence to ricin A chain significantly increases its cytotoxicity to mammalian cells. *J. Biol. Chem.* 268: 23986–23990.

44. Lambert, J. M., G. McIntyre, N. Gauthier, D. Zullo, V. Rao, R. M. Steeves, V. S. Goldmacher, W. A. Blattler. 1991. The galactose-binding sites of the cytotoxic lectin ricin can be chemically blocked in high yields with reactive ligands prepared by chemical modification of glycopeptides containing triantennary N-linked oligosaccharides. *Biochemistry* 30: 3234–3247.

45. Lambert, J. M., V. S. Goldmacher, A. R. Collinson, L. M. Nadler, W. A. Blattler. 1991. An immunotoxin prepared with blocked ricin: a natural plant toxin adapted for therapeutic use. *Cancer Res.* 51: 6236–6242.

46. Frankel, A. E., C. Burbage, T. Fu, E. Tagge, J. Chandler, M. C. Willingham. 1996. Ricin toxin contains at least three galactose-binding sites located in B chain subdomains 1 alpha, 1 beta, and 2 gamma. *Biochemistry* 35: 14749–14756.

47. Collier, R. J. 1975. Diphtheria toxin: mode of action and structure. *Bact. Rev.* 39: 54–85.

48. Chung, D. W., R. J. Collier. 1977. Enzymatically active peptide from the adenosine diphosphate-ribosylating toxin of Pseudomonas aeruginosa. *Infect. and Immun.* 16: 832–841.

49. Iglewski, B. H., D. Kabat. 1975. NAD-Dependent inhibition of protein synthesis by Pseudomonas aeruginosa toxin. *Proc. Natl. Acad. Sci. USA* 72: 2284–2288.

50. Endo, Y., K. Mitsui, M. Motizuki, K. Tsurugi. 1987. Mechanism of action of ricin and related toxic lectins on eukaryotic ribosomes: the site and characteristics of the modification in 28S rRNA caused by the toxins. *J. Biol. Chem.* 262: 5908–5912.

51. Chaudhary, V. K., C. Queen, R. P. Junghans, T. A. Waldmann, D. J. FitzGerald, I.

Pastan. 1989. A recombinant immunotoxin consisting of two antibody variable domains fused to Pseudomonas exotoxin. *Nature* 339: 394–397.

52. Reiter, Y., U. Brinkmann, K. O. Webber, S. H. Jung, B. Lee, I. Pastan. 1994. Engineering interchain disulfide bonds into conserved framework regions of Fv fragments: improved biochemical characteristics of recombinant immunotoxins containing disulfide-stabilized Fv. *Protein Eng.* 7: 697–704.

53. Reiter, Y., U. Brinkmann, R. J. Kreitman, S. H. Jung, B. Lee, I. Pastan. 1994. Stabilization of the Fv fragments in recombinant immunotoxins by disulfide bonds engineered into conserved framework regions. *Biochemistry* 33: 5451–5459.

54. Reiter, Y., U. Brinkmann, S. H. Jung, B. Lee, P. G. Kasprzyk, C. R. King, I. Pastan. 1994. Improved binding and antitumor activity of a recombinant anti-erbB2 immunotoxin by disulfide stabilization of the Fv fragment. *J. Biol. Chem.* 269: 18327–18331.

55. Siegall, C. B., S. L. Gawlak, D. Chace, E. A. Wolff, B. Mixan, H. Marquardt. 1994. Characterization of ribosome-inactivating proteins isolated from Bryonia dioica and their utility as carcinoma-reactive immunoconjugates. *Bioconjug. Chem.* 5: 423–429.

56. Gawlak, S. L., M. Neubauer, H. E. Klei, C. Y. Chang, H. M. Einspahr, C. B. Siegall. 1997. Molecular, biological, and preliminary structural analysis of recombinant bryodin 1, a ribosome-inactivating protein from the plant Bryonia dioica. *Biochemistry* 36: 3095–3103.

57. Francisco, J. A., S. L. Gawlak, C. B. Siegall. 1997. Construction, expression, and characterization of BD1-G28-5 sFv, a single-chain anti-CD40 immunotoxin containing the ribosome-inactivating protein bryodin 1. *J. Biol. Chem.* 272: 24165–24169.

58. Studier, F. W., A. H. Rosenberg, J. J. Dunn, J. W. Dubendorff. 1990. Use of T7 RNA polymerase to direct expression of cloned genes. *Methods Enzymol.* 185: 60–89.

59. Buchner, J., I. Pastan, U. Brinkmann. 1992. A method for increasing the yield of properly folded recombinant fusion proteins: single-chain immunotoxins from renaturation of bacterial inclusion bodies. *Anal. Biochem.* 205: 263–270.

60. Robinson, D. F., I. H. Maxwell. 1995. Suppression of single and double nonsense mutations introduced into the diphtheria toxin A-chain gene: a potential binary system for toxin gene therapy. *Hum. Gene Ther.* 6: 137–143.

61. Garabedian, E. M., L. J. Roberts, M. S. McNevin, J. I. Gordon. 1997. Examining the role of Paneth cells in the small intestine by lineage ablation in transgenic mice. *J. Biol. Chem.* 272: 23729–23740.

62. Chen, S. Y., A. G. Yang, J. D. Chen, T. Kute, C. R. King, J. Collier, Y. Cong, C. Yao, X. F. Huang. 1997. Potent antitumour activity of a new class of tumour-specific killer cells. *Nature* 385: 78–80.

63. Tagge, E. P., J. Chandler, B. Harris, M. Czako, L. Marton, M. C. Willingham, C. Burbage, L. Afrin, A. E. Frankel. 1996. Preproricin expressed in Nicotiana tabacum cells in vitro is fully processed and biologically active. *Protein Expr. Purif.* 8: 109–118.

64. Francisco, J. A., S. L. Gawlak, M. Miller, J. Bathe, D. Russell, D. Chace, B. Mixan, L. Zhao, H. P. Fell, C. B. Siegall. 1997. Expression and characterization of bryodin

1 and a bryodin 1-based single-chain immunotoxin from tobacco cell culture. *Bioconjug. Chem.* 8: 708–713.

65. Heimbrook, D. C., S. M. Stirdivant, J. D. Ahern, N. L. Balishin, D. R. Patrick, G. M. Edwards, J. D. Defeo, D. J. FitzGerald, I. Pastan, A. Oliff. 1990. Transforming growth factor alpha-Pseudomonas exotoxin fusion protein prolongs survival of nude mice bearing tumor xenografts. *Proc. Natl. Acad. Sci. USA* 87: 4697–4701.
66. Brinkmann, U., L. H. Pai, D. J. FitzGerald, M. C. Willingham, I. Pastan. 1991. B3(Fv)-PE38KDEL, a single chain immunotoxin that causes complete regression of a human carcinoma in mice. *Proc. Natl. Acad. Sci. USA* 88: 8616–8620.
67. Kreitman, R. J., R. K. Puri, I. Pastan. 1995. Increased antitumor activity of a circularly permuted interleukin 4-toxin in mice with interleukin 4 receptor-bearing human carcinoma. *Cancer Res.* 55: 3357–3363.
68. Benhar, I., Y. Reiter, L. H. Pai, I. Pastan. 1995. Administration of disulfide-stabilized Fv-immunotoxins B1(dsFv)-PE38 and B3(dsFv)-PE38 by continuous infusion increases their efficacy in curing large tumor xenografts in nude mice. *Int. J. Cancer* 62: 351–355.
69. Ghetie, M. A., J. Richardson, T. Tucker, D. Jones, J. W. Uhr, E. S. Vitetta. 1991. Antitumor activity of Fab' and IgG-anti-CD22 immunotoxins in disseminated human B lymphoma grown in mice with severe combined immunodeficiency disease: effect on tumor cells in extranodal sites. *Cancer Res.* 51: 5876–5880.
70. Fulton, R. J., J. W. Uhr, E. S. Vitetta. 1988. In vivo therapy of the BCL1 tumor: effect of immunotoxin valency and deglycosylation of the ricin A chain. *Cancer Res.* 48: 2626–2631.
71. Francisco, J. A., G. J. Schreiber, C. R. Comereski, L. E. Mezza, G. L. Warner, T. J. Davidson, J. A. Ledbetter, C. B. Siegall. 1997. In vivo efficacy and toxicity of a single-chain immunotoxin targeted to CD40. *Blood* 89: 4493–4500.
72. Frankel, A. E., D. FitzGerald, C. Siegall, O. W. Press. 1996. Advances in immunotoxin biology and therapy: a summary of the Fourth International Symposium on Immunotoxins. *Cancer Res.* 56: 926–932.
73. Pai, L. H., R. Wittes, A. Setser, M. C. Willingham, I. Pastan. 1996. Treatment of advanced solid tumors with immunotoxin LMB-1: an antibody linked to Pseudomonas exotoxin. *Nat. Med.* 2: 350–353.
74. Avramoglu, R. K., J. Nimpf, R. S. McLeod, K. W. Ko, Y. Wang, D. Fitzgerald, and Z. Yao. 1998. Functional expression of the chicken low density lipoprotein receptor-related protein in a mutant chinese hamster ovary cell line restores toxicity of Pseudomonas exotoxin A and degradation of alpha2-macroglobulin. *J. Biol. Chem.* 273(11): 6057–6065.

5

$F(ab')_2$ FUSION PROTEINS AND BISPECIFIC $F(ab')_2$

J. YUN TSO
Protein Design Labs, Inc.
Fremont, CA 94555

5.1 STRUCTURE AND FUNCTION OF $F(ab')_2$

$F(ab')_2$ is the largest proteolytic fragment that retains the bivalent binding sites of an antibody. It can serve as a targeting module to bring toxins, enzymes, or effector cells to antigen-bearing cells. Recent developments in genetic engineering have produced recombinant $F(ab')_2$ fragments possessing novel properties. The properties of $F(ab')_2$ and its fusion proteins, including bispecific $F(ab')_2$, are discussed in detail in this chapter.

5.1.1 Proteolytic Cleavage of IgG

Immunoglobulin G (IgG) molecules are susceptible to proteolytic cleavage. The most susceptible sites are located at the hinge, which is the most exposed and extended portion of the molecule. Cleavage of IgG at the upper hinge region above the interheavy chain disulfide bonds splits IgG into two monovalent Fab fragments and one Fc fragment. Digestion of the molecule below these disulfide bonds splits the molecule into one bivalent $F(ab')_2$ fragment and one Fc fragment. The $F(ab')_2$ fragment is the NH_2 portion of the molecule containing the two antigen-binding sites of an antibody linked at the hinge by

Antibody Fusion Proteins, Edited by Steven M. Chamow and Avi Ashkenazi
ISBN 0471-18358-X

interheavy chain disulfide bonds. The IgG domains retained in F(ab')$_2$ are: the V_H, the C_H1 and the hinge segment of the heavy chain; and the V_L and C_κ or C_λ regions of the light chain. The truncated heavy chains in F(ab')$_2$ are sometimes referred to as Fd fragments. The two symmetrical halves of F(ab')$_2$ are named Fab', which can be obtained by reduction of the interheavy chain disulfide bonds. Once they are reduced, two different Fab' can associate via interheavy chain disulfide bonds to form bispecific F(ab')$_2$. As a result, a bispecific F(ab')$_2$ has two different antigen-binding sites, each of which interacts with its antigen monovalently.

Since each IgG subclass has a unique protease susceptibility, digestion conditions to generate F(ab')$_2$ are mostly empirical. Methods for murine monoclonal IgG fragmentation have been described by P. Parham.[1] F(ab')$_2$ fragments can be obtained from mouse IgG1, IgG2a, and IgG3 by pepsin digestion. The susceptibility of these antibodies to pepsin varies and the digestion conditions used to generate F(ab')$_2$ have to be modified accordingly. IgG2b is also susceptible to pepsin, but no F(ab')$_2$ can be obtained from this subclass because the cleavage sites are located in the upper hinge region. More recently, Yamaguchi et al.[2] have tested digestion conditions using several proteases with narrower substrate specificity than the traditional papain and pepsin. They found lysyl endopeptidase could be used on IgG2b to generate F(ab')$_2$. In human IgG subclasses, IgG1, IgG2, and IgG4 are susceptible to pepsin digestion to generate F(ab')$_2$.[3] Fragmentation conditions for human IgG3 are not available.

5.1.2 Properties of F(ab')$_2$

The lack of Fc in F(ab')$_2$ renders it unable to mediate antigen-nonspecific mechanisms such as complement fixation and antibody-dependent cell-mediated cytotoxicity. Depending on the application, the lack of Fc in F(ab')$_2$ can be viewed as an advantage or a disadvantage. If a reagent targeting to the antigen without the interference of the Fc is desired, F(ab')$_2$ may be the molecule of choice. In vitro, F(ab')$_2$ is often used instead of whole antibody for tissue and cell staining to avoid Fc-directed binding. In in vivo imaging, radioisotope-labeled F(ab')$_2$ often achieves a better target-to-blood ratio than that achieved by IgG. F(ab')$_2$ is frequently used as a control to assess the contribution of the Fc in an antibody-mediated reaction. Sometimes the reaction is absolutely Fc-dependent, whereas other times the Fc contributes to side effects. A good example of the latter is the case of anti-CD3. Anti-CD3 antibodies such as OKT3 (mouse antihuman CD3) and 145.2C11 (hamster antimouse CD3) have dual activities in vivo: T-cell activation and T-cell immunosuppression. The first activity is Fc dependent and contributes to the toxicity[4,5] and immunogenicity of the antibody,[6] whereas the second activity is not Fc dependent and can be achieved by the F(ab')$_2$ without Fc-dependent side effects.[7,8] Another example is the murine antihuman gpIIb/IIIa antibody C4G1, which blocks the binding of fibrinogen to platelets. The humanized version of this antibody causes severe thrombocytopenia and death in monkeys, whereas the $F(ab')_2$ and Fab versions have less effect on platelet counts but still inhibit platelet aggregation.[9]

5.1.3 Purification

Since Fc is absent from $F(ab')_2$, techniques based on Fc binding cannot be used for the purification of $F(ab')_2$. Staphylococcal protein A affinity chromatography, which is Fc-dependent, is not generally used to purify $F(ab')_2$. It is, however, quite frequently used to separate whole or partially digested IgG from $F(ab')_2$ after proteolytic cleavage. In addition to Fc binding, protein A also has some affinity for the variable region in the human heavy chain subgroup III.[10] $F(ab')_2$ derived from this subgroup can sometimes be purified by protein A affinity chromatography. While protein G from streptococci also has affinity for the Fc of immunoglobulins, it can also bind to C_H1 domains of heavy chains bound to C_κ light chains.[11] Protein G affinity chromatography is therefore the method of choice for purifying $F(ab')_2$ and $F(ab')_2$ fusion proteins. Human or mouse $F(ab')_2$ molecules bind to protein G Sepharose in 0.1 M sodium acetate buffer, pH 5.0, and can be eluted with 0.1 M glycine-HCl at pH less than 3.0. Interestingly, $F(ab')_2$ containing λ light chains cannot be purified using these conditions.

When $F(ab')_2$ is the major protein in the mixture, hydrophobic interaction chromatography works very well for purification. Most $F(ab')_2$ fragments bind to Phenyl Sepharose or Phenyl-5-PW in 1 M ammonium sulfate/0.1 M sodium phosphate buffer, pH 7.0, and can be eluted using a 1.0–0.0 M ammonium sulfate gradient in 0.1 M sodium phosphate buffer, pH 7.0. Ion-exchange chromatography can also be used to purify $F(ab')_2$, but empirical knowledge is often required, due to the variation in pI of different $F(ab')_2$. Furthermore, the requirement of low salt buffer for binding often makes ion-exchange chromatography less convenient to use with spent culture supernatants. For example, the binding of IgG or $F(ab')_2$ fragments to BAKERBOND ABx requires the salt concentration to be less than 20 mM; antibodies or their fragments often come out of solution under these conditions.

5.1.4 Avidity

The affinity of an antibody measures the strength of binding of a single antigenic epitope to a single antigen-binding site. Bivalent IgG sometimes uses both of its binding sites to react with multiple identical epitopes on the cell surface. In this case, there is an increase of the apparent strength of binding since both antigen-binding sites must release simultaneously in order for the antigen and antibody to dissociate. The overall strength of binding of an antibody to an antigen is called its avidity. Since $F(ab')_2$ contains both antigen-binding sites, its avidity for antigen is generally not much different from that of its parent IgG. In the case of bispecific $F(ab')_2$, however, the situation may be different. Some antibodies exhibit a significant decrease in apparent binding strength when they are converted from a bivalent molecule to a monovalent molecule. The avidity effect can be as much as 50-fold.[12] Two different antibodies that bind to different epitopes of the same protein antigen with similar apparent avidity may still have very different affinity in monovalent form. If affinity is an important concern, it may be best to determine the affinity of each half of a bispecific antibody, in the form of Fab, before the bispecific molecule is constructed.

5.1.5 Pharmacokinetics and Biodistribution of $F(ab')_2$ in vivo

The size of $F(ab')_2$ is about 100 kD, too big to be filtered rapidly through the kidney. It therefore remains in the blood much longer than the smaller antibody fragments such as Fab or Fv. In human subjects, the serum half-life of $F(ab')_2$ is about 12 to 27 h.[13] Fv and Fab, on the other hand, can have a half-life ranging from 2.8 to 7.5 h.[14] Compared to IgG, with a serum half-life of 21 days in humans,[15] the half-life of $F(ab')_2$ is nevertheless still very short. The property of long serum half-life has been mapped by mutagenesis to the C_H2 and C_H3 domains of IgG (Box 5.1).[16–20]

Since Fc is missing in $F(ab')_2$ fragments, it is expected that $F(ab')_2$ or its fusion proteins would have a short serum half-life. This property would make $F(ab')_2$ unsuitable for treating chronic diseases that require constant blocking of the target sites. Even in acute situations, such as transplant rejection and cancer, frequent dosing of $F(ab')_2$ may be required to achieve efficacy.[7,21–23]

Box 5.1 Molecular Basis for the Long Serum Half-Life of IgG.

It has long been known that the property of long serum half-life of IgG resides in its Fc fragment, but detailed molecular information was lacking. Recently, Ward and colleagues[16] have localized by mutagenesis a site on murine IgG1 Fc that is involved in the control of catabolism. The essential residues are in close proximity to each other at the C_H2 and C_H3 domain interface. These investigators[17] also showed that the same residues are needed for binding to FcRn, the receptor involved in transferring IgG from mother to young by neonatal transcytosis. The importance of these residues is in agreement with the X-ray crystallographic structure of a rat FcRn:Fc complex.[18] Results in two recent papers[19,20] show that the function of FcRn may be more than just transferring IgG to fetus or neonate. In transgenic mice lacking FcRn, the catabolism of IgG was drastically increased, indicating that FcRn is the protector for IgG, responsible for the latter's long serum half-life. The proposed mechanism is that FcRn retains and rescues IgG intracellularly in the endosome, the site of protein catabolism, and recycles it to the cell surface where it dissociates into the extracellular fluid.

On the other hand, the faster clearance and lack of Fc receptor interaction make it a superior reagent for targeting and imaging. In radioimmunotherapy, $F(ab')_2$ fragments were found to accumulate at the tumor sites faster and deliver a higher percentage of the injected dose than intact antibodies.[24,25] Because of their faster clearance, isotope-labeled $F(ab')_2$ fragments also irradiate blood and normal tissues much less than intact antibodies. The only exception to this rule is that labeled $F(ab')_2$ fragments tend to accumulate more in kidney, which may have the undesirable effect of kidney toxicity.

Healthy individuals have natural antibodies against $F(ab')_2$ fragments that are generated in vivo by pepsin digestion.[26,27] These antibodies are potent immunoregulatory molecules and are part of the physiologic immune repertoire. The levels of these anti-$F(ab')_2$ are inversely correlated with the severity of certain autoimmune diseases.[28] These anti-$F(ab')_2$ antibodies react only with self-$F(ab')_2$ rather than $F(ab')_2$ from other species. For this reason, they are sometimes referred to as homoreactive antibodies. They recognize a conformational epitope that is buried in the lower hinge region of IgG.[29] How these antibodies regulate the immune repertoire is not entirely clear. Whatever normal physiological functions these antibodies perform, their presence might compromise the efficacy of $F(ab')_2$ if they are derived from the same or similar species as the host. Yano et al.[30] found that the levels of homoreactive antibody activity in monkeys correlated with the intensity of platelet depletion when the animals were injected with the $F(ab')_2$ fragment of humanized C4G1 (antihuman gpIIb/IIa antibody). In this treatment, thrombocytopenia is considered as an undesirable side effect. When using humanized $F(ab')_2$ or its fusion proteins in human subjects, it is important to keep in mind that the homoreactive epitope may compromise the desired therapeutic effect, and it is best to have it eliminated by mutagenesis.

5.2 PRODUCTION OF $F(ab')_2$ AND $F(ab')_2$ FUSION PROTEINS BY GENETIC ENGINEERING

$F(ab')_2$ fusion proteins are produced genetically by gene fusion. They are then expressed in host cells appropriate for the desired products. Even in generating $F(ab')_2$, it may be advantageous to produce them genetically to bypass the ill-controlled proteolytic process. In both cases, genetic engineering methods allow investigators the freedom to design $F(ab')_2$ with desirable properties such as humanized sequences, optimal linkers to the fused protein, and appropriate mutations to mask immunogenic epitopes. Mammalian and bacterial systems are both widely used for the expression of genetically engineered $F(ab')_2$.

5.2.1 Expression in Mammalian Cells

Genetically manipulated antibody genes are usually expressed in mammalian cells. The most commonly used cells are mouse myeloma cells derived from

MOPC21, which is a plasmacytoma line obtained by injection of a Balb/c mouse with mineral oil. This cell line makes an antibody (IgG1/kappa) of an unknown specificity. Several sublines of MOPC21 have lost the ability to synthesize the endogenous antibody chains. These sublines, such as Sp2/0, NS1, and NS0, are ideal for use as fusion partners for hybridomas; moreover, they are good host lines for the expression of exogenously introduced antibody genes. The other frequently used cell line for antibody expression is the Chinese hamster ovary (CHO) cell line.[31] Expression vectors for the heavy and light chains are transfected into these cells by electroporation, and stable transfectants are isolated using a selection marker that is cotransfected with the antibody genes. It usually takes two to three weeks to obtain stable transfectants and another two to three weeks to obtain transfectant subclones. Most of these transfectants can grow in serum-free medium and can be scaled up for production in large bioreactors. In general, the antibody yields are very good in these systems, producing about 2–40 mg/L in tissue culture flasks and 200–1000 mg/L in bioreactors. Since antibodies are the major protein secreted by these cells, they are often straightforward to purify, and are substantially endotoxin free.

The first attempt to express $F(ab')_2$-like molecules was performed in mammalian cells using a method similar to that described above for antibody expression. Neuberger et al.[32] deleted the C_H2 and C_H3 regions from a heavy chain gene and expressed the truncated gene fused at the 3′ end with two different genes. Each modified gene was then introduced to J558L cells, which secretes only an endogenous light chain, to make antibody fragments. In the first version, the hinge exon of the heavy chain was spliced into the $C_{\delta s}$ exon that encodes the COOH-terminal 21 amino acid residues of the secreted IgD heavy chain. J558L cells harboring this gene expressed $F(ab')_2$-like molecules consisting of two Fab′ molecules disulfide-linked via the hinge, and also Fab′ molecules that failed to dimerize. Neuberger et al.[32] also fused a *Staphylococcus aureus* nuclease (SNase) gene 3′ to the first five codons of the C_H2 exon to make a Fab′-nuclease fusion protein. Again, both $F(ab')_2$-SNase and Fab′-SNase were made. At the time, the fact that not all of these molecules were uniformly $F(ab')_2$-like was somewhat puzzling. It was not clear whether the extra domains inhibited crosslinking of the hinge regions, or that Fab′ molecules are intrinsically difficult to dimerize in vivo. Subsequently, Co et al.[33] expressed the Fab′ of a humanized antibody, C4G1 (antihuman gpIIb/IIIa) in Sp2/0 cells, and found that only about 30% of the antibody fragments were expressed in the $F(ab')_2$ form, while the rest were Fab′. Similarly, Fab′ expressed in *Esherichia coli* also failed to dimerize completely into $F(ab')_2$ (see Section 5.5.2 below). These data indicated that the hinge sequences have little, if any, affinity for each other to hold the interhinge sulfhydryl groups in place to form disulfide bonds. Hence, the formation of $F(ab')_2$ molecules is dependent on mass action rather than on specific protein–protein interaction. In order to facilitate the formation of $F(ab')_2$, Kostelny et al.[34] introduced the dimer-forming sequences, Fos or Jun, 3′ to the hinge exon of the heavy chain gene. When expressed, the fused proteins [$Fab'-Fos+Fab'-Jun \rightarrow F(ab')_2-Fos/Jun$] were entirely in the form of $F(ab')_2$ linked by interhinge disulfides. The fused sequences were therefore needed to aid the formation of $F(ab')_2$ dimer. These results suggest that, in designing $F(ab')_2$-fusions, it is useful to employ sequences or domains that are likely to interact to form a dimer. The nature of the interaction can be further exploited to yield heterodimer, as in the case of a bispecific antibody. The formation of bispecific $F(ab')_2$ by Fos and Jun is discussed in further detail in Section 5.3.

5.2.2 Expression in Bacteria

The pioneering work of Skerra and Plückthun,[35] and Better et al.[36] demonstrated that it is possible to express antibody Fv or Fab fragments in *E. coli*. In both cases, genes for the light and heavy chain fragments are coexpressed in a dicistronic operon under the control of an inducible promoter. Upon induction, these fragments are secreted into the periplasmic space where the chains are assembled and disulfide bonds are formed. Compared to mammalian expression systems, the generation of transformants expressing antibody fragments takes only a few days, and scale-up in fermenters can achieve cell growth at very high density. The bacterial system is also the preferred system to use for the expression of immunotoxins since the end products can be toxic to mammalian cells. The production levels in *E. coli* are modest in shaker flasks, ranging from 0.1 mg/L to 10 mg/L. If high cell density can be achieved, the production level can be as high as 1–2 g/L.[37,38] The expression level in bacteria is inversely related to folding problems of antibody fragments and the stress they impose on the bacteria.[39] These problems depend on the variable region sequences of the antibody. Because soluble antibody fragments are a minor portion of the cell extract, their purification is more complicated, and removal of contaminating endotoxin is an important consideration.

Although Fab can be expressed in *E. coli*, extending the heavy chain fragment to include hinge sequences does not ensure formation of $F(ab')_2$ in the bacterial system. Upon expressing a humanized Fab′ fragment (anti-HER-2) containing a single hinge cysteine in *E. coli*, Carter et al.[37] found that despite the high Fab′ titers, only trace quantities of the antibody fragments were recovered as $F(ab')_2$. Expressing Fab′ fragments containing two hinge cysteine residues yielded partial formation of $F(ab')_2$ (up to 30%), whereas fusion of an additional cysteine residue to the hinge increased the yield of $F(ab')_2$ further to 70%.[40] Similar results were reported by Better et al.[41] when they expressed anti-human CD5 antibody fragments in *E. coli*. As with the mammalian expression system, these results confirmed that Fab′ fragments are intrinsically difficult to dimerize in vivo. The yield of $F(ab')_2$ depends not only on the concentration of Fab′ expressed, but also the number of hinge cysteine residues that are available. In contrast, when dimer-forming peptides were fused to the hinge, the yield of bivalent, disulfide-linked antibody fragments was almost 100%, despite their lower expression in *E. coli*.[42]

The Fab' fragments produced in *E. coli* can be made into F(ab')$_2$ by a disulfide exchange reaction in vitro or coupled as thioether-bridged F(ab')$_2$ by chemical modification (see Section 5.3.3.1 below). The latter F(ab')$_2$ is no longer sensitive to reduction. Rodrigues et al.[40] have compared the pharmacokinetics of various humanized F(ab')$_2$ fragments in mice. The *E. coli*-expressed F(ab')$_2$ fragments used had one hinge disulfide bond, one hinge thioether-bridged bond, or three hinge disulfide bonds. In addition, F(ab')$_2$ generated by pepsin digestion was also used. These authors found that F(ab')$_2$ with one hinge disulfide had the fastest clearance time (permanence time 0.7 h), and the pepsin generated-version the slowest (permanence time 3.3 h). The thioether version and the three disulfide version each had an equivalent intermediate clearance time (permanence time about 2.3 h). It appears that stable inter-hinge bonding is important for F(ab')$_2$ to have a favorable serum half-life. It was not clear why the pepsin-generated F(ab')$_2$ had the longest serum half-life. It is quite possible that some F(ab')$_2$ from *E. coli* are linked in a tail-to-tail (antiparallel) fashion, and that such a configuration is intrinsically less stable.

5.2.3 F(ab')$_2$ Fusions

Fusion of the antigen-binding portion of the immunoglobulin to different protein domains can yield recombinant antibodies possessing novel properties. Protein domains can be fused either to the heavy chain or to the light chain portion of F(ab')$_2$, and the fusion site can be either the NH_2- or the COOH-terminus of each chain. In choosing the fusion site, one must consider whether the protein domain or the antigen-binding domain requires a free NH_2-terminus for activity. If the protein domain is fused to the NH_2-terminus of the antigen-binding domain, a linker is needed to separate them so that each domain can fold separately. For protein domains that do not require a free NH_2-terminus, it is most convenient to fuse them distal to the hinge. In these cases, the hinge serves not only as a linker but also provides interheavy chain disulfide bonds to link the fused molecules into F(ab')$_2$. For these two reasons, hinge fusion is the most widely used method to join protein domains to antigen-binding domains. Quite often the protein domain replaces the Fc portion of the antibody, and if it has any affinity for itself, it may dimerize the fused protein into a bivalent F(ab')$_2$-like fragment (for example, see Section 5.3.3.2 below).

5.3 APPLICATIONS

F(ab')$_2$ fused with various protein domains may be used as diagnostic antibodies, as immunotoxins or as building blocks of bispecific F(ab')$_2$.

5.3.1 Diagnostic $F(ab')_2$

In immunoassays, antibodies are frequently labeled with enzymes by chemical procedures. Such approaches involve multiple steps and are quite labor intensive. Very often the antibody conjugates are chemically heterogeneous and are subjcct to batch-to-batch variations in potency. A genetic approach is an attractive alternative to bypass these problems. Although it takes time to clone the variable region sequences of a particular antibody in preparation for genetic fusion with an enzyme, the effort is compensated by having a product with a consistent stoichiometry of $F(ab')_2$ to enzyme. In addition, the only manipulation needed to obtain the labeled protein is to purify it from the producing cells, thereby omitting the conjugation steps. In immunoassays where standardization is important, having a labeled antibody that is consistent in potency is key to the assay.

Ducancel et al.[43] showed that a $F(ab')_2$-alkaline phosphatase fusion protein could be made in *E. coli*. The fused protein retains the same affinity for antigen as its parent antibody, and it possesses full alkaline phosphatase activity. The expression vector they designed contains a dicistronic operon encoding a gene for the truncated heavy chain (Fd) of an IgG inserted between residues +6 and +7 of bacterial alkaline phosphatase (PhoA), and a gene for the light chain of the same IgG. The Fd used in this construct is without the hinge, but the fused protein still dimerizes as $F(ab')_2$-alkaline phosphatase due to the interaction of the enzyme molecules. These investigators[44] made a recombinant $F(ab')_2$-alkaline phosphatase directed against human IgG and used it as a tracer for quantitative analysis of human IgG. In an enzyme immunoassay, this recombinant tracer was able to detect human IgG specific for hepatitis B antigen with a very high sensitivity.

5.3.2 $F(ab')_2$ Immunotoxins

Immunoconjugates containing antibody or antibody fragment linked to cytotoxic proteins have been widely described for their ability to seek and destroy specific cells. These proteins have great potential to be effective therapeutic reagents. Most of these conjugates are made by chemical cross-linking. The process has two disadvantages: It is labor intensive and difficult to yield homogeneous products. Recombinant immunotoxins, on the other hand, require only a single train of production and are generally easier to purify to homogeneity. Because cytotoxic proteins are toxic to mammalian cells, most recombinant immunotoxins are expressed in *E. coli*. Most toxins are fused to antibody fragments such as single-chain Fv, disulfide stabilized Fv, or Fab (see Chapters 3 and 4). Fusions to $F(ab')_2$ are rare, possibly due to difficulty of dimerizing Fab′-toxin. Better et al.[41] have fused the ribosomal inactivating protein gelonin to the NH_2-terminus of a light chain and expressed it with a Fd fragment containing the entire hinge in *E. coli* to obtain humanized anti-CD5 $F(ab')_2$-toxin. Most of the other expressed products are in the form of Fab′-toxin. In a comparison among single-chain Fv, Fab, and F(ab′)$_2$ fusions, the bivalent molecule was the most potent and had the longest serum half-life.[41]

5.3.3 Bispecific F(ab′)$_2$

Bispecific antibodies are molecules with dual antigen specificities. With the ability to link two different molecules together, they can mediate cytotoxicity against cancer cells, initiate signal transduction events in target cells, and bind to target cells with exquisite specificity. Of the three functions mentioned, the most studied is the retargeted cytotoxicity mediated by a bispecific antibody (see Chapter 7). By crosslinking cytotoxic effector cells to target cells, bispecific antibodies can trigger lethal destruction of unwanted cells. To be effective, the antigen on the effector cells must be able to trigger signal transduction for cytotoxicity upon binding the antibody. Such antigens include the α, β and CD3 polypeptide chains of the T-cell receptor on cytotoxic T cells, the FcγRIII (CD16) on natural killer cells, and the FcγRI receptor (CD64) on monocytes.

There are two methods for the production of whole bispecific antibodies: chemical conjugation and hybrid hybridoma. The first method utilizes chemical linkers to conjugate two different antibodies together.[45] Like all chemical methods for protein modification, these procedures are labor intensive and are difficult to scale up. Moreover, these methods give heterogeneous products that are not suitable for use as therapeutic agents. The second method produces bispecific antibodies by fusing two different hybridomas into one cell.[46] The resulting hybrid-hybridoma secretes antibodies with randomly assembled heavy and light chains. Of the ten possible combinations of heavy and light chain pairings, only one yields the desired bispecific antibody. The yield of bispecific antibody by this method is low, and it requires extensive purification from hybrid-antibody side products. To lessen the problem of random association of antibody chains, a method called "knobs-into-holes" was introduced to engineer the heavy chain domains for heterodimerization.[47,48] In this approach a "knob" was made in the C_H3 domain of antibody I to insert into a "hole" in the C_H3 domain of antibody II. The knob variant was created by judicious replacement of a small residue in the C_H3 domain with a large one, and the hole variant by the reverse procedure. Coexpression of the light chain-less CD4-IgG immunoadhesin knob variant with the anti-CD3 hole variant in mammalian cells yields high percentage (92%) of the bispecific molecule.

In some bispecific IgG antibodies, such as those that are based on anti-CD3, the presence of Fc in these molecules often causes problems in vivo. Because of the Fc-mediated interaction involving Fc-receptors, these molecules can cross-link CD3 on T cells independent of the tumor antigen, leading to side effects similar to those elicited by intact anti-CD3 antibodies.[22,49,50] These side effects are caused by nonspecific T cell activation, and the resulting toxicity can be dose limiting. The Fc-mediated reactions may also cause T cell depletion and long-term immunosuppression,[51] which are counterproductive to cancer treatment.

Due to the difficulties of manufacturing intact bispecific antibodies and the possible interference of their activities by Fc-mediated reactions, many investigators have turned to $F(ab')_2$ as the building blocks of bispecific antibodies. The basic production method is the disulfide exchange reaction described by Nisonoff and Mandy[52] and is illustrated in Figure 5.1. In this method, two different $F(ab')_2$ fragments are reduced to Fab', mixed and then reoxidized to form parent $F(ab')_2$ as well as bispecific $F(ab')_2$. Since Fab' have little, if any, affinity for each other, very high protein concentrations are required for the formation of the inter-Fab' disulfide bonds. Among the side products in the disulfide exchange reaction are parent $F(ab')_2$ and overreacted products such as $F(ab')_3$. This method has been improved by two recent developments: the chemical-mediated disulfide exchange reaction and the leucine zipper-mediated bispecific $F(ab')_2$ formation.

5.3.3.1 The Chemical-mediated Disulfide Exchange Reaction Brennan et al.[53] introduced a chemical modification step to the disulfide exchange reaction to aid the formation of bispecific $F(ab')_2$. As illustrated in Figure 5.2, one of the Fab' is first modified by 5, 5'-dithiobis(2-nitrobenzoic acid) (Ellman's reagent) to form a thionitrobenzoate derivative. The modified Fab' is then mixed with the other Fab' to form bispecific $F(ab')_2$. The advantages of this method are: (1) the modified Fab' cannot form homodimeric $F(ab')_2$, and (2) the thionitrobenzoate derivative stabilizes the modified Fab' and facilitates the disulfide exchange reaction. Fab' can also be modified with N.N'-1,2-phenylenedimaleimide.[54] Nucleophilic attack by the other unmodified Fab' leads to the formation of $F(ab')_2$ linked at the hinge by thioether bonds that cannot be reduced. Side products not linked by thioether bonds can then be reduced to Fab', and removed by gel filtration chromatography. Other side products such as $F(ab')_3$ and $F(ab')_4$ can also be removed by the gel filtration step.[54]

The chemical-aided scheme for the production of bispecific $F(ab')_2$ is somewhat labor intensive. It involves proteolytic cleavage to generate $F(ab')_2$, reduction and modification of Fab', a coupling reaction to make bispecific $F(ab')_2$, a gel filtration step to remove side products, and a final chromatography step to purify the bispecific $F(ab')_2$. An alternative approach, which is equally labor intensive, is to purify the intact bispecific IgG from hybrid-hybridoma first, and then to generate the $F(ab')_2$ molecules by pepsin digestion. Despite multiple steps and low yield, several bispecific $F(ab')_2$ have been made by these two approaches, some of them in quantities of hundreds of miligrams, for clinical investigations. Bispecific $F(ab')_2$ currently under preclinical and clinical evaluations include humanized anti-CD3 × anti-HER-2[55,56] and anti-CD64 × anti-HER-2[57] for breast cancer, anti-CD3 × anti-NCAM[58] for brain tumor, anti-CD3 × anti-EGP-2[59] for renal carcinoma, and anti-CD3 × anti-folate binding protein[60,61] for ovarian cancer. Anti-CD3 ×

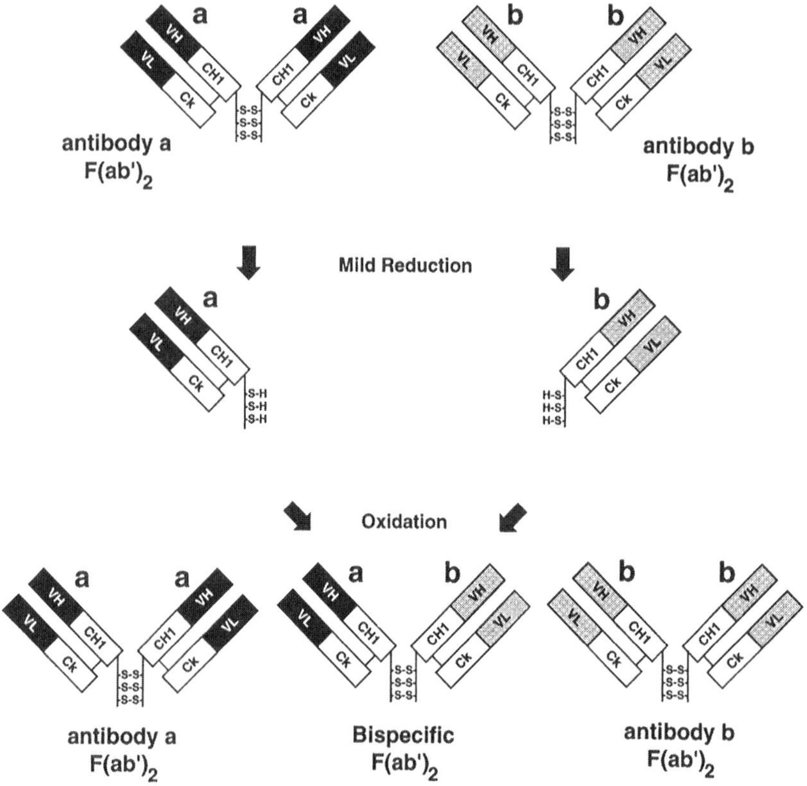

Figure 5.1 The disulfide exchange reaction to generate bispecific F(ab')$_2$.

anti-NCAM and anti-CD3 × anti-folate binding protein were introduced to patients in situ in conjunction with activated autologous T cells. These drugs were well tolerated and some clinical responses were observed in Phase I trials. Single infusions of anti-CD64 × anti-HER-2 were also well tolerated in patients up to 17 mg/patient. In contrast, intravenous administration of anti-CD3 × anti-EGP-2 or anti-CD3 × anti-folate binding protein to patients caused severe morbidity, with dose limited to only a few hundred μg of the bispecific F(ab')$_2$ per patient. The side effects, which were probably caused by tumor antigen-induced T cell activation, were similar to thosc caused by OKT3 in nonimmunosuppressed patients.[62,63] These observations indicated that, even when nonspecific, Fc-mediated, T cell activation is eliminated from bispecific antibody by the use of F(ab')$_2$, specific T cell activation by tumor antigen-bearing cells can still cause systemic morbidity in patients. The choice of tumor antigen is therefore very important when anti-CD3-based bispecific F(ab')$_2$ is to be used systematically. It is expected that the bispecific F(ab')$_2$ may cause less morbidity if the tumor antigen is more specific to the tumor and the tumor load is small.

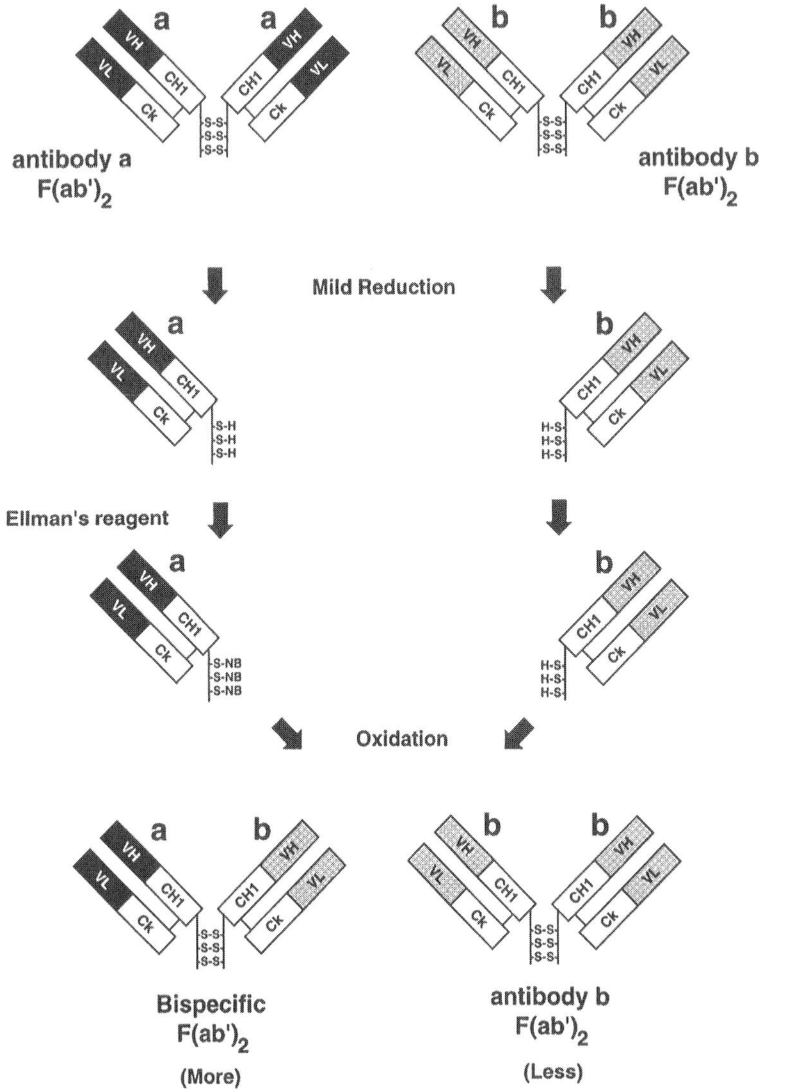

Figure 5.2 The chemical-mediated disulfide exchange reaction to generate bispecific $F(ab')_2$.

5.3.3.2 The Formation of Bispecific $F(ab')_2$ by Leucine Zippers Kostelny et al.[34] introduced a genetic method, which involves $F(ab')_2$ fusion, for the production of bispecific $F(ab')_2$. The method is illustrated in Figure 5.3. In this procedure, antibody fragments, such as Fab′, are genetically fused at the hinge to leucine zipper peptides, which are derived from the COOH-terminal region of the transcription factors Fos and Jun (Box 5.2).[64–69] These peptides are

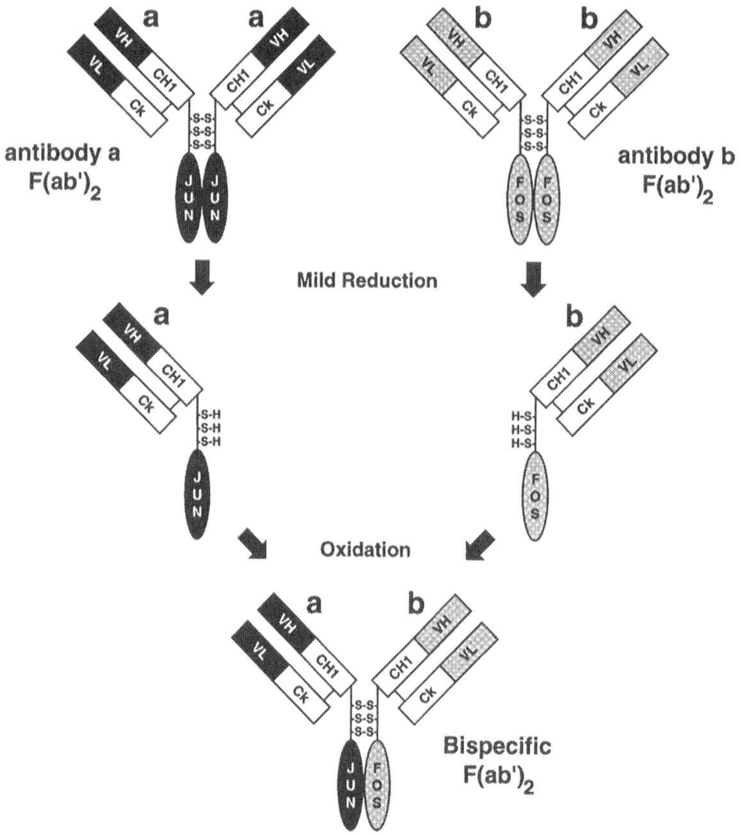

Figure 5.3 The leucine zipper-mediated formation of bispecific $F(ab'\text{-zipper})_2$. (Figure courtesy of *J. Hematotherapy*, Mary Ann Liebert, Inc.)

specific amino acid sequences about 30 residues long with leucine occurring at every seventh residue. Such sequences form amphipathic α-helices, with the leucine residues aligned on the hydrophobic side for dimer formation. Because they have the propensity to dimerize, antibody fragments fused to them are expressed as $F(ab')_2$. Two Fab′ fragments separately fused to the leucine zippers Fos and Jun can form bispecific $F(ab')_2$ preferentially in vitro. The pairing of the two different Fab′ is ensured because the paring of Fos/Jun is thermodynamically more stable than that of Jun/Jun or Fos/Fos. Subsequent disulfide bond formation at the hinge region covalently locks the two Fab′ fragments into a bispecific $F(ab')_2$. Since the formation of the bispecific $F(ab')_2$ is catalyzed by the Fos/Jun interaction, the yield of the desired product is high. In several preparations, more than 80% of the final $F(ab')_2$ molecules are in the form of bispecific $F(ab')_2$.[70] The side products are mostly parental $F(ab')_2$. Few aggregates are formed in this method.

Box 5.2 Structure and Function of Leucine Zippers.

The leucine zipper was originally identified in a transcription factor called C/EBP. It is a stretch of 30 amino acid residues with leucine occurring at every seventh residue. McKnight and colleagues[64] proposed that these leucine residues are aligned on one side of an amphipathic α-helix and these residues on two such helices would interdigitate, leading to dimerization. Subsequently similar sequences were found in protooncogene products Fos, Jun, Myc, and the yeast transcription factor GCN4. O'Shea et al.[65] showed that the leucine zipper regions of GCN4 associate in parallel and the leucine residues are lined up side by side on opposed helices. Furthermore, they showed the joined helices form a classical coiled coil—a well-known structure in dimers of many fibrous proteins. These zippers could either join two identical proteins, as in GCN4, or join different proteins, as in Fos and Jun.[66] The mechanism of specificity in the Fos–Jun heterodimer has been investigated by O'Shea et al.[67,68]

Most of these leucine zipper-containing molecules mentioned above are DNA-binding proteins. The DNA-binding regions in these proteins precede the zipper regions in the primary structure, and they are rich in basic amino acid residues. The basic region plus the leucine zipper motif is often abbreviated as bZIP. Upon dimerization, the bZIP elements form a Y shape molecule, a structure that is reminiscent of IgG. X-ray crystal structure of the bZIP elements from GCN4 complexed with DNA has been determined.[69] The bZIP dimer is a pair of continuous alpha helices that form a parallel coiled coil over their carboxy-terminal 30 residues in the leucine zipper region and gradually diverge toward their amino termini to pass through the major groove of the DNA-binding site. The coiled-coil dimerization interface is oriented almost perpendicular to the DNA axis.

Different investigators have also expressed antibody fragments fused to leucine zippers in *E. coli.* The form of antibody fragment used is single-chain Fv fused to the hinge and then the leucine zipper.[38,42,71] The end product is a bivalent molecule containing two single-chain Fv fragments, which can be made into bispecific antibody fragments using the method outlined above. Since V_L is covalently linked to V_H in this system, it should be possible to express two single-chain Fv fragments in one cell, one linked to Fos and the other linked to Jun, and to obtain bispecific antibody fragments in one production line. This method may be a good alternative to the bispecific diabody method,[72] or the single-chain bispecific $(sFv)_2$ method[73,74] for the expression of bispecific single-chain Fv in a single-host system.[38]

Using a mammalian expression system, small amounts (up to 20 mg) of several different bispecific $F(ab')_2$ have been made by the leucine zipper method. Most of these are anti-CD3-based bispecific $F(ab')_2$ used in mouse models for cancer treatment.[22,23,75] Results of these experiments have indicated that bispecific $F(ab')_2$ fragments were efficacious in eliminating tumor cells in immunocompetent mice. In contrast to the severe side effects caused by anti-CD3-based bispecific $F(ab')_2$ in patients, these bispecific $F(ab')_2$ fragments did not seem to cause any serious morbidity in animals. In addition, anti-CD3-based bispecific $F(ab')_2$ fragments were also made to study signal transduction in T cells. Anti-CD3 × anti-CD4 bispecific $F(ab')_2$ can trigger $CD4^+CD8^+$ thymocytes to mature into $CD4^+CD8^-$ cells.[76] The same fragments were also shown to recruit CD4 to the CD3 complex in T cell clones to stimulate signal transduction events that lead to proliferation.[77]

Scaled-up procedures for the production of several hundred milligrams of one humanized bispecific $F(ab')_2$ have also been described. In designing a humanized $F(ab')_2$, Tso et al.[70] introduced a modification to the hinge of human IgG1 to provide three cysteine residues instead of two for interheavy chain disulfide bond formation. Two residues (K—C) from the hinge of mouse IgG2a were inserted to the lower hinge of human IgG1 for the additional bond (Figure 5.4). The modification was introduced to maximize the linking of Fab′ arms by disulfide bonds and to render the $F(ab')_2$ unrecognizable by natural antibodies (homoreactive antibodies). The modified hinge is susceptible to proteolytic cleavage for the removal of the leucine zippers to yield authentic $F(ab')_2$. The pepsin cleavage site in the modified human IgG1 hinge has not

(A) **A-E-P-K-S-(C)-D-K-T-H-T-C-P-P-C-K-C-P-A-G-G-Zipper**

------------------Modified Human IgG1 Hinge-----------------

(B) **A-E-P-K-S-(C)-D-K-T-H-T-C-P-P-C-P-**

-------------------Human IgG1 Hinge--------------------

(C) **E-P-R-G-P-T-I-K-P-C-P-P-C-K-C-P-**

------------------Mouse IgG2a Hinge--------------------

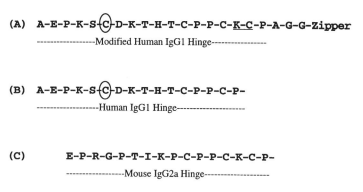

Figure 5.4 *A*. The sequence of the modified human IgG1 hinge used in the hinge-zipper fusion. Two residues Lys-Cys (underlined) were inserted in the modified hinge. The first Cys (circled) in this modified hinge forms a disulfide bond with the light chain, and the last three Cys residues form interheavy chain disulfides. For comparison, hinge sequences of the human IgG1 (*B*) and the mouse IgG2a (*C*) are also shown. All three Cys residues in the mouse IgG2a hinge are used for interheavy chain disulfides. After the insertion of Lys-Cys, the modified hinge and the mouse IgG2a hinge have extensive sequence homology near the COOH-terminus. (Figure courtesy of *J. Hematotherapy*, Mary Ann Liebert, Inc.)

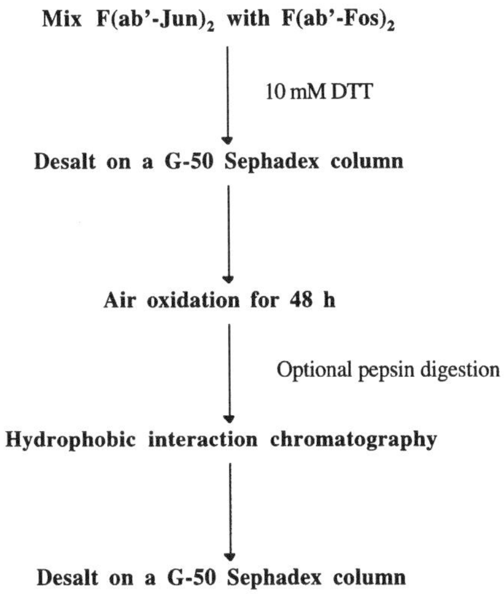

Figure 5.5 Redox reaction and purification steps used to make bispecific $F(ab')_2$.

been determined. Presumably it is located between the inserted K—C residues, the site that is susceptible to a variety of proteases in mouse IgG2a.[2] In the production scheme, a pepsin digestion step can be introduced to remove the leucine zippers once they have been used to aid heterodimer formation. The newly devised procedures for the production of humanized bispecific $F(ab')_2$ are summarized in Figure 5.5. Briefly, production cell lines expressing $F(ab'\text{-}Fos)_2$ and $F(ab'\text{-}Jun)_2$ are created by transfections. Both $F(ab'\text{-}Fos)_2$ and $F(ab'\text{-}Jun)_2$ are purified from culture media by protein G affinity chromatography, mixed in equimolar amounts, and reduced with 10 mM dithiothreitol (DTT). The use of 10 mM DTT is excessive; DTT reduces not only all interheavy chain disulfides, but also the disulfide bonds between the heavy and light chains. However, the noncovalent interactions between the heavy and light chains are strong enough to avoid chain dissociation. After reduction, DTT is removed by gel filtration, and the Fab′-zipper fragments are allowed to air-oxidize to form bispecific $F(ab'\text{-zipper})_2$. A hydrophobic interaction chromatography step is introduced to further purify the bispecific $F(ab')_2$. The optional pepsin digestion step for the removal of the leucine zippers can be introduced prior to hydrophobic interaction chromatography. SDS-PAGE analysis of products from each step is shown in Figure 5.6. All of these steps can be easily performed in the laboratory. The only drawback in such a scheme is that it requires two separate production trains prior to the coupling step. Using these procedures, a bispecific $F(ab')_2$, which has affinity for both the α and β subunits of the human IL-2 receptor, was made in sufficient quantity and purity for preclinical evaluation.[70]

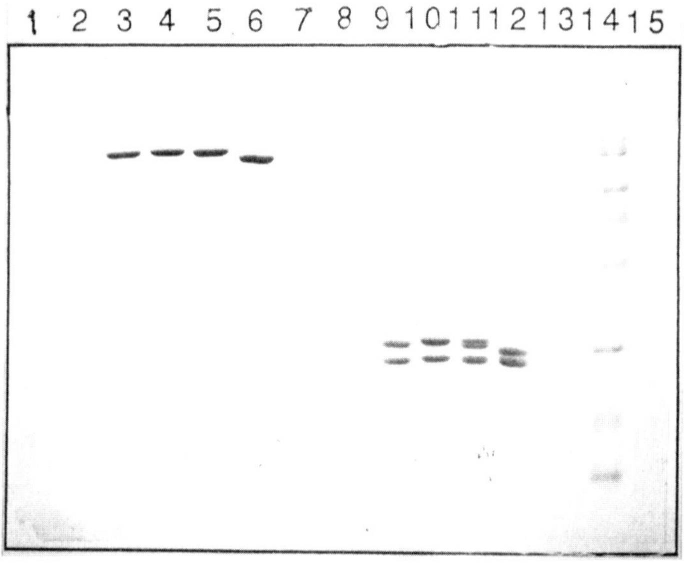

Figure 5.6 SDS-PAGE analysis of bispecific F(ab′)$_2$ generated by pepsin digestion. Lanes 3–6 were run under nonreducing conditions, and lanes 9–12 under reducing conditions. Lane 3, F(ab′-Fos)$_2$; lane 4, F(ab′-Jun)$_2$; lane 5, bispecific F(ab′-zipper)$_2$; lane 6, pepsin-generated bispecific F(ab′)$_2$; Lane 9, F(ab′-Fos)$_2$; lane 10, F(ab′-Jun)$_2$; lane 11, bispecific F(ab′-zipper)$_2$; lane 12, pepsin-generated bispecific F(ab′)$_2$; and lane 14, molecular weight markers. Molecular weight markers are: myosin 200 kDa; phosphorylase b, 97 kDa; bovine serum albumin, 69 kDa; ovalbumin, 46 kDa; carbonic anhydrase, 30 kDa; soy bean trypsin inhibitor, 21 kDa; and lysozyme, 14 kDa. (Figure courtesy of *J. Hematotherapy*, Mary Ann Liebert, Inc.)

5.4 CONCLUSION

Recent developments in antibody engineering have focused on using small antibody fragments such as Fab and single-chain Fv rather than whole antibodies. While the smaller size fragments offer great advantages in expression and in repertoire cloning, they may not be ideal molecules for immunotherapy. These monovalent fragments have short serum half-life and are sometimes unstable upon storage.[78] F(ab′)$_2$ fragments, on the other hand, are stable, easy to purify, retain the avidity of parent antibodies, and have the longest serum half-life among all antibody fragments. Because of these properties, they are ideal building blocks for many antibody fusions. Historically, F(ab′)$_2$ molecules have not been favored as potential therapeutic agents because production was difficult, and expression of Fab′ genes failed to produce uniformly bivalent F(ab′)$_2$. Fusions of dimer-forming protein domains to Fab′ molecules have solved several of these problems. Furthermore, the introduced domains can offer additional properties that are not present in whole antibodies.

5.5 SUMMARY

$F(ab')_2$ is the most stable antibody fragment that retains the bivalent binding sites of an antibody. The dimeric structure of an antibody is maintained in $F(ab')_2$ by interheavy chain disulfide bonds. These interheavy chain disulfide bonds, once reduced, cannot be readily reoxidized to form the bivalent $F(ab)_2$. This is because dimer-forming sequences are needed to aid interheavy chain disulfide formation, and these sequences have already been removed from $F(ab')_2$. For the same reason, recombinant Fab′ fragments expressed in mammalian cells or bacteria also have difficulty in forming $F(ab')_2$. Fusion of Fab′ with other protein domains sometimes restores the propensity of the fused proteins to form $F(ab')_2$. These protein domains include enzymes, toxins, or simply dimer-forming peptides. If these protein domains have biological activities, the antigen-binding sites of the fused $F(ab')_2$ deliver these activities to target sites in vitro or in vivo. Even though the dimer-forming peptides have no biological activities, they can be used to aid formation of bispecific $F(ab')_2$, a molecule with two different antigen-binding sites. Bispecific $F(ab')_2$ can be used to recruit and trigger effector cells to destroy target cells or to induce signal transduction in target cells. Both of these functions are potentially useful for treating human diseases.

ACKNOWLEDGMENTS

I thank Sheri Kostelny, Michael Cole, and Randy Robinson for reviewing this chapter.

REFERENCES

1. Parham, P. 1983. Preparation and purification of active fragments from mouse monoclonal antibodies. In *Cellular Immunology*, Vol. 1, ed. E. M Weir, Palo Alto: Blackwell Scientific Press, pp. 14.1–14.23.
2. Yamaguchi, Y., H. H. Kim, K. Kato, K. Masuda, I. Shimada, Y. Arata. 1995. Proteolytic fragmentation with high specificity of mouse immunoglobulin G. *J. Immunol. Methods* 181:259–267.
3. Buchegger, F., A. Pelegrin, N. Hardman, C. Heusser, J. Lukas, W. Dolci, J. P. Mach. 1992. Different behavior of mouse-human chimeric antibody $F(ab')_2$ fragments of IgG1, IgG2 and IgG4 subclass in vivo. *Int. J. Cancer* 50: 416–422.
4. Alegre, M., M. Depierreux, S. Florquin, D. Abramowicz, T. Najdovski, V. Flamand O. Leo, M. Deshodt-Lanckman, M. Goldman. 1990. Acute toxicity of an anti-CD3 monoclonal antibody in mice: a model of OKT3 first dose reactions. *Transplant. Proc.* 22: 1920–1921.
5. Ferran, C., K. Sheehan, M. Dy, R. Schreiber, S. Merite, P. Landais, L. H. Noel, G. Grau, J. A. Bluestone, J. F. Bach, L. Chatenoud. 1990. Cytokine-related syndrome

following injection of anti-CD3 monoclonal Ab: further evidence for transient in vivo T cell activation. *Eur. J. Immunol.* 20: 509–515.

6. Alegre M-L., J. Y. Tso, H. A. Sattar, J. Smith, M. S. Cole, J. A. Bluestone. 1995. An anti-murine CD3 monoclonal antibody with a low affinity for Fcγ receptor supresses transplantation responses while minimizing acute toxicity and immunogenicity. *J. Immunol.* 155: 1544–1555.
7. Hirsch, R., J. A. Bluestone, L. Denenno, R. E. Gress. 1990. Anti-CD3 F(ab′)$_2$ fragments are immunosuppressive in vivo without evoking the strong humoral response or morbidity associated with whole mAb. *Transplantation* 49: 1117–1123.
8. Woodle, E.S., J. R. Thistlewaite, I. A. Ghobrial, L. K. Jolliffe, F. P. Stuart, J. A. Bluestone. 1991. OKT3 F(ab′)$_2$ fragments-retention of the immunosuppressive properties of whole antibody with marked reduction in T cell activation and lymphokine release. *Transplantation* 52: 354–360.
9. Kaku, S., S. Yano, T. Kawasaki, Y. Sakai, K. Suzuki, K. Kawamura, Y. Masuho, N. Sato, T. Takenaka, N. Landofi, M. S. Co. 1996. Comparison of the antiplatelet agent potential of the whole molecule, F(ab′)$_2$ and Fab fragments of humanized anti-gpIIb/IIIa monoclonal antibody in monkeys. *Gen. Pharmac.* 27: 435–439.
10. Randen, I., K. N. Potter, Y. Li, K. M. Thompson, V. Pascual, O. Forre, J. B. Natvig, J. D. Capra. 1993. Complementarity-determining region 2 is implicated in the binding of staphylococcal protein A to the human immunoglobulin VHIII variable regions. *Eur. J. Immunol.* 23: 2682–2686.
11. Derrick, J.P., D. B. Wigley. 1992. Crystal structure of a streptococcal protein G domain bound to an Fab fragment. *Nature* 359: 752–754.
12. Chamow, S. M., D. Z. Zhang, X. Y. Tan, S. M. Mhatre, S. A. Marsters, D. H. Peers, R. A. Byrn, A. Ashkenazi, R. P. Junghans. 1994. A humanized, bispecific immunoadhesin-antibody that targets CD3+ effectors to kill HIV-1-infected cells. *J. Immunol.* 153: 4268–4280.
13. Trang, J. M. 1992. Pharmacokinetics and metabolism of therapeutic and diagnostic antibodies. In *Protein Pharmacokinetics and Metabolism*, ed. B. L. Ferraiolo, New York: Plenum Press, pp. 223–270.
14. Colcher, D., R. Bird, M. Roselli, K. D. Hardman, S. Johnson, S. Pope, S. W. Dodd, M. W. Pantoliano, D. E. Milenic, J. Schlom. 1990. In vivo tumor targeting of a recombinant single-chain antigen-binding protein. *J. Natl. Cancer Inst.* 82: 1191–1197.
15. Morell, A., W. D. Terry, T. A. Waldmann. 1970. Metabolic properties of IgG subclasses in man. *J. Clin. Invest.* 49: 673–680.
16. Kim, J-K., M-F. Tsen, V. Ghertie, E. S. Ward. 1994. Identifying amino acid residues that influence plasma clearance of murine IgG1 fragments by site-directed mutagenesis. *Eur. J. Immunol.* 24: 542–548.
17. Kim, J-K., M-F. Tsen, V. Ghertie, E. S. Ward. 1994. Localization of the site of murine IgG1 molecule that is involved in binding to the murine intestinal Fc receptor. *Eur. J. Immunol.* 24: 2429–2434.
18. Burmeister, W. P., A. H. Huber, P. Bjorkman. 1994. Crystal structure of the complex of rat neontal Fc receptor with Fc. *Nature* 372: 379–383.
19. Ghertie, V., J. G. Huboard, J-K. Kim, M-F. Tsen, Y. Lee, E. S. Ward. 1996. Abnormally short serum half-lives of IgG in β2-microglobulin-deficient mice. *Eur.*

J. Immunol. 26: 690–696.

20. Junghans, R. P., C. L. Anderson. 1996. The protection receptor for IgG catabolism is the β2-microglobulin-containing neonatal intestinal transport receptor. *Proc. Natl. Acad. Sci. USA* 93: 5512–5516.
21. Warmerdam, P. A., J. G. van de Winkel, A. Vlug, N. A. Westerdaal, P. J. Capel. 1991. A single amino acid in the second Ig-like domain of the human Fc gamma receptor II is critical for human IgG2 binding. *J. Immunol.* 147: 1338–1343.
22. Weiner, G., S. A. Kostelny, J. R. Hillstrom, M. S. Cole, B. K. Link, S. L. Wang, J. Y. Tso. 1994. The role of T cell activation in anti-CD3 × anti-tumor bispecific antibody therapy. *J. Immunol.* 152: 2385–2392.
23. Bikacs, T., J. Lee, M. B. Moreno, C. M. Zacharchuk, M. S. Cole, J. Y. Tso, C. H. Paik, J. M. Ward, D. M. Segal. 1995. A bispecific antibody prolongs survival in mice bearing lung metastases of syngenic mammary adenocarcinoma. *Internatl. Immunol.* 7: 947–955.
24. Pedley, R. B., J. A. Boden, R. Boden, R. Dale, R. H. Begent. 1993. Comparative radioimmunotherapy using intact or $F(ab')_2$ fragments of ^{131}I anti-CEA antibody in a colonic xenograft model. *Br. J. Cancer* 68: 69–73.
25. Behr, T., W. Becker, E. Hannappel, D. M. Goldenberg, F. Wolf. 1995. Targeting of liver metastases of colorectal cancer with IgG, $F(ab')_2$, and Fab′ anti-carcinoembryonic antigen antibodies labeled with ^{99m}Tc: the role of metabolism and kinetics. *Cancer Res.* 55: 5777s–5585s.
26. Shinomiya, T., J. Koyama. 1974. Naturally occurring antibodies reacting with some buried determinants of homologous IgG exposed by spleen protease digestion. *J. Immunochem.* 11: 467–473.
27. Nsu, H., D. S. Chia, D. W. Knutson, E. V. Barnette. 1980. Naturally occurring human antibodies to the $F(ab')_2$ portion of IgG. *Clin. Exp. Immunol.* 42: 378.
28. Silvestris, F., A. D. Bankhurst, R. P. Searles, R. C. Williams. 1984. Studies of the anti-$F(ab')_2$ antibodies and possible immunologic control mechanisms in systemic lupus erythematosus. *Arthritis Rheum.* 27: 1387–1396.
29. Terness P., I. Kohl, G. Hubener, R. Battistutta, L. Moroder, M. Welshof, C. Dufter, M. Finger, C. Hain, M. Jung, G. Opelz. 1995. The natural human IgG anti-$F(ab')_2$ antibody recognizes a conformational IgG1 hinge epitope. *J. Immunol.* 154: 6446–6452.
30. Yano, S., S. Kaku, K. Suzuki, C. Terazaki, T. Sakayori, T. Kawasaki, K. Kawamura, Y. Sugita, K. Hoshino, Y. Masuho. 1995. Natural antibodies against the immunoglobulin $F(ab')_2$ fragment cause elimination of antigens recognized by the $F(ab')_2$ from the circulation. *Eur. J. Immunol.* 25: 3128–3133.
31. Wood, C.R., A. J. Dorner, G. E. Mottis, E. M. Alderman, D. Wilson, R. M. O'Hara, R. J. Kaufman. 1990. High level synthesis of immunoglobulin in Chinese hamster ovarian cells. *J. Immunol.* 145: 3011–3016.
32. Neuberger, M. S., G. T. Willams, R. O. Fox. 1984. Recombinant antibodies possessing novel effector functions. *Nature* 312: 604–608.
33. Co, M. S., S. Yano, R. Hsu, N. F. Landolfi, M. Vasquez, M. Cole, J. Y. Tso, T. Bringman, W. Larid, D. Hudson, K. Kawamura, K. Suzuki, K. Furuichi, C. Queen, Y. Masuho. 1994. A humanized antibody specific for the platelet integrin, gpIIb/IIIa. *J. Immunol.* 152: 2968–2976.

34. Kostelny, S. K., M. S. Cole, J. Y. Tso. 1992. Formation of a bispecific antibody by the use of leucine zippers. *J. Immunol.* 148: 1547–1553.

35. Skerra, A., A. Plückthun. 1988. Assembly of functional immunoglobulin Fv fragment in *Escherichia coli. Science* 249: 1038–1039.

36. Better, M., C. P. Chang, R. R. Robinson, A. H. Horwitz. 1988. *Escherichia coli* secretion of an active chimeric antibody fragment. *Science* 240: 1041–1043.

37. Carter, P., R. F. Kelly, M. L. Rodrigues, B. Snedecor, M. Covarrubias, M. D. Velligan, W. L. T. Wong, A. M. Rowland, C. E. Kotts, M. E. Carver, M. Yang, J. H. Bourell, M. SH, D. Henner. 1992. High level Escherichia coli expression and production of a bivalent humanized antibody fragment. *Bio/Technology* 10: 163–167.

38. Carter, P., J. Ridgway, Z. Zhu. 1995. Toward the production of bispecific antibody fragments for clinical applications. *J. Hematotherapy* 4: 463–470.

39. Knappik, A., A. Plückthun. 1995. Engineering turns of a recombinant antibody improve its in vivo folding. *Protein Engineering* 81: 81–89.

40. Rodrigues, M. L., B. Snedecor, C. Chen, W. L. T. Wong, G. Shaily, G. S. Blank, D. Maneval, P. Carter. 1993. Engineering Fab′ fragments for efficient F(ab′)$_2$ formation in *Escherichia coli* and for improved in vivo stability. *J. Immunol.* 151: 6954–6961.

41. Better, M., S. L. Bernhard, R. E. Williams, S. D. Leigh, R. J. Bauer, A. H. C. Kung, S. F. Carroll, D. M. Fishwild. 1995. T cell-targeted immunofusion protein from *Escherichia coli. J. Biol. Chem.* 25: 1451–1457.

42. Pack, P., A. Plückthun. 1992. Miniantibodies: use of amphipathic helices to produce functional, flexibly linked dimeric Fv fragments with high avidity in *Escherichia coli. Biochemistry* 31: 1579–1584.

43. Ducancel, F., D. Gillet, A. Carrier, E. Lajeunesse, A. Menez, J-C. Boulain. 1993. Recombinant colorimetric antibodies: construction and characterization of a bifunctional F(ab)$_2$/alkaline phosphatase conjugate produced in *Escherichia coli. Bio/Technology* 11: 601–605.

44. Carrier, A., F. Ducancel, N. B. Settiawan, L. C, B. Maillère, M. Léonetti, P. Drevet, A. Ménez, J.-C. B. 1995. Recombinant antibody-alkaline phosphatase conjugates for diagnosis of human IgGs: application to antiHBsAg detection. *J. Immunol. Methods* 181: 177–186.

45. Segal, D. M., D. P. Snider. 1989. Targeting and activation of cytotoxic lymphocytes. *Chem. Immunol.* 47: 179–213.

46. Milstein, C., A. C. Cuello. 1984. Hybrid hybridomas and the production of bi-specific monoclonal antibodies. *Immunol. Today* 5: 299–304.

47. Ridgway, J. B., L. Presta, P. Carter. 1996. 'Knobs-into-holes' engineering of antibody CH3 domains for heavy chain heterodimerization. *Protein Eng.* 9: 617–627.

48. Atwell, S., J. B. Ridgway, J. A. Wells, P. Carter. 1997. Stable heterodimers from remodeling the domain interface of a homodimer using a phage display library. *J. Mol. Biol.* 270: 26–35.

49. Chappel, M. S., D. E. Isenamn, M. Everett, Y. Y. Xu, K. J. Dorrington, M. H. Klein. 1991. Identification of the Fcγ receptor class I binding site in human IgG through

the use of recombinant IgG1/IgG2 hybrid and point mutated antibodies. *Proc. Natl. Acad. Sci. USA* 88: 9036–9040.

50. Belani, R., G. J. Weiner. 1996. T cell activation and cytokine production in anti-CD3 bispecific antibody therapy. *J. Hematotherapy* 4: 395–402.

51. Demanet, C., J. Brissinck, J. DeJong, K. Thielemans. 1996. Bispecific antibody-mediated immunotherapy of BCL1 lymphoma: increased efficacy with multiple injections and CD28-induced costimulation. *Blood* 87: 4390–4398.

52. Nisonoff, A., W. J. Mandy. 1962. Quantitative estimation of the hybridization of the rabbit antibodies. *Nature* 194: 355–357.

53. Brennan, M., P. F. Davison, H. Paulus. 1985. Preparation of bispecific antibodies by chemical recombination of monoclonal immunoglobulin G1 fragments. *Science* 229: 81–83.

54. Glennie, M. J., H. M. McBride, A. T. Worth, G. T. Stevenson. 1987. Preperation and performance of bispecfic $F(ab'\gamma)_2$ antibody containing thiother-linked $Fab'\gamma$ fragments. *J. Immunol.* 139: 2367–2375.

55. Shalaby, M. R., H. M. Shepard, M. L. Presta, M. L. Rodrigues, P. C. L. Beverley, P. Carter. 1992. Development of humanized bispecific antibodies reactive with cytotoxic lymphocytes and tumor cells overexpressing the HER-2 proto-oncogene. *J. Exp. Med.* 175: 217–225.

56. Shalaby, M. R., P. Carter, D. Maneval, D. Gillinan, C. Kotts. 1995. Bispecific HER-2 × CD3 antibodies enhance T-cell cytotoxicity in vitro and localize to HER2-overexpressing xenograft in nude mice. *Clin. Immunol. Immunopathol.* 74: 185–192.

57. Valone, F. H., P. A. Kaufman, P. M. Guyre, L. D. Lewis, V. Memoli, M. S. Ernstoff, W. Wells, R. Barth, Y. Deo, J. Fisher, K. Phipps, R. Graziano, L. Meyer, M. Mrozek-Orlowski, K. Wardwell, V. Guyer, T. L. Morley, C. Arvizu, P. Wallace, M. W. Fanger. 1995. Clinical trials of bispecific antibody MDX-210 in women with advanced breast or ovarian cancer that overexpresses HER-2/neu. *J. Hematotherapy* 4: 471–475.

58. Nitta, T., K. Sato, H. Yagita, K. Okomura, S. Ishi. 1990. Preliminary trial of specific targeting therapy against malignant glioma. *Lancet* 335: 368–371.

59. Kroesen, B-J., R. J. Janssen, J. Buter, J. Nieken, D. T. Sleijfer, N. O. Mulder, L. De Leij. 1995. Bispecific monoclonal antibodies for intravenous treatment of carcinoma patients: immunobiologic aspects. *J. Hematotherapy* 4: 409–414.

60. Canevari, S., D. Mezzanzanica, A. Mazzoni, D. V. R. Negri, R. L. H. Bolhuis, M. I. Colnaghi, G. Bolis. 1995. Bispecific antibody targeted T cell therapy of ovarian cancer: clinical results and future directions. *J. Hematotherapy* 4: 423–427.

61. Tibben, J. G., O. C. Boerman, L. F. Massuger, C. P. Schijf, R. A. Claessen, F. H. Corstens. 1996. Pharmacokinetics, biodistribution and biological effects of intravenously administered bispecific monoclonal antibody OC/TR $F(ab')_2$ in ovarian carcinoma patients. *Int. J. Cancer* 66: 477–483.

62. Urba, W. J., C. Ewel, W. Kopp, J. W. D. Smith, R. G. Steis, J. D. Ashwell, S. P. Creekmore, J. Rossio, M. Sznol, W. Sharfman, R. Fenton, J. Janik, T. Watson, J. Beveridge, D. Longo. 1992. Anti-CD3 monoclonal antibody treatment of patients with CD3-negative tumors: a phase IA/B study. *Cancer Res.* 52: 2394–2401.

63. Hank, J. A., M. Albertini, O. H. Wesley, J. H. Schiller, A. Borchert, K. Moore, R. Bechhofer, B. Storer, J. Gan, C. Gambacorti, J. Sosman, P. M. Sondel. 1995. Clinical and immunological effects of treatment with murine anti-CD3 monoclonal antibody along with interleukin 2 in patients with cancer. *Clin. Cancer Res.* 1: 481–491.

64. Landschulz, W. H., P. F. Johnson, S. L. McKnight. 1988. The leucine zipper: a hypothetical structure common to a new class of DNA binding proteins. *Science* 240: 1759–1764.

65. O'Shea, E. K., R. Rutkowski, P. S. Kim. 1989. Evidence that the leucine zipper is a coiled coil. *Science* 243: 538–542.

66. Curran, T., B. R. Franza. 1988. Fos and Jun: The AP-1 connection. *Cell* 55: 395–397.

67. O'Shea, E. K., R. Rutkowski, W. Stafford, P. S. Kim. 1989. Preferential heterodimer formation by isolated leucine zippers from Fos and Jun. *Science* 245: 646–648.

68. O'Shea, E. K., R. Rutkowski, P. S. Kim. 1992. Mechanism of specificity in the Fos-Jun oncoprotein heterodimer. *Cell:* 68: 699–708.

69. Ellenberger, T. E., C. J. Brandi, K. Struhl, S. C. Harrison. 1992. The GCN4 basic region leucine zipper binds DNA as a dimer of uninterrupted alpha helices: crystal structure of the protein-DNA complex. *Cell* 7: 1223–1237.

70. Tso, J. Y., S. L. Wang, W. Levin, J. Hakimi. 1995. Preparation of a bispecific $F(ab')_2$ targeted to the human IL-2 receptor. *J. Hematotherapy* 4: 389–394.

71. de Kruif, J., T. Logtenberg. 1996. Leucine zipper dimerized bivalent and bispecific scFv antibodies from a semi-synthetic antibody phage display library. *J. Biol. Chem.* 271: 7630–7634.

72. Holliger, P., T. Prospero, G. Winter. 1993. Diabodies: Small bivalent and bispecific antibody fragments. *Proc. Natl. Acad. Sci. USA* 90: 6444–6449.

73. Gruber, M., E. R. Schodin, E. R. Wilson, D. M. Kranz. 1994. Efficient tumor cell lysis mediated by a bispecific single-chain antibody expressed in *Escherichia coli*. *J. Immunol.* 152: 5368–5374.

74. Mallender, W.D., E. W. J. Voss. 1994. Construction, expression, and activity of a bivalent bispecific single-chain antibody. *J. Biol. Chem.* 269: 199–206.

75. Moreno, M. B., J. A. Titus, J. Y. Tso, C. M. Zacharchuk, D. M. Segal, J. R. Wonderlich. 1995. Bispecific antibodies retarget murine T cell cytotoxicity against syngeneic breast cancer in vitro and in vivo. *Cancer Immunol. Immunother.* 40: 182–190.

76. Bommhardt, U., M. S. Cole, J. Y. Tso, R. Zamoyska. 1997. Signal through CD8 or CD4 can induce commitment to the CD4 lineage in the thymus. *Eur. J. Immunol.* 27: 1152–1163.

77. Smith, J. A., J. Y. Tso, M. R. Clark, M. S. Cole, J. A. Bluestone. 1997. Nonmitogenic anti-CD3 monoclonal antibodies deliver a partial TCR signal and influence clonal anergy. *J. Exp. Med.* 185: 1413–1422.

78. Kurucz, I., J. Titus, C. R. Jost, C. M. Jacobus, D. M. Segal. 1995. Retargeting of CTL by an efficiently refolded bispecific single chain Fv dimer produced in bacteria. *J. Immunol.* 154: 4576–4582.

6

MONOVALENT PHAGE DISPLAY OF Fab AND scFv FUSIONS

DAVID B. POWERS AND JAMES D. MARKS
San Francisco General Hospital
San Francisco, CA 94110

6.1 INTRODUCTION

The preceding chapters described theoretical and practical aspects of immunofusions, molecules that use the heavy (V_H) and light (V_L) chain variable domains of an antibody to impart specific binding properties to the fusion protein. Until recently, antigen-binding variable domains for use in immunofusions have come from murine monoclonal antibodies produced using hybridoma technology. In fact, monoclonal antibodies (and their engineered derivatives) have been important tools to demonstrate "proof of concept" for almost all immunofusions to date. Despite their utility, however, monoclonal antibodies suffer from a number of key drawbacks that can limit their application to immunofusions.

Monoclonals are time consuming to make—initial injection and periodic boosts can require weeks to months, and hybridoma construction and screening require additional time. In addition, the immune response is often biased toward certain "immunodominant" epitopes, making it difficult or impossible to produce monoclonals with the desired specificity for a particular aim. Furthermore, production of antibodies against proteins conserved between species may be difficult or impossible. The maximum binding affinities of

Antibody Fusion Proteins, Edited by Steven M. Chamow and Avi Ashkenazi
ISBN 0471-18358-X

monoclonal antibodies—seldom more than nanomolar[1]—may also be inadequate for many targeting applications.

Once an appropriate hybridoma has been generated, DNA encoding the antibody V_H and V_L genes must be cloned from the hybridoma cell for expression as an Fv, Fab, or single chain Fv (scFv) antibody fragment. However, mutations introduced by the somatic hypermutation machinery into regions where primers anneal may make PCR amplification difficult or impossible, necessitating another amplification approach such as RACE or oligoligation PCR.[2–4] PCR itself may also introduce mutations coding for stop codons or destabilizing amino acids, necessitating the sequencing of multiple clones. Cloning the correct V_H and V_L can also be complicated by the presence of several immunoglobulin transcripts, some of them arising from the fusion partner.[5] Furthermore, V genes may be cut internally by restriction enzymes, complicating cloning strategies that utilize hexanucleotide recognition sites.[6,7]

Once the V genes have been successfully cloned, the antibody fragment should be expressed and characterized biophysically and biochemically prior to use as an immunofusion. Although the easiest method for expression is in *Escherichia coli*, expression levels in bacteria vary considerably[8] due to antibody fragment toxicity and poor folding kinetics.[9] These differences are sequence dependent, differ dramatically between antibodies, and in many instances result in failure to produce adequate quantities of antibody fragment for further in vitro characterization.[9] Thus, despite the large number of well-characterized hybridoma cell lines, very few of these antibodies can be successfully reconstructed as antibody fragments that express at high levels in *E. coli.* Secretion in a eukaryotic cell line is necessary to generate the antibody fragments that are functional and can be used for biochemical and biophysical characterization.[10,11]

Finally, murine antibodies are immunogenic when administered to humans for therapeutic applications, resulting in the HAMA (human antimouse antibodies) response.[12,13] This response leads to an increase in clearance of the foreign antibody, a decrease in efficacy with repeat administration, and the potential for allergic reactions or serum sickness. While murine monoclonals can be "chimerized" (by grafting the murine V-domains onto human constant domains)[14–16] or "humanized" (by grafting the murine complementarity-determining regions onto human framework regions);[17] the process is laborious and not necessarily straightforward. Unfortunately, adaptation of hybridoma technology to produce human monoclonals has been largely unsuccessful.[18]

All of the foregoing limitations can be overcome by taking advantage of recent advances in biotechnology which permit production of antibody fragments directly in bacteria (reviewed in Refs. 19–22). Antigen-specific antibody fragments are directly selected from antibody fragment gene repertoires expressed on the surface of bacteriophage, viruses that infect bacteria (phage biology is discussed in Box 6.1).[23,24] The V-genes are already cloned

Box 6.1 Biology of Filamentous Bacteriophage.

The filamentous bacteriophages M13, fd, and f1 are closely related phages that infect *E. coli* cells via the F-pili. Under the electron microscope they appear as elongated filaments of approximately 7 nm in width by 1000 nm in length. They have a single-stranded, covalently closed circular DNA genome of ~6500 bases, coding for a total of 11 genes (called genes I through XI; the protein products of these genes are known as p1 through p11).

The phage DNA genome is encased in a protein coat primarily consisting of approximately 3000 copies of the major coat protein p8. Several other minor coat proteins (p3, p6, p7, and p9) are found in a few copies at one end or the other of the filament. The most important of these minor coat proteins is p3, which is found in ~5 copies at the tip of the phage filament. The gene 3 protein comprises three domains (N1, N2, and CT) connected by glycine-rich linkers. The CT domain anchors the p3 protein in the protein coat, while the N1 and N2 domains are responsible for phage binding and entry into the host cell.

Infection of *E. coli* cells begins with binding of the N2 domain of the p3 protein to the end of an F-pilus. After binding, the F-pilus retracts, bringing the phage particle into contact with the bacterial outer membrane. Entry into the cell is mediated by binding of the N1 domain of p3 to the TolA protein, causing the phage protein coat to dissolve into the cell membrane concurrently with the phage genome being extruded into the host cytoplasm.

Once inside, the single-stranded phage genome (designated the + strand) is converted by host enzymes into a double-stranded closed circle (the RF, or replicative form). The RF reproduces by a rolling circle mechanism: the + strand is nicked by the phage p2 protein, then this + strand is displaced by host enzymes as they continue to produce additional copies of the + strand, using the − strand as template. These + strands are then covalently closed by p2 and converted into additional RFs. Meanwhile, the accumulating RFs serve as the template for the transcription of the remaining phage genes. The newly synthesized coat protein p8 and the minor coat proteins (p3, p6, p7, and p9) are transferred to the cell membrane.

RF reproduction continues until a sufficient concentration of the single-stranded DNA binding protein p5 accumulates in the cytosol. At this point, newly synthesized + strands are no longer converted into RFs, instead they become coated with the p5 protein, and assemble into phage particles. Phage assembly occurs at the inner surface of the cell membrane. Viral + strands, coated with p5, attach to the membrane, and the phage genome is extruded from the cell in the reverse of entry. As the genome is extruded through the cell membrane, the p5 protein is stripped from the genome and replaced by the major coat protein p8 along with the minor coat proteins. When the end of the phage genome is reached, the phage particle is released into the medium. Approximately 10^{12} phage particles per milliliter are produced in cultures of infected *E. coli* cells.

and almost invariably express at high levels in bacteria.[25,26] Higher affinity antibody fragments can be selected from phage antibody libraries created by mutating the antibody fragment genes of initial isolates.[27,28] Moreover, the approach can be used to produce human antibodies, which are difficult to produce using conventional hybridoma technology, and to produce antibodies without immunization.

6.2 OVERVIEW OF ANTIBODY PHAGE DISPLAY

Phage display mimics the strategy used by the humoral immune system to produce antibodies in vivo. This strategy has three key components: (1) the generation of millions of different antibody molecule genes through recombination and combinatorial pairing; (2) expression of the antibody gene repertoire on the surface of B lymphocytes, where the antibody functions as an antigen receptor; and (3) antigen-driven selection of antigen-binding B lymphocytes for proliferation and differentiation (Fig. 6.1A). The B lymphocyte serves an essential role by providing a physical linkage between surface antibody (phenotype) and the gene encoding the antibody (genotype).

Three technical achievements make it possible to reproduce the process of immune selection in vitro (Fig. 6.1B). First, it has proved possible to express the antigen-binding V_H and V_L domains of antibodies in *E. coli*, either as Fab, Fv, or scFv antibody fragments. Second, large and diverse repertoires of Fab or scFv genes can be generated using the polymerase chain reaction. The antibody fragment repertoires can be built from V_H and V_L genes obtained from B lymphocytes, either before or after immunization, or from cloned V gene segments rearranged in vitro. Third, the antibody fragments can be expressed on the surface of phage. This is accomplished by cloning the antibody gene into a phage vector so that it is genetically fused with a gene that encodes a protein expressed on the phage surface. The resulting phage has the antibody displayed on its surface, anchored to the phage via the coat protein, and the phage contains the gene encoding the antibody inside. Thus, the phage mimics the function of the B lymphocyte, providing a physical linkage between phenotype on the surface and genotype within.

Antibody fragment gene repertoires can be cloned into phage vectors, resulting in the creation of phage antibody libraries. Phage antibodies binding a specific antigen can be separated from nonbinding phage by selection on antigen. (Phage are incubated with immobilized antigen, nonbinding phage are removed by washing, then bound phage are eluted). A single round of selection will result in a 20- to 1000-fold enrichment for binding phage. Eluted phage are used to infect *E. coli*, which produce more phage for the next round of selection. Repetition of the selection process makes it possible to isolate binding phage present at frequencies of less than one in a billion.

Antibody phage display has resulted from concurrent progress in prokaryotic expression of antibody fragments, PCR cloning of antibody gene repertoires, and display of peptides and proteins on filamentous bacteriophages. These three areas are reviewed in detail in the following sections. Subsequent sections describe specific applications of antibody phage display, including its use to (1) bypass hybridoma technology; (2) bypass immunization; and (3) affinity mature antibodies.

6.3 PROKARYOTIC EXPRESSION OF ANTIBODY FRAGMENTS

Initial efforts to express full-length antibodies in the cytoplasm of *E. coli* by simultaneous coexpression of heavy and light chains were met with limited success. When a mouse heavy (μ) and light (λ) chain on two compatible plasmids were introduced into *E. coli* and expressed off the trp promoter, the proteins were produced as insoluble inclusion bodies in the cytoplasm. Antigen-binding activity was only recovered after solubilizing and refolding the aggregated material.[29] Similarly, expression of the heavy and light chains of a murine anti-CEA antibody in the cytoplasm of *E. coli* yielded only insoluble aggregates that required solubilization and refolding to detect antigen binding.[30] Presumably, these results reflect the difference between the reducing environment of the prokaryotic cytoplasm versus the oxidizing environment of the eukaryotic secretion pathway where disulfide bonds can form.

More favorable results were achieved when researchers expressed antibody fragments fused to bacterial signal sequences that directed their secretion into the periplasmic compartment. Better et al. constructed a discistronic operon that expressed genes encoding the two chains of a human–mouse chimeric Fab; each polypeptide chain was directed to the periplasm by a bacterial signal sequence which was subsequently cleaved by signal peptidase. In the periplasm, the two chains folded correctly and formed a fully functional Fab.[31] Likewise, Skerra and Plukthun engineered *E. coli* to secrete into the periplasm the V_H and V_L polypeptide chains of a mouse monoclonal antibody. The two chains correctly folded to form a fully functional Fv fragment.[32]

While Fvs are the smallest antibody subunit that retain antigen binding, their utility is limited by instability. At typically utilized concentrations, Fv dissociate into V_H and V_L domains, with the K_d for dissociation ranging between 10^{-4} to 10^{-8} M.[33–37] Differences in K_d values result from differences in residues composing the β-sheets that make up the V_H-V_L interface.[38] While many of these interface residues are conserved, 25% of the interface residues reside in the hypervariable CDRs.[38] Several strategies have been utilized to stabilize the Fv fragment, including engineering a disulfide bond into the V_H-V_L interface.[36,39,40] The most commonly employed strategy has been to physically link the V_H and V_L chains into a continuous polypeptide to form a single-chain Fv (scFv).[41,42] Both V_H-linker-V_L and V_L-linker-V_H orientations have been shown to yield functional scFv and as long as the linker is of sufficient length, many different linker sequences are tolerated[8,41,43,44] scFv generally retain the specificity and affinity of the antibody from which they were derived.[45]

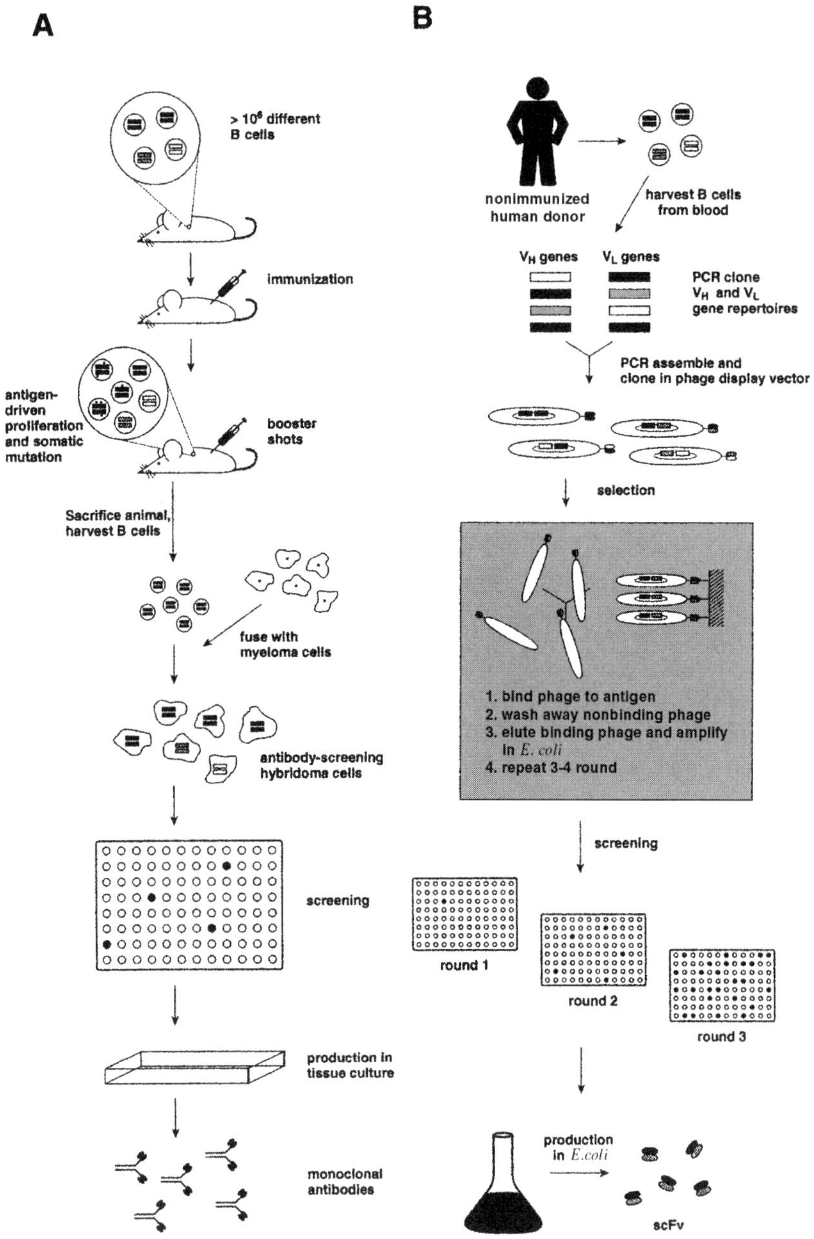

Figure 6.1 A comparison of monoclonal antibodies from murine hybridomas (panel A) vs. phage display (panel B). In panel A, a naive mouse contains more than 10^6 rearranged V genes (shaded bars) in B cells, coding for antibodies that are displayed as membrane-bound molecules. Immunization causes antigen-driven proliferation and

Furthermore, the crystal structure of a Fab and its related scFv were recently solved, and only minor differences in the antigen-binding sites were observed.[46]

6.4 GENERATION OF ANTIBODY GENE REPERTOIRES USING THE POLYMERASE CHAIN REACTION

Before the advent of the polymerase chain reaction (PCR), cloning antibody genes was a laborious process that required creation and screening of genomic or cDNA libraries. PCR has streamlined the process considerably, permitting the construction of large antibody fragment gene repertoires. For PCR amplification, first strand cDNA is obtained by reverse transcription of mRNA with primers specific to the constant region (primers that anneal in the C_H gene ($C_\gamma 1$, $C_\gamma 2$, $C_\gamma 3$, $C_\gamma 4$, C_μ, C_ε, C_α, or C_δ) or to the C_L gene (C_κ or C_λ)). Since sequences of the constant domain exons are known,[47] primer design is straightforward.

To create Fabs, the heavy chain V_H-C_H1 (Fd) and light chain (V_L-C_L) genes are amplified. Design of primers that anneal to the 3′ end of these genes is straightforward, because the constant regions have been sequenced.[47] For Fv or scFv, only the rearranged V_H and V_L genes are amplified. Design of PCR primers for the 3′ end of rearranged murine[48] or human[25] V_H and V_L genes is also straightforward because primers can be based on sequence in the J gene segments. These primers can also be used for first strand cDNA synthesis instead of constant region primers.

Design of primers for the 5′ end of the V-gene is less straightforward due to the sequence variability of different V-genes. In the earliest attempt to use PCR

Figure 6.1 *continued*

somatic hypermutation (stars next to the V genes). The B cells are harvested from the immunized mouse and fused with immortal myeloma cells (wrinkled edges) to generate immortalized, antibody secreting hybridomas. Hybridomas are screened by ELISA for antigen specificity, and the monoclonal antibodies produced in tissue culture. In panel B, the phage display method begins with isolation of B cells from either the spleen of an immunized mouse (as in A) or from the peripheral blood of a naive or immunized human donor. Heavy and light chain V-gene regions (shaded bars) are amplified by PCR and assembled as a single-chain Fv region (scFv). (Note that the original heavy and light chain pairings become scrambled during scFv assembly). The repertoire of scFv genes are cloned and inserted into filamentous bacteriophage, where the encoded scFv are displayed in a functional form on the phage surface (colored ovals represent scFv). Multiple rounds of selection with a solid-phase antigen allow isolation of even rare phage from the original library. The selected scFv will generally have affinities similar to monoclonal antibodies (~ nanomolar dissociation constants) and can be expressed in *E. coli*. The original binding scFv can be subjected to repeated rounds of mutagenesis and selection to affinity mature the scFv to ~ 10 picomolar affinities.

to amplify V-genes, N-terminal protein sequencing was carried out on purified antibody from a hybridoma, and the sequence was used to assign the V_H and V_L (V_λ or V_κ) gene families. These gene assignments were used to design degenerate primers.[49] A generally applicable approach was taken by Orlandi et al.[50] The nucleotide sequences of murine V_H and V_L genes were extracted from the Kabat database, aligned, and the frequency of the commonest nucleotide plotted for each position. Conserved regions identified at the 5′ region of the V_H and V_L genes were then used to design degenerate oligonucleotide primers containing restriction sites for directional cloning. Additional groups have designed sets of universal V-gene primers containing internal or appended restriction sites suitable for amplification of murine,[51–55] human,[25,56–58] chicken,[59] and rabbit[60] V-genes.

Fv, Fab, and scFv genes are typically constructed by sequential cloning of V_H (V_H-C_H1) and V_L (V_L-C_L) genes.[41,42] Sequential cloning, however, requires four restriction sites which increases the chances of restriction enzymes cutting internally.[61,62] Alternatively, PCR splicing by overlap extension (PCR assembly) of V_H and V_L genes has been used to construct scFv genes[25,48] (Fig. 6.2). With PCR assembly, only two restriction sites are required to clone the scFv. These sites can be appended to the 5′ and 3′ end of the scFv gene cassette. Furthermore, expression systems have been described where octanucleotide cutters can be utilized, markedly decreasing the likelihood that a restriction enzyme will cut the V-gene internally.

6.5 ANTIBODY PHAGE DISPLAY

Phage display technology exploits several unique features of the filamentous *E. coli* bacteriophages M13, fd, and f1. Phage are single-stranded DNA viruses with genomes of approximately seven thousand base pairs; the viral particle is a filament containing one copy of the viral genome encased in a protein coat. This coat is composed primarily of 2700 copies of the gene 8 protein pVIII. In addition there are several minor coat proteins; including the gene 3 product pIII which is found in 3–5 copies at one end of the viral particle. Phage replication begins with binding of the pIII protein to the F-pilus of a male (F^+) *E. coli* bacterium, and entry of the viral genome into the bacterial cytoplasm. Once inside the bacterium, the viral genome is copied into a double-stranded replicative form (RF), which serves as a template to make more copies of the single-stranded viral genome for packaging. The intragenic region of the phage genome contains a signal for the newly created genomes to be packaged and extruded from the host cell. During extrusion, the phage DNA becomes coated with pVIII and pIII, forming a new phage particle, which can begin a new round of infection.

In 1985, George Smith showed that when DNA encoding a peptide was cloned as a gene fusion with gene III, the resulting phage "displayed" the peptide on its surface as a fusion with pIII.[63] Phage displaying the peptide were

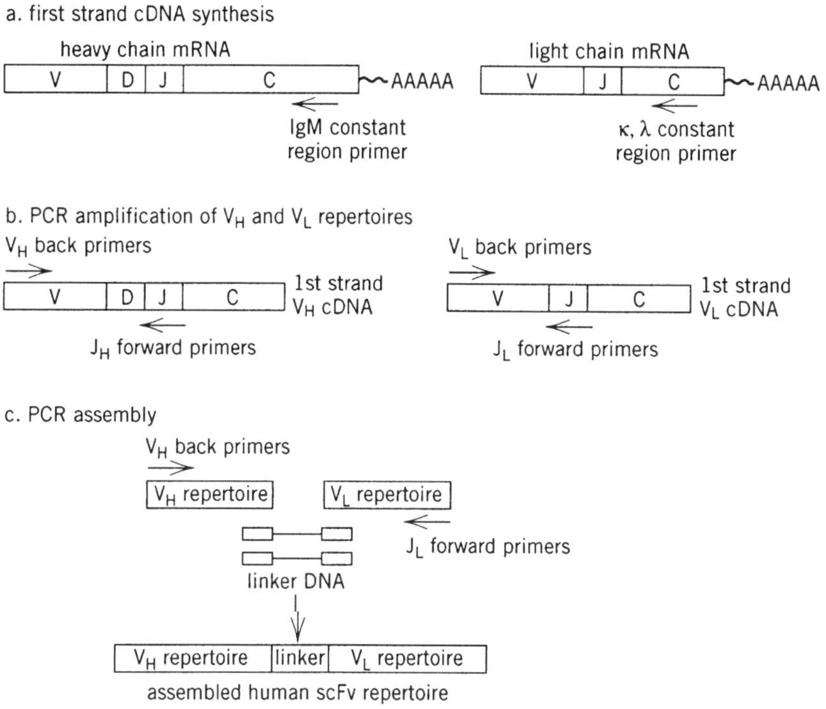

Figure 6.2 PCR cloning of antibody V-gene repertoires as single-chain Fvs (scFv). (*a*) mRNA is isolated from peripheral blood, spleen, or bone marrow, and antibody genes are transcribed (by reverse transcriptase) using IgG, IgM, κ, and λ constant-region specific primers, creating first strand cDNA. (*b*) The V_H and V_L variable regions are amplified in a series of PCR reactions, using primers specific for each of the known heavy and light chain V-gene families. (*c*) Assembly by PCR. PCR-amplified V_H and V_L genes are pooled, together with a short linker DNA that overlaps the 3′ and 5′ ends of the V_H and V_L regions respectively, and PCR amplified to yield one continuous DNA fragment. A final PCR reaction (not shown) adds flanking restriction sites to the assembled repertoire for cloning into the phage display vector.

enriched from populations of wild-type phage by affinity chromatography using a monoclonal antibody specific for that peptide. Following infection of bacteria with the enriched, eluted phage, more phage were grown and subjected to further selection. Repetition of the selection process enabled the isolation of binding phage present at frequencies of less than one in a billion. The same technique could be used to enrich peptide phage with high affinity for ligand from medium or low affinity phage.

In 1990, McCafferty et al. demonstrated that it was possible to display a functional antibody fragment on the surface of filamentous phage, using a scFv fragment of the antilysozyme MAb D1.3.[23] They constructed a scFv-pIII fusion in the phage fd vector fdCAT1 and showed that the displayed scFv retained antigen binding and specificity. Phage displaying the scFv were enriched from a mixture with wild-type phage by affinity chromatography on a lysozyme column. Enrichment factors of 10^3 for one round of selection and 10^6 for two rounds of selection were achieved.

6.5.1. Vectors for the Display of Antibody Fragments on Filamentous Phage

A large number of vector systems have subsequently been described for the display of antibody fragments. Some of the more important ones are summarized in Figure 6.3 and Table 6.1. The vectors differ primarily with respect to the type of antibody fragment displayed (Fab[24,55,64–67] or scFv[23,48,68]), the fusion partner (pIII[23,24,48,55,65,67,68] or pVIII[64,66]) and the type of vector (phage[23,48,66] or phagemid[24,55,64,65,67,68]). Vectors for display of scFv have a single leader sequence and either two or four restriction sites, depending on whether the V-genes are cloned sequentially[69] or are first spliced together using PCR to create an scFv gene repertoire.[48] Vectors for display of Fv or Fab have two leader sequences and two pairs of restriction sites for sequential cloning of V_H-C_H1 and V_L-C_L genes. Fv and Fab fragments are displayed by fusing one of the chains to a phage coat protein and secreting the other chain into the bacterial periplasm, where the heterodimer forms.[24]

The most widely used phage vectors create fusions with the minor coat protein pIII.[23,24,65,69–71] Alternatively, antibody fragments can be displayed as pVIII fusions.[64,66,72] These fusions result in many more copies of antibody fragment per phage (up to 24 copies/phage[64]). Display of antibody fragments in phage vectors also leads to multicopy display, with each of the three to five pIII having an antibody fragment fusion. Multivalent display leads to an increase in the apparent affinity (avidity) when the phage antibody is selected on antigen immobilized on a solid support. Increased avidity enables selection of very low affinity antibody fragments. However, multivalent display makes discrimination difficult between phage with only minor differences in affinity.[65] This can hinder selection of the highest affinity binders.

Use of a phagemid vector results in "monovalent" antibody fragment display. Phagemids are plasmids that contain the phage f1 intragenic region and therefore can be packaged in a phagelike particle, but lack other essential phage genes and therefore cannot independently produce phage. Essential phage genes are provided in trans by infection of phagemid containing *E. coli* with a helper phage. The helper phage has a "weakened" packaging signal, and thus the phagemid genome is preferentially packaged into the phage particle. In phagemid vectors, antibody fragment-pIII fusion expressed from the phagemid DNA competes with wild-type pIII expressed from the helper phage genome. The resulting phage have on average less than one copy of antibody-pIII fusion/phage. Monovalent phage display leads to more efficient selection of phage on the basis of binding affinity. For example, in a single round of selection using a mixture of two Fab that differed 100-fold in affinity

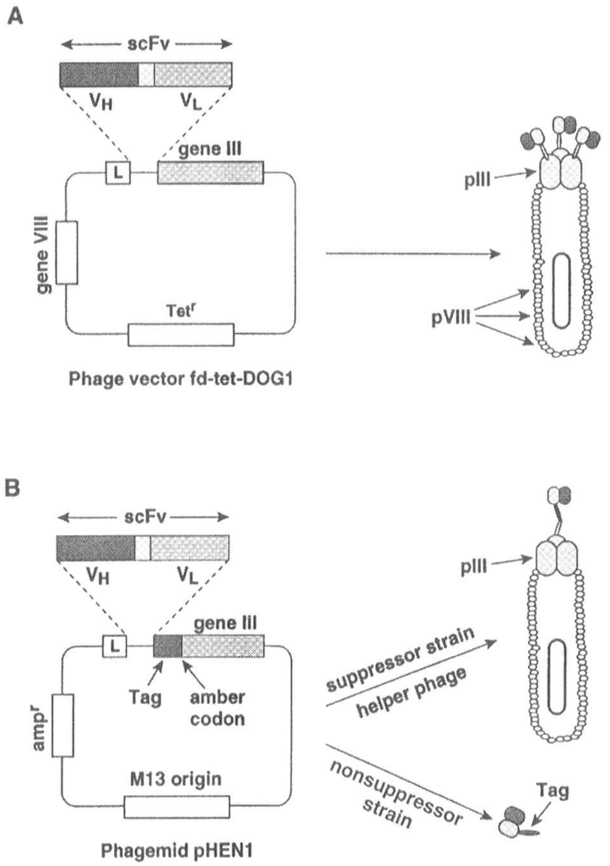

Figure 6.3 Features of phage display vectors. A representative phage and phagemid vector are shown. (*A*) Phage vector fd-tet-DOG1 and (B) phagemid vector pHEN1. Both vectors display scFv (V_H-linker-V_L) as fusions to the amino terminus of the pIII protein. In phage fd-tet-DOG1, all copies of pIII are scFv fusions, leading to 3–5 copies displayed per phage particle. With phagemid pHEN1, expression in supressor strains of *E. coli* allows the amber codon following the scFv-tag to be read as a glutamine, causing the scFv to be fused to the pIII protein. In phagemids, both wild-type pIII (from the helper phage) and fusion pIII (from the phagemid) compete for inclusion in the viral particle. In nonsuppressor *E. coli* strains, the scFv is expressed as a soluble protein. Amp^r = ampicillin resistance gene, tet^r = tetracycline resistance gene.

(K_d 10^{-7} M vs. 10^{-9} M), the enrichment ratio was 253-fold for monovalent display on pIII vs. only 5.5-fold for multivalent pVIII display.[65] Phagemid vectors also yield higher transformation efficiencies, making construction of large phage antibody libraries easier. As a result of these two features, the most widely used systems for antibody phage display use phagemid vectors with display on pIII.

TABLE 6.1 Commonly Used Vectors for Phage and Phagemid Display of Antibody Fragments

Vector	Form	Fusion	Marker	Tag	Reference	Comments
pHEN1	phagemid	pIII	Amp	myc	24	For cloning scFv either NcoI-NoTI or SfiI-NotI, followed by c-myc tag for detection. Amber codon between myc tag and pIII for expressing soluble scFv in nonsuppressor strain.
fd-tet-DOG1	phage	pIII	Tet	none	24	For cloning scFv ApaLI-NotI as pIII fusions.
pCANTAB 5E	phagemid	pIII	Amp	E tag	(Pharmacia)	scFv are cloned SfiI to NotI, fused to E tag and then to pIII, amber codon after E tag allows expression of soluble scFv in nonsupressor strains.
pCOCK	phagemid	pIII	Amp	myc	69	Basically pHEN1 with an AscI site between SfiI and NotI which allows independent cloning of V_H and V_L genes as SfiI-AscI and AscI-NotI, respectively; the scFv linker has to be recreated by PCR cloning; the scFv linker is GGGGSGGRASGGGGS to accommodate the AscI site.
pCOMB3	phagemid	pIII	Amp	none	65	For displaying Fabs: heavy chain Fd fragments are cloned XhoI to SpeI fused to pIII (198–406); light chains are cloned SacI to XbaI and associate in the periplasm; SpeI + NheI digestion and religation removes pIII sequences for soluble Fab production.
M131XL604	phage	pVIII	none	none	66	For pVIII display of Fab; the heavy chain Fd is cloned in XhoI-SpeI, and the light chain is cloned NcoI to XbaI. Heavy chain is fused through amber codon to pVIII, light chain associates noncovalently in periplasm. Amber codon allows expression of soluble Fab in nonsupressor strains.

Analysis for phage binding to antigen can be performed in an ELISA format using antiphage antibodies.[48] Analysis for binding is simplified by including an amber codon between the antibody fragment gene and gene III[24] (Fig. 6.3). This makes it possible to easily switch between displayed and soluble antibody fragment simply by changing the host bacterial strain. When phage is grown in a supE supressor strain of *E. coli*, the amber stop codon between the antibody gene and gene III is read as glutamine and the antibody fragment is displayed on the surface of the phage. When eluted phage is used to infect a nonsupressor strain, the amber codon is read as a stop codon and soluble antibody is secreted from the bacteria into the periplasm and culture media. Binding of soluble scFv to antigen can be detected by ELISA using epitope tags, such as c-myc[25] or E-tag[73] incorporated into the vectors 5′ to the amber codon. Inclusion of a hexahistidine tag makes it possible to easily purify native scFv using immobilized metal affinity chromatography (IMAC)[74] without the need for subcloning.[26]

6.6 USE OF PHAGE DISPLAY TO BYPASS HYBRIDOMA TECHNOLOGY

Phage display can be used to produce monoclonal Fab or scFv antibody fragments from immunized mice and humans. The primer systems described previously in Section 6.4 can be used to produce diverse murine or human V_H (V_H-C_H1) and V_L (V_L-C_L) gene repertoires from spleen, bone marrow, or peripheral blood lymphocytes.[25,48,56,75] The V_H and V_L genes can be randomly spliced together using PCR,[25,48] or cloned sequentially to create scFv or Fab gene repertoires displayed on the surface of phage.[55,64–67,69]

In the first example, a scFv phage antibody library was created from the V-genes of a mouse immunized with the hapten phenyloxazolone (phOx).[48] After cloning, a 2×10^5 member scFv phage antibody library was created in a phage (fd) vector. Binding phage antibodies were selected on a phOx-BSA column. After two rounds of selection, the majority of clones analyzed produced phOx binding scFv and more than twenty unique scFv were isolated. The K_d of the highest affinity phage antibodies (1.0×10^{-8} M) were comparable to the affinities of IgG from hybridomas constructed from mice immunized with the same hapten. Similar panels of scFv have been obtained using phage display and mice immunized with EGF receptor[76] and Botulinum neurotoxin type A.[75] This approach has also been used to produce monoclonal chicken[59] and rabbit[60] antibody fragments using species-specific primers.

A greater number of examples exist of phage libraries constructed from the V-genes of immunized humans or patients with diseases where a humoral immune response is generated. In most of these examples, Fab libraries were constructed. Variable region genes were obtained either from peripheral blood lymphocytes, bone marrow, lymph nodes, or rarely, from spleen. In the area of infectious disease, human monoclonal scFv or Fab antibody fragments against

Tetanus toxin,[65] HIV-1 gp120,[77] HIV-1 gp41,[78] hepatitis B surface antigen,[79] hepatitis C,[80] respiratory syncytial virus,[81,82] and hemophilus influenza,[83] have been isolated from immunized volunteers or infected patients. Autoimmune antibodies have been isolated from patients with systemic lupus erythematosus (anti-DNA),[84] Hashimoto's disease (antithyroid),[85] myasthenia gravis (anti-acetylcholine receptor),[86] as well as other diseases.

Immune libraries allow exploitation of the affinity maturation process, which occurs by somatic hypermutation. In addition, immunization amplifies the number of lymphocytes producing binding antibodies, as well as their RNA content. This is particularly true for IgG-producing plasma cells. Thus libraries are enriched for V-genes from antigen-binding antibodies. However, it is important to keep in mind that this may be offset by the fact that the original V_H and V_L pairings are lost during library construction when the lymphocytes are lysed for RNA isolation.[87] With achievable library sizes, the probability of recovering the original V_H and V_L pairings is small. Techniques for "in-cell" PCR have been devised for retaining the original V-gene pairings,[88] but this approach has not been applied to library construction. That immune libraries yield high affinity binders is partly due to the fact that in many cases, a binding light chain can be replaced by a homologous light chain with recovery of antigen binding. Such chain promiscuity has been observed in immune libraries[48,77] and during chain shuffling, where the binding light chain is replaced by a library of light chains.[27,48,89,90] After light chain shuffling, the frequency of binding antibody fragments can be as high as 10%.

Antibodies from immune phage antibody libraries share a number of characteristics. Typically, a panel of antibodies are obtained. Some of these are clonally related, being derived from the same V-D-J rearrangement with differences due to somatic hypermutation. Others are clearly derived from different V-D-J rearrangements. The antibodies are highly specific for the antigen used for selection, and they bind to antigen with affinities typical of IgG produced from hybridomas derived from immunized mice or humans. Moreover, the ability to select rather than screen for binding properties (see Section 6.5.1) permits the isolation of antibodies to rare epitopes,[91–93] cell surface markers[94] or unstable antigens,[95] which have proved difficult to produce using hybridoma technology. The antibody fragments can be used for the same purposes as IgG derived from hybridomas, including Western blotting,[96] epitope mapping,[78] cell agglutination assays,[97,98] cell staining, FACS,[99] and immunohistochemistry.[100]

The use of immune phage antibody libraries allows the isolation of a large number of scFv or Fab of high specificity and affinity. This approach, however, is subject to the limitations associated with immunization. First, with many antigens the immune response is directed to only a few immunodominant epitopes.[75,78] Second, production of antibodies against phylogenetically conserved proteins is problematic. Third, for each desired specificity a new library must be constructed.

6.7 USE OF PHAGE DISPLAY TO BYPASS IMMUNIZATION

As an alternative to immune phage antibody libraries, monoclonal antibody fragments can be produced without prior immunization by displaying very large and diverse scFv or Fab gene repertoires on phage.[25] This theoretically results in the ability to isolate antibodies to any desired specificity from a single phage antibody library. In the first example, rearranged human V_H, V_κ, and V_λ gene repertoires were amplified from the mRNA of peripheral blood lymphocytes obtained from two nonvaccinated volunteers.[25] Diverse scFv gene repertoires were obtained by randomly joining the V_H and V_L gene repertoires with DNA encoding the 15 amino acid linker sequence $(G_4S)_3$. The human scFv gene repertoire was cloned into the phagemid pHEN1 to create a nonimmune phage antibody library of 3.0×10^7 members. This is similar to the size of a primary human or mouse B-cell repertoire.

From this single nonimmune library, one to several scFv were isolated against more than 20 different antigens, including a hapten, 3 different polysaccharides, and 16 different proteins.[25,97,101–103] The scFv were highly specific for the antigen used for selection and had affinities typical of the primary immune response (1 mM to 15 nM range).[25,101,102] Using this approach, it was possible to isolate scFv that bound self proteins; for example human scFv which bound human thyroglobulin, immunoglobulin, tumor necrosis factor, CEA, c-erbB-2 and blood group antigens.[25,97,101–103] The ability to isolate antiself specificities may result from the generation of specificities that had been deleted from the B-cell repertoire. These new specificities would arise from the random reshuffling of V_H and V_L genes that occurs during library construction. Alternatively, antiself specificities could arise from the 10–30% of human B cells which may be making low affinity antiself antibodies at any time.[104]

Larger or more diverse phage antibody libraries should theoretically provide higher affinity antibodies against a greater number of epitopes on all antigens used for selection.[105,106] One approach to increase diversity is to rearrange V-genes in vitro using cloned V-gene segments and random oligonucleotides encoding part of the antigen combining site.[96,99,107–110] Using this approach, it is possible to ensure that the potential diversity is large enough so that a library will only contain a single member of each sequence. In contrast, there may be considerable duplication of sequences using V-gene segments rearranged in vivo. The design of semisynthetic libraries is based on structural, genetic, and mutagenesis data demonstrating that the determinates of antibody specificity reside in the six complementarity determining regions (CDRs) (reviewed in Ref. 111). Five of the 6 CDRs are encoded in a small number of germline gene segments[112–114] and have limited sequence variability and main chain conformations (canonical structures).[114–116] The exception is V_H CDR3, which is generated by recombination of three gene segments (V, D, and J). Addition of random nucleotides (N-segment addition) at the V-D and D-J junctions leads to tremendous sequence and structural diversity. Not surprisingly, this CDR contributes a disproportionate number of antigen contacting amino acid side chains and binding energy.[117–119] Thus synthetic antibody libraries have focused the diversity in V_H CDR3.

In the first example of a semisynthetic phage antibody library, a V_H gene repertoire was constructed from 51 cloned germline human V_H gene segments[113] and oligonucleotides designed to yield a V_H CDR3 of five or eight residues in which five residues were of random sequence. The repertoire was cloned into a vector containing a single rearranged V_λ light chain gene to create an scFv phage library of 2.2×10^7 members.[20] This library yielded large panels of monoclonal scFv to hapten antigens, but not to protein antigens.[20] Increasing the length of the synthetic V_H CDR3 to between 4 and 12 residues resulted in a library from which antibodies could be isolated to protein antigens as well.[96] The affinities of the antibodies, however, were not reported. Subsequent examples of semisynthetic antibody libraries have used a larger number of V_L genes.[99,109,110]

The greatest improvements in the utility of nonimmune libraries has resulted from increasing library size several orders of magnitude. This yields libraries capable of yielding panels of high affinity antibody fragments against any antigen. Two published examples of such libraries exist, a 6.5×10^{10} member Fab library constructed using semisynthetic V-genes,[110] and a 1.3×10^{10} member scFv library constructed from human V-genes rearranged in vivo (obtained from healthy human volunteers).[26] In the case of the Fab library, combinatorial infection was used to overcome limitations imposed on library size by transformation efficiency.[120] From these two libraries, panels of antigen-specific antibody fragments were isolated against 43 different antigens, including haptens and proteins. Affinities of the antibody fragments were typical of the secondary immune response (K_d ranging between 4.1×10^{-8} M and 3.0×10^{-10} M for antibody fragments directed against 11 different antigens). Similarly, we have constructed a 6.7×10^9 member scFv phage antibody library from the V_H and V_L genes of healthy humans.[121] An average of 9.2 scFv with K_d as high as 3.7×10^{-10} M were isolated to 10 different protein antigens. These libraries represent the current state of the art: the ability to generate high-affinity human monoclonal scFv within weeks and without immunization.

6.8 A COMPARISON OF DIFFERENT LIBRARY TYPES AND THEIR APPLICATIONS

6.8.1 Immune vs. Nonimmune Libraries

The choice of whether to generate immune or nonimmune phage antibody libraries will depend on the uses intended and the skills of the investigator. Compared to nonimmune libraries, immune libraries should yield a larger number of antibody fragments that have higher average affinity. These libraries are also easier to construct, since significantly smaller libraries can be utilized. In addition, immune libraries are the only possibility for probing the in vivo immune response to a particular pathogen or disease. (It is also possible to obtain human antibodies, either where there is an immune response to a pathogenic state, or when it is possible to immunize.) Immune libraries, however, suffer from the drawback that a new library has to be constructed for each new antigen desired. In addition, for many antigens the immune response is directed to only a few immunodominant epitopes.[75,78] This may result in failure to isolate the precise specificity desired for a particular aim. Furthermore, production of antibodies against proteins conserved between species may prove difficult or impossible. Production of non-immunogenic human antibodies is also limited by the inability to immunize humans with many antigens, especially those of human origin.

In contrast, nonimmune libraries need to be made only once, can provide antibodies to any antigen, including those that are conserved or self and can serve as a source of therapeutic human antibodies. Compared to immune libraries, however, fewer antibodies will be obtained and the affinities may not be as high. Moreover, useful libraries are of a size that has proven difficult to produce except in a limited number of laboratories. Once this last limitation is overcome, large nonimmune libraries are likely to assume increasing importance for antibody generation. Antibodies from nonimmune libraries have been used successfully for Western blotting,[96] epitope mapping,[75] cell agglutination assays,[97,98] cell staining, and FACS.[99]

It is presently unclear whether nonimmune libraries are best constructed from V-genes rearranged in vivo (natural or naive libraries) or in vitro (semisynthetic). A theoretical advantage of semisynthetic libraries is that you can ensure that only a single copy of each antibody exists in a library. In addition, V-gene segments can be selected that express well in *E. coli*, providing a higher number of functional antibodies and reducing selection biases based on expression efficiency. The use of synthetic CDRs, however, assumes that the V_H CDR is a random peptide, which is not the case.[122] It is therefore likely that a significant number of synthetic CDR3 do not fold properly, or do not make useful binding pockets, compared to those constructed in vivo from V, D, and J gene segments. Semisynthetic libraries are also more difficult to construct, requiring the cloned V-gene segments. At present, it appears that library size is a more important parameter than where the V-genes are derived from.

6.8.2 Fab vs. scFv Libraries

The choice of whether to produce scFv or Fab libraries depends partly on the intended use. For construction of fusion proteins, such as immunoadhesins or immunotoxins, the single gene format of the scFv is an advantage. This is especially true for targeted gene therapy approaches where the scFv gene is fused to a viral envelope protein gene[123,124] or the gene for a DNA binding protein.[125] scFv also appear to be the preferred antibody fragment for intracellular immunization, a technique where the antibody gene is delivered intracellularly to achieve phenotypic knockout.[126,127] In theory, Fab libraries may be preferred where the final product will be a complete antibody, since in some instances removal of the scFv linker might alter antigen-binding properties. In practice, too few examples exist where either Fab or scFv have been retroengineered into complete antibodies to draw conclusions.

Technical issues also influence the choice of antibody fragment type for library construction. In general, scFv libraries are easier to construct. The genes are smaller, amplify more easily, and clone with higher efficiency. scFv are also generally less toxic to *E. coli* and fold more efficiently. This leads to a lower deletion rate of insert from phage libraries and higher expression levels of native antibody fragment. This makes subsequent characterization easier. In fact, the only disadvantage of scFv is the tendency of some to dimerize and aggregate. Dimerization and aggregation occur from the V_H domain of one scFv molecule pairing with the V_L domain of a second scFv molecule.[128,129] The tendency of scFv to dimerize is sequence dependent, with some scFv existing as stable monomer[98,101,102,128] and others as mixtures of monomeric and oligomeric scFv.[96,98,102,129,130] Our bias is that the advantages of working with scFv libraries greatly exceeds this single disadvantage. We therefore work exclusively with scFv.

6.9 STRATEGIES FOR SELECTION OF PHAGE ANTIBODIES

In its earliest implementation, selection involved: (1) the binding of phage expressing a relevant antibody fragment to antigen bound on a solid surface; (2) removal of irrelevant (nonbinding) phage by washing; and (3) elution of specifically bound phage. The basic selection process is first described in some detail. The remainder of the section is devoted to more complicated selection strategies (Fig. 6.4).

The first phage antibody libraries were selected in columns using antigen covalently linked to a matrix such as cyanogen bromide activated Sepharose.[23,48] The easiest technique, however, is merely to adsorb the antigen of interest onto a polystyrene surface, such as an ELISA plate well[58] or immunotube.[25] Antigens are typically adsorbed at concentrations of 10 to 50 μg/ml in either PBS or carbonate buffer, however specific conditions may need to be optimized for a particular antigen. After antigen is bound to the surface, the remaining binding capacity of the surface is blocked by an excess of irrelevant protein, e.g., BSA, casein, or gelatin. A 2% solution of nonfat powdered milk in PBS (2% MPBS) has proved particularly effective. Subsequent incubations are done in 2% MPBS to prevent isolation of milk protein-specific phage. If the antigen was produced as a fusion protein, for example to GST or maltose binding protein, the addition of soluble GST or maltose protein to the 2% MPBS will reduce the selection of antibodies that bind the fusion partner. After blocking, approximately 10^{11}–10^{13} phage are added and allowed to bind for one to several hours. Nonbinding phage are removed by washing and specifically bound phage eluted by addition of agents that disrupt the antibody–antigen interaction such as extremes of pH or chaotropic agents. Even with extensive washes, however, considerable nonspecific phage can remain bound to antigen and are eluted by these general eluants. This leads to lower apparent enrichment ratios. One way to increase enrichment is to use a specific eluate such as a high concentration of soluble antigen,[28,48] competitive elution with a known ligand, for example an antibody,[131] or cleavage of a susceptible linker in the antigen-support linkage.[132]

6.9.1 Selection Using Soluble Antigen

Adsorption of protein antigens to plastic may lead to partial or complete antigen denaturation, which can lead to the disappearance of certain epitopes and the exposure of other cryptic epitopes. Thus antibodies selected on surface-absorbed antigens may not react with the native antigen, for example in solution or on the surface of cells.[101,133] The problem of antigen denaturation can be overcome by selecting on soluble antigen in solution. The antigen can be chemically tagged, for example by biotinylation, and captured along with bound phage using avidin or streptavidin magnetic beads.[28,90] Alternatively, the antigen can be genetically tagged with a hexahistidine tag with capture on Ni-NTA agarose,[75] or expressed as GST or maltose-binding fusions with capture on either glutathione or maltose-columns. Not only will this approach increase the likelihood of obtaining antibodies that recognize native antigen, it is also extremely useful for limiting the antigen concentration for selection of antibodies on the basis of affinity. Furthermore, selecting in solution reduces the avidity effect of dimeric scFv, thus biasing for the selection of stable monomeric scFv.[90]

When selecting in solution, measures must be taken to prevent selection of phage that bind to the molecules used to capture antigen-bound phage, for example selection of streptavidin-binding antibodies when selecting on biotinylated antigen. Two general strategies have been used to prevent the selection of binders to the capture molecule: (1) rotation and (2) preclearance. For rotation strategies, alternating rounds of selection are carried out with different capture molecules. This works exceedingly well in the case of biotinylated antigen, where capture is alternated between avidin and streptavidin magnetic beads. For preclearance, the phage stock is depleted of binders by preincubation with the capture molecule: e.g., streptavidin-coated magnetic beads, glutathione resin, or maltose-binding protein. Preclearance does not work as well as rotation, since it is possible to remove only a fraction of the binding phage. Final screening for binding antibodies should include proper controls to detect binders to the capture molecules.

Specific elution strategies can also be used to reduce the frequency of binders to the capture molecule (see Section 6.9). For example, antigen can be labeled

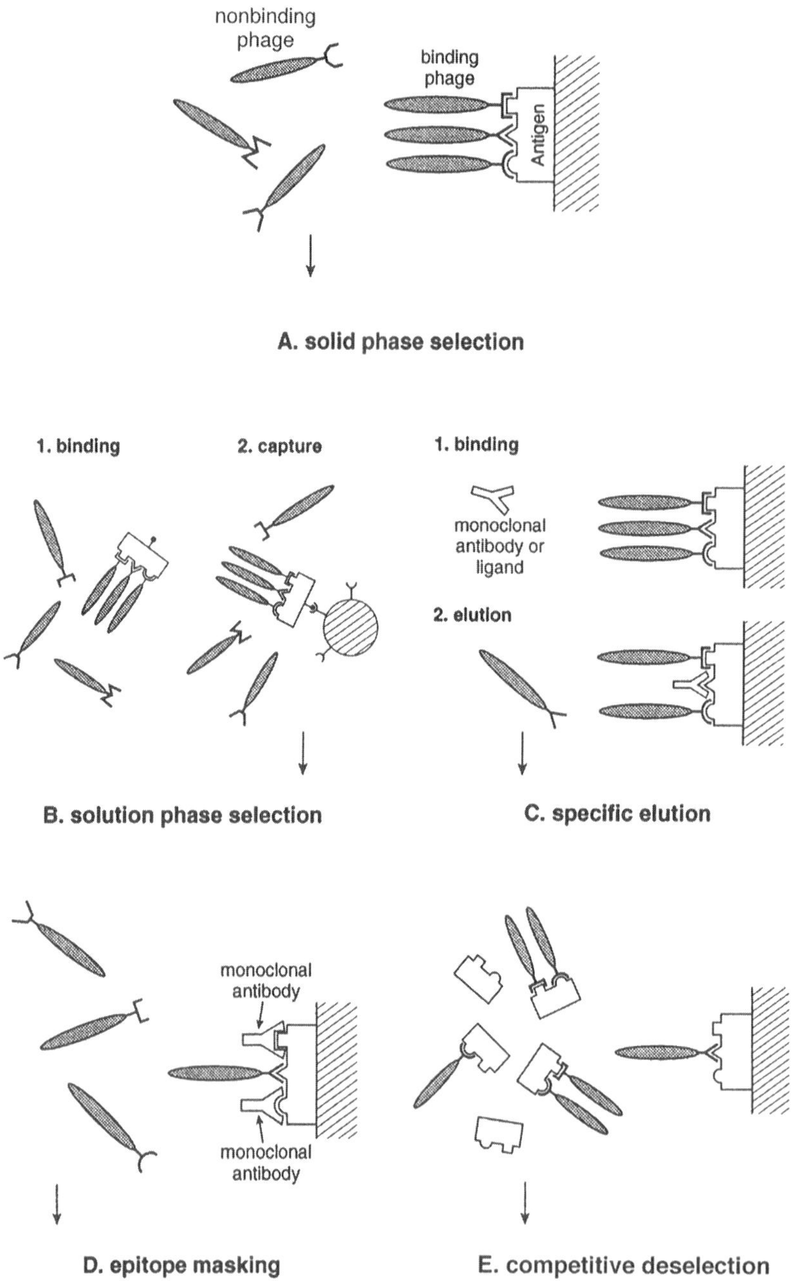

Figure 6.4 Some variations on selection. (*A*) solid phase selection, with binding phage recognizing three different epitopes on the target antigen and nonbinding phage being washed away. (*B*) Solution phase selection. A biotinylated antigen is allowed to bind phage in solution, then the phage-biotinylated antigen complex is separated from

with biotin containing a reducible disulfide bond. Phage are then eluted from streptavidin or avidin using a reducing agent. Alternatively, Balass et al.[134] have described the use of a modified form of streptavidin created by reaction of a tyrosine in the biotin-binding site with tetranitromethane to generate a nitrotyrosine-containing streptavidin (nitrostreptavidin).[135] Nitrostreptavidin has a weaker affinity for biotin, so a biotinylated antigen can be eluted with free biotin.

Soluble antigen is also employed for an elegant selection technique where phage antibody binding to antigen results directly in infection of *E. coli*. For this approach, phage antibodies are generated that lack functional pIII.[136–138] This requires the use of a gene III deleted helper phage, for phagemid systems[136] or use of a phage vector.[138] Antigen-binding phage are rendered infectious by binding to an antigen-pIII fusion, constructed chemically or genetically. For genetic fusions, placing the phage antibody gene and antigen-pIII gene on the same replicon permits the coselection of antibody–antigen pairs that bind to each other.[138]

6.9.2 Guiding Selections to Specific Epitopes

Various strategies have been used to direct the selection of phage antibodies to a particular epitope on an antigen, or at least to prevent selection of antibodies that bind to undesired epitopes. One strategy is to mask undesired epitopes during phage antibody selection, using monoclonal antibodies.[92,139,140] Alternatively, a specific eluant that binds to the desired epitope can be employed, such as a monoclonal antibody.[131,141] A related approach is to use solution phase competition to direct the selection to epitopes that differ between two related molecules. Ames et al. described "competitive biopanning," where activated complement C5a was immobilized on a tube, and selections done in the presence of unactivated C5 in solution to select two Fab that specifically bound only activated C5a.[142] This approach can thus be employed to select for "neoepitopes," which are formed upon protein activation.

Figure 6.4 *continued*
nonbinding phage with a streptavidin- or avidin-coated magnetic bead. (*C*) Specific elution. Antibody-phage bind to multiple epitopes on the target antigen. Phage binding to one particular epitope are eluted with a monoclonal antibody that binds the same epitope. (*D*) Epitope masking. Monoclonal antibodies are used to mask undesired epitopes on the target antigen, driving the selection of phage, which bind to the unmasked epitopes. (*E*) Competitive deselection. A target antigen with multiple epitopes is immobilized on a solid support. During selection, excess of a related antigen with some of the same epitopes as the target antigen is included in solution, absorbing all the phage specific for the common epitopes (half-squares and triangles), thereby driving the selection toward the unique epitopes on the solid phase antigen.

6.9.3 Direct Selection on Cells

Selections can be performed on more complex mixtures of antigen, including intact cells, as long as measures are taken to prevent enrichment of phage antibodies that bind to nonrelevant cell-surface antigens. Cell-surface antigen-specific antibodies have been isolated by selecting phage antibody libraries on erythrocytes,[97] lymphocytes,[99] and melanoma cells.[94,100] For highly expressed antigens, such as blood group B, specific antibodies were isolated by direct selection on erythrocytes.[97] For antigens expressed at lower density, it is necessary to deplete the library of binders to other antigens using a closely related cell type.[97] For example, de Kruif et al.[99] used a fluorescence-activated cell sorter (FACS) to select phage specific for a subset of a cell population by (1) binding the phage library to the entire cell population and (2) separating out the desired subpopulation with a fluorescently tagged antibody to a known antigen. Presumably cells not expressing the known antigen absorbed phage, which bound to common antigens widely distributed in the entire population of cells.

Antigens can also be displayed as fusion proteins on the surface of bacteria, resulting in the creation of "living columns" which could be used for selection of phage antibodies.[143] In this example, an epitope of p21*ras* was fused to the outer membrane protein LamB of *E. coli.* Enrichments of 10^8-fold could be achieved after three rounds of panning when a monoclonal anti-*ras* phage antibody was mixed with irrelevant phage. This approach could be expanded to antigen display on eukaryotic cells, for example by fusion of the antigen gene to a gene encoding a transmembrane domain.[10]

Using living cells, it should also be possible to select for antibodies with specific functional characteristics such as induction of receptor mediated endocytosis[144] or apoptosis. Ultimately it may prove possible to select for desired properties in vivo. For example, Pasqualini et al. isolated tissue specific phage peptides after intravenous injection of a peptide phage library into mice.[145] The widespread applicability of this approach remains to be demonstrated.

6.9.4 Monitoring Selections

Several techniques exist for monitoring the success of phage antibody selections. The titer of eluted phage and the ratio of eluted/input phage typically increases from round to round. (This may not be true, however, if the antigen concentration is decreased or if the library is taken over by deletion mutants as can occur if too many rounds of selection are performed.) To determine the effectiveness of washing techniques, mock selections can be performed using tubes not coated with antigen. The earliest appearance of binding phage antibodies can be detected by analysis of polyclonal phage for antigen binding by ELISA. Signals are detectable when the frequency of binding clones is approximately 1/100. An additional round of selection should lead to a significant increase in ELISA signal and should trigger the analysis of individual clones for binding.

6.10 SCREENING AND INITIAL CHARACTERIZATION OF SELECTED PHAGE ANTIBODIES

Antigen-binding phage antibodies are typically identified by ELISA. Single bacterial colonies can be grown in 96 well micro-titre plates resulting in the production of either phage antibodies[48] or native antibody fragment,[25] depending on the vector system and *E. coli* strain used. Bacterial supernatant containing either phage or native antibody is used for ELISA, with antigen binding detected using either anti-phage antibodies,[48] anti-tag antibodies (for scFv)[19,73] or anti-light chain antibodies (for Fab).[64]

Once binders have been identified, the number of unique antibody fragments can be estimated by PCR screening, followed by restriction digest fingerprinting.[25,48,146] This is accomplished by PCR amplifying ELISA positive clones with primers that flank the antibody gene. After amplification, the PCR fragment is digested with a frequently cutting restriction enzyme, e.g., BstN1. Each unique antibody typically has a unique "BstN1 fingerprint". This is confirmed by DNA sequencing several clones of each fingerprint pattern. The V_H and V_L gene sequences can be compared to databases of germline V-gene segment sequences to determine the family and germline gene of origin and extent of somatic mutation.

6.11 PRODUCTION AND PURFICATION OF ANTIBODIES FROM PHAGE LIBRARIES

For subsequent characterization, such as measurement of affinity, it is necessary to obtain purified native (nonfusion) antibody fragment. For expression systems in which the antibody fragment is fused to gene III or VIII via an amber codon (UAG), native scFv (Fab) can be expressed by transferring the plasmid to a nonsupressor strain of *E. coli.*[24] For other vector systems, it will be necessary to subclone the scFv (Fab) gene. Native antibody fragments are secreted into the bacterial periplasm and can then be harvested by osmotic shock.[68] The optimal duration of induction prior to harvesting from the periplasm may vary considerably, depending on the antibody fragment expression level and extent of toxicity to *E. coli.* As induction continues, the *E. coli* becomes leaky and the antibody fragment appears in the bacterial supernatant. While antibodies can be purified from the supernatant, working volumes are much larger than for periplasmic preparations, and the likelihood of proteolysis is greater.

scFv can be purified using monoclonal antibody affinity columns that bind an epitope tag.[25] We prefer, however, to subclone the scFv gene into a vector that results in the fusion of a C-terminal hexahistidine tag.[101] scFv is then purified using immobilized metal affinity chromatography (IMAC).[74] Use of a phage display vector that incorporates a hexahistidine tag eliminates the need for subcloning.[26] When antibody fragments are used for affinity measurements, we utilize a gel filtration step after IMAC to remove aggregated or dimeric antibody fragments.[101]

6.12 INCREASING ANTIBODY AFFINITY USING PHAGE DISPLAY

Phage display is a powerful tool for increasing antibody affinity to values not achievable using hybridoma technology.[1] To increase affinity, the antibody sequence is diversified, a phage antibody library is constructed, and higher affinity binders are selected on antigen.[27,147,149] The two major considerations that must be addressed to successfully apply this approach are: (1) where to introduce mutations into the antibody fragment gene and (2) how to efficiently select rare higher affinity antibodies from more frequent lower affinity antibodies. These questions are addressed in the following sections. For convenience, we generally refer to scFv antibody fragments, but the approaches described could also be used to increase Fab affinity.

6.12.1 Where and How Mutations Should Be Introduced

When designing a mutant phage antibody library, decisions must be made as to how and where to introduce mutations. One approach is to randomly introduce mutations, thus apparently mimicking the process of somatic hypermutation in vivo. Random mutations can be introduced using chain shuffling,[48,49] where the V_H or V_L gene of a binding scFv is replaced with a V-gene repertoire. For example, Marks et al. used chain shuffling to increase the affinity of a scFv specific for the hapten phenyloxazolone (phOx) 300-fold.[27] Light chain shuffling resulted in a 20-fold increase in affinity to 16 nanomolar. A further 15-fold increase in affinity, to 1.1 nM, was achieved by conserving the new light chain and V_H CDR3, then shuffling the V_H gene segment. As with in vivo affinity maturation, the kinetic basis for the increase in affinity[27] was largely due to a decrease in the dissociation rate constant (k_{off}).[150] Mutations can also be randomly introduced using error prone PCR[28] or mutator strains of *E. coli*.[151]

The random introduction of mutations is simple and easy to perform, and has resulted in large increases in affinity (greater than 100-fold) for hapten binding antibody fragments.[27,151] Results with protein binding antibody fragments, however, have been more modest ($<$10-fold).[28,90] Moreover, the relatively random distribution of mutations in higher affinity clones provides little useful information about where to direct additional mutations to further increase affinity.

As an alternative to random mutagenesis, mutations can be specifically introduced into the CDRs that form the contact interface between antibody and antigen. Targeting mutations to the CDRs has been shown to be a very effective approach for increasing antibody affinity, including the affinity of protein-binding antibodies. Yang et al. increased the affinity of an anti-HIV gp120 Fab 420-fold (to a $K_d = 1.5 \times 10^{-11}$ M) by mutating four CDRs in five libraries and combining independently selected mutations.[148] Similarly, Schier et al. increased the affinity of an anti-ErbB-2 scFv more than 1200-fold (to a $K_d = 1.3 \times 10^{-11}$ M) by sequentially mutating V_H and V_L CDR3.[149]

Complete randomization of an amino acid position typically requires the use of the nucleotide sequence NNS, where N = A, G, C, or T and S = G or C. Since this generates 32 possible nucleotide sequences, randomization of only 5 amino acid positions will generate $(32)^5$ or 3.4×10^7 possible sequences. This number is at the limit of the size of phage libraries that can be conveniently produced. Thus only 4 to 5 amino acids can be mutated at a time if the entire sequence space is to be sampled. This is only a fraction of the number of amino acids in the CDRs, which can exceed 50 residues. One approach to reduce the number of CDR residues that need to be sampled is to perform molecular modeling on a homologous Fab or Fv crystal structure.[149] For all of the CDRs except V_H CDR3, modeling permits identification of CDR residues that have a structural role, either in maintaining the main chain conformation or packing at the V_H-V_L interface.[38,114,115] Mutation of residues with predicted structural roles results in reselection of the wild-type amino acid residue.[73,149] Thus, these residues should not be randomized. Instead, the amino acids selected for mutagenesis should be those whose side chains are predicted to be solvent accessible. We have also observed complete conservation of glycine and tryptophan residues within CDRs.[73,149] Glycine residues are typically key residues in turns, and the chemical properties of tryptophan make it a frequent structural or high energy contact residue.[152] Thus conservation of these two residues when randomizing CDRs should be considered. With respect to the initial CDRs to select for mutagenesis, mutation of the V_H and V_L CDR3 has yielded greater increases in affinity than mutation of other CDRs.[148,153]

Affinity maturation can be performed sequentially, with each subsequent library constructed from the sequence of the highest affinity mutant. Alternatively, to save time a parallel strategy can be used where different regions of the molecule are randomized independently, higher affinity mutants are isolated and mutations are combined. The combined effect of two mutations on affinity can be described by the equation:

$$\Delta\Delta G_{A+B} = \Delta\Delta G_A + \Delta\Delta G_B + \Delta\Delta G_I$$

where $\Delta\Delta G_A$ is the change in free energy due to mutation A, $\Delta\Delta G_B$ is the change in free energy from mutation B, $\Delta\Delta G_I$ is the free energy change due to interaction between mutations A and B, and $\Delta\Delta G_{A+B}$ is the total free energy change in the double mutant. The value of the interaction energy term may be positive, negative, or zero; and this value determines whether combining mutations leads to a net increase in affinity.[154] With respect to results from phage antibody libraries, some mutations are additive,[148,149] while others are not.[90,149] Therefore, a sequential strategy may be more prudent. Alternatively, it is possible to "scan" many residues at a low mutation frequency (parsimonious mutagenesis[155]) to identify residues that modulate affinity and structural and functional residues that are conserved.[73] Residues identified as modulating affinity could then be completely randomized in a subsequent library.

6.12.2 Selection of Higher Affinity Phage Antibodies

Efficient selection of higher affinity phage antibodies is less than straightforward due to sequence-dependent differences in phage antibody expression and in toxicity to *E. coli*. These differences can lead to selection for increased expression levels, or decreased toxicity, rather than for higher affinity. In the case of scFv phage antibodies, selection is also complicated by the tendency of some scFv to dimerize.[127,128] Dimeric scFv can form on the phage surface by noncovalent association of the V-domains of the scFv-pIII fusion with the V-domains of native scFv in the periplasm. Native scFv appears in the periplasm both from incomplete suppression of the amber codon between the scFv gene and gene III, as well as by proteolysis. Dimeric scFv exhibit increased apparent affinity due to avidity, and are preferentially enriched over monomeric scFv when selections are performed on antigen immobilized on a solid phase.[90,120] Thus, selections must be carefully designed to ensure enrichment based on affinity rather than expression level, toxicity to *E. coli*, or avidity.

Selection for monomeric higher affinity antibody fragments is optimal when selections are performed in solution, for example using biotinylated antigen with subsequent capture on streptavidin-coated magnetic beads.[28,90] For the initial round of selection, an antigen concentration greater than the K_d of the wild-type antibody is used in order to capture rare or poorly expressed phage antibodies. In subsequent rounds, the antigen concentration is reduced to significantly less than the desired K_d.[90] Failure to adequately reduce the antigen concentration results in failure to sort on the basis of affinity, while a reduction that is too large results in loss of binding phage.[90,156] Moreover, the optimal antigen concentration cannot be predicted a priori, due to variability in phage antibody expression levels and uncertainty regarding the highest affinities present in the mutant phage antibody library. An elegant technique for monitoring the stringency of selections is to measure the concentration and percentage of binding phage present in polyclonal phage prepared after each round of selection using surface plasmon resonance in a BIAcore.[156] The results can then be used to determine the antigen concentration needed for the next round of selection. Alternatively, selection can be monitored using the titre of eluted phage.[156]

It is also necessary to ensure that all specifically bound phage are eluted for selection of the highest affinity antibodies. Solutions used for elution include competition with soluble antigen;[28,48] 100 mM triethylamine;[25] glycine, pH 2.2;[65] 100 mM NaOAc, pH 2.8 containing 500 mM NaCl;[30] or 76 mM citric acid, pH 2.8.[97] Alternatively, antigen-bound phage can be directly used to infect *E. coli*.[157] However, it has recently been shown that eluting with soluble antigen or by adding *E. coli* directly to antigen bound phage leads to selection of scFv phage antibodies with the fastest dissociation rate constant (k_{off}) (lower affinity).[156] This is because the phage must dissociate from antigen before infection can occur. As affinity increases, more stringent eluants are required to ensure that the highest affinity phage are eluted.[156] Choice of eluent can be determined by measuring their ability to dissociate polyclonal phage antibodies from antigen using surface plasmon resonance in a BIAcore.[156] Efficient elution can be ensured by using a cleavable linker between antigen and the affinity matrix, for example using sulfo-NHS-SS-biotin (Pierce) which can be cleaved by reduction with DTT.

6.13 CONCLUSIONS AND FUTURE PROSPECTS

Phage display technology is already replacing hybridoma technology for the production of therapeutic antibodies. Completely human antibodies can be isolated from immune or nonimmune libraries, and these should prove to be less immunogenic than humanized rodent monoclonal antibodies. For preexisting murine monoclonal antibodies, phage display can be used to obtain antibodies that are entirely human in sequence, but which bind to the same epitope as the murine monoclonal.[157,158] Furthermore, the technique can be used to increase antibody affinity to values not obtainable by immunization. The availability of these improved antibody binding domains should open up new therapeutic and diagnostic applications for immunofusions, for example by allowing the detection of antigen concentrations not previously possible.

Nonimmune phage antibody libraries will also prove tremendously powerful as drug discovery technology. They will allow the rapid production of antibodies to any antigen for use in the research laboratory. It may also prove possible to develop approaches to pan phage antibody libraries against cDNA expression libraries and simultaneously isolate cDNA and antibodies that bind the gene product. Similarly, expansion of selection technologies to whole cells will facilitate identification of novel antigens. The antibodies can then be used for determination of the location of cellular and tissue gene expression, as well as for study of function—for example using intracellular immunization to achieve phenotypic knockout (reviewed in Refs. 126, 127). This will vastly facilitate functional genomics projects. Ultimately, the use of phage antibody libraries will be limited only by the investigator's imagination.

REFERENCES

1. Foote, J., H. N. Eisen. 1995. Kinetic and affinity limits on antibodies produced during immune responses. *Proc. Natl. Acad. Sci. USA* 92: 1254–1256.
2. Edwards, J. B., J. Delort, J. Mallet. 1991. Oligodeoxyribonucleotide ligation to single stranded cDNAs: a new tool for cloning 5′ ends of mRNAs and for constructing cDNA libraries by in vitro amplification. *Nucl. Acids Res.* 19: 5227–5232.
3. Heinrichs, A., C. Milstein, E. Gherardi. 1995. Universal cloning and direct sequencing of rearranged antibody V genes using C region primers, biotin captured cDNA and one sided PCR. *J. Immunol. Meth.* 178: 241–251.
4. Ruberti, F., A. Cattaneo, A. Bradbury. 1994. The use of the RACE method to clone hybridoma cDNA when V region primers fail. *J. Immunol. Meth.* 173: 33–39.
5. Kipriyanov, S. M., O. A. Kupriyanova, G. Moldenhauer. 1996. Rapid detection of recombinant antibody fragments directed against cell-surface antigens by flow cytometry. *J. Immunol. Meth.* 13: 51–62.
6. Ge, L., A. Knappik, P. Pack, C. Freund, A. Pluckthun. 1995. Expressing antibodies in *Escherichia coli*. In *Antibody Engineering*, Vol. 2, ed. C. A. K. Borrebaeck, Oxford: Oxford University Press, pp. 229–266.
7. Persic, L., A. Roberts, J. Wilton, A. Cattaneo, A. Bradbury, H. R. Hoogenboom. 1997. An integrated vector system for the eukaryotic expression of antibodies or their fragments after selection from phage display libraries. *Gene* 187: 9–18.
8. Knappik, A., C. Krebber, A. Pluckthun. 1993. The effects of folding catalysts on the in vivo folding process of different antibody fragments expressed in *Escherichia coli*. *Bio/Technology* 11: 77–83.
9. Knappik, A., A. Pluckthun. 1995. Engineered turns of a recombinant antibody improve its in vivo folding. *Protein Eng.* 8: 81–89.
10. Chesnut, J. D., A. R. Baytan, M. Russell, M. P. Chang, A. Bernard, I. H. Maxwell, J. P. Hoeffler. 1996. Selective isolation of transiently transfected cells from a mammalian cell population with vectors expressing a membrane anchored single-chain antibody. *J. Immunol. Meth.* 193: 17–27.
11. Kretzschmar, T., L. Aoustin, O. Zingel, M. Marangi, B. Vonach, H. Towbin, M. Geiser. 1996. High-level expression in insect cells and purification of secreted monomeric single-chain Fv. *J. Immunol. Meth.* 195: 93–101.
12. Lamers, C. H. J., J. W. Gratama, S. O. Warnaar, G. Stoter, R. L. H. Bolhuis. 1995. Inhibition of bispecific monoclonal antibody (bsAb)-targeted cytolysis by human anti-mouse antibodies in ovarian carcinoma patients treated with bsAb-targeted activated T-lymphocytes. *Int. J. Cancer.* 60: 450–457.
13. Jaffers, G. J., T. C. Fuller, A. B. Cosimi, P. S. Russel, H. J. Winn, R. B. Colvin. 1986. Monoclonal antibody therapy. Anti-idiotypic and non-anti-idiotypic antibodies to OKT3 arise despite intense immunosuppression. *Transplantation* 41: 572–578.
14. Morrison, S. L., M. J. Johnson, L. A. Herzenberg, V. T. Oi. 1984. Chimeric human antibody molecules: mouse antigen-binding domains with human constant region domains. *Proc. Natl. Acad. Sci. USA* 81: 6851–6855.

15. Boulianne, G. L., N. Hozumi, M. J. Shulman. 1984. Production of functional chimaeric mouse/human antibody. *Nature* 312: 643–646.

16. Neuberger, M. S., G. T. Williams, E. B. Mitchell, S. S. Jouhal, J. G. Flanagan, T. H. Rabbitts. 1985. A hapten-specific chimaeric IgE antibody with human physiological effector function. *Nature* 314: 268–270.

17. Winter, G., W. J. Harris. 1993. Humanized antibodies. *Trends Pharm. Chem.*14: 139–143.

18 James, K., G. T. Bell. 1987. Human monoclonal antibody production: current status and future prospects. *J. Immunol. Meth.* 100: 5–40.

19. Marks, J. D., H. R. Hoogenboom, A. D. Griffiths, G. Winter. 1992. Molecular evolution of proteins on filamentous phage: mimicking the strategy of the immune system. *J. Biol. Chem.* 267: 16007–16010.

20. Hoogenboom, H. R., J. D. Marks, A. D. Griffiths, G. Winter. 1992. Building antibodies from their genes. *Immunol. Rev.* 130: 41–68.

21. Winter, G., A. Griffiths, R. Hawkins, H. Hoogenboom. 1994. Making antibodies by phage display technology. *Ann. Rev. Immunol.* 12: 433–455.

22. Marks, C., J. D. Marks. 1996. Phage libraries: a new route to clinically useful antibodies. *N. Engl. J. Med.* 335: 730–733.

23. McCafferty, J., A. D. Griffiths, G. Winter, D. J. Chiswell. 1990. Phage antibodies: filamentous phage displaying antibody variable domains. *Nature* 348: 552–554.

24. Hoogenboom, H. R., A. D. Griffiths, K. S. Johnson, D. J. Chiswell, P. Hudson, G. Winter. 1991. Multi-subunit proteins on the surface of filamentous phage: methodologies for displaying antibody (Fab) heavy and light chains. *Nucl. Acids Res.* 19: 4133–4137.

25. Marks, J. D., H. R. Hoogenboom, T. P. Bonnert, J. McCafferty, A. D. Griffiths, G. Winter. 1991. By-passing immunization: Human antibodies from V-gene libraries displayed on phage. *J. Mol. Biol.* 222: 581–597.

26. Vaughan, T. J., A. J. Williams, K. Pritchard, J. K. Osbourn, A. R. Pope, J. C. Earnshaw, J. McCafferty, R. A. Hodits, J. Wilton, K. S. Johnson. 1996. Human antibodies with sub-nanomolar affinities isolated from a large non-immunized phage display library. *Nature Biotech.* 14: 309–314.

27. Marks, J. D., A. D. Griffiths, M. Malmqvist, T. Clackson, J. M. Bye, G. Winter. 1992. Bypassing immunisation: high affinity human antibodies by chain shuffling. *Bio/Technology* 10: 779–783.

28. Hawkins, R. E., S. J. Russell, G. Winter. 1992. Selection of phage antibodies by binding affinity: mimicking affinity maturation. *J. Mol. Biol.* 226: 889–896.

29. Boss, M. A., J. H. Kenten, C. R. Wood, J. S. Emtage. 1984. Assembly of functional antibodies from immunoglobulin heavy and light chains synthesised in E. coli. *Nucl. Acids Res.* 12: 3791–3806.

30. Cabilly, S., A. D. Riggs, H. Pande, J. E. Shively, W. E. Holmes, M. Rey, L. J. Perry, R. Wetzel, H. L. Heyneker. 1984. Generation of antibody activity from immunoglobulin polypeptide chains produced in *Escherichia coli*. *Proc. Natl. Acad. Sci. USA* 81: 3273–3277.

31. Better, M., C. P. Chang, R. R. Robinson, A. H. Horwitz. 1988. Escherichia coli secretion of an active chimeric antibody fragment. *Science* 240: 1041–1043.

32. Skerra, A., A. Pluckthun. 1988. Assembly of a functional immunoglobulin Fv fragment in *Escherichia coli*. *Science* 240: 1038–1041.

33. King. 1992. *Antibody and Immunoconjugate Radiopharm.* 5: 159–170.

34. Horne, C., M. Klein, I. Polidoulis, K. J. Dorrington. 1982. Noncovalent association of heavy and light chains of human immunoglobulins. III. Specific interactions between V_H and V_L. *J. Immunol.* 129: 660–664.

35. Hammel, P. A., M. H. Klein, S. J. Smith-Gill, K. J. Dorrington. 1987. Relative noncovalent association constant between immunoglobulin H and L chains is unrelated to their expression or antigen-binding activity. *J. Immunol.* 139: 3012–3020.

36. Glockshuber, R., M. Malia, I. Pfitzinger, A. Pluckthun. 1990. A comparison of strategies to stabilize immunoglobulin Fv-fragments. *Biochemistry* 29: 1362–1367.

37. Rodrigues, M. L., L. G. Presta, C. E. Kotts, C. Wirth, J. Mordenti, G. Osaka, W. L. T. Wong, A. Nuijens, B. Blackburn, P. Carter. 1995. Development of a humanized disulfide-stabilized anti-p185^{HER2} Fv-β-lactamase fusion protein for activation of a cephalosporin doxorubicin prodrug. *Cancer. Res.* 55: 63–70.

38. Chothia, C., J. Novotny, R. Bruccoleri, M. Karplus. 1985. Domain association in immunoglobulin molecules. The packing of variable domains. *J. Mol. Biol.* 186: 651–663.

39. Reiter, Y., U. Brinkmann, K. O. Webber, S.-H. Jung, B. Lee, I. Pastan. 1994. Engineering interchain disulfide bonds into conserved framework regions of Fv fragments: improved biochemical characteristics of recombinant immunotoxins containing disulfide-stabilized Fv. *Protein Eng.* 7: 697–704

40. Reiter, Y., U. Brinkmann, S. Jung, I. Pastan, B. Lee. 1995. Disulfide stabilization of antibody Fv: computer predictions and experimental evaluation. *Protein Eng.* 8: 1323–1331.

41. Huston, J. S., D. Levinson, H. M. Mudgett, M. S. Tai, J. Novotny, M. N. Margolies, R. J. Ridge, R. E. Bruccoleri, E. Haber, R. Crea, H. Oppermann. 1988. Protein engineering of antibody binding sites: recovery of specific activity in an anti-digoxin single-chain Fv analogue produced in *Escherichia coli*. *Proc. Natl. Acad. Sci. USA* 85: 5879–5883.

42. Bird, R. E., K. D. Hardman, J. W. Jacobson, S. Johnson, B. M. Kaufman, S. M. Lee, T. Lee, S. H. Pope, G. S. Riordan, M. Whitlow. 1988. Single-chain antigen-binding proteins. *Science* 242: 423–426.

43. Huston, J. S., M. Mudgett-Hunter, M. Tai, J. McCartney, F. Warren, E. Haber, H. Opperman. 1991. Protein engineering of single-chain Fv analogs and fusion proteins. *Meth. Enzymol.* 203: 46–88.

44. Whitlow, M., D. Filpula. 1991. Single-chain Fv proteins and their fusion partners. Methods: A Companion to *Meth. Enzymol.* 2: 97.

45. Bird, R. E., B. W. Walker. 1991. Single chain antibody variable regions. *Trends Biotech.* 9: 132–138.

46. Zdanov, A., Y. Li, D. R. Bundle, S.-J. Deng, C. R. MacKenzie, S. A. Narang, N. M. Young, M. Cygler. 1994. Structure of a single-chain antibody variable domain (Fv) fragment complexed with a carbohydrate antigen at 1.7 A resolution. *Proc. Natl. Acad. Sci. USA* 91: 6423.

47. Kabat, E. A., T. T. Wu, H. M. Perry, K. S. Gottesman, C. Foeller. 1991. *Sequences*

of Proteins of Immunological Interest. Washington, D.C.: US Department of Health and Human Services, US Government Printing Office.

48. Clackson, T., H. R. Hoogenboom, A. D. Griffiths, G. Winter. 1991. Making antibody fragments using phage display libraries. *Nature* 352: 624–628.
49. Larrick, J. W., Y. L. Chiang, R. Sheng-Dong, G. Senck, P. Casali. 1988. Generation of specific human monoclonal antibodies by in vitro expansion of human B cells: a novel recombinant DNA approach. In *In vitro Immunisation in Hybridoma Technology*, ed. C. A. K. Borrebaeck, Amsterdam: Elsevier Science Publishers B.V., pp. 231–246.
50. Orlandi, R., D. H. Gussow, P. T. Jones, G. Winter. 1989. Cloning immunoglobulin variable domains for expression by the polymerase chain reaction. *Proc. Natl. Acad. Sci. USA* 86: 3833–3837.
51. LeBoeuf, R. D., F. S. Galin, S. K. Hollinger, S. C. Peiper, J. E. Blalock. 1989. Cloning and sequencing of immunoglobulin variable-region genes using degenerate oligodeoxyribonucleotides and polymerase chain reaction. *Gene* 82: 371–377.
52. Iverson, S. A., L. Sastry, W. D. Huse, J. A. Sorge, S. J. Benkovic, R. A. Lerner. 1989. A combinatorial system for cloning and expressing the catalytic antibody repertoire in Escherichia coli. *Cold Spring Harb. Symp. Quant Biol.* 1: 273–281.
53. Sastry, L., M. M. Alting, W. D. Huse, J. M. Short, J. A. Sorge, B. N. Hay, K. D. Janda, S. J. Benkovic, R. A. Lerner. 1989. Cloning of the immunological repertoire in Escherichia coli for generation of monoclonal catalytic antibodies: construction of a heavy chain variable region-specific cDNA library. *Proc. Natl. Acad. Sci. USA* 86: 5728–5732.
54. Kettleborough, C. A., J. Saldanha, K. H. Ansell, M. M. Bendig. 1993. Optimization of primers for cloning libraries of mouse immunoglobulin genes using the polymerase chain reaction. *Eur. J. Immunol.* 23: 206–211.
55. Orum, H., P. S. Andersen, A. Oster, L. K. Johansen, E. Riise, M. Bjornvad, I. Svendsen, J. Enberg. 1993. Efficient method for constructing comprehensive murine Fab antibody libraries displayed on phage. *Nucl. Acids Res.* 21: 4491–4498.
56. Marks, J. D., M. Tristrem, A. Karpas, G. Winter. 1991. Oligonucleotide primers for polymerase chain reaction amplification of human immunoglobulin variable genes and design of family-specific oligonucleotide probes. *Eur. J. Immunol.* 21: 985–991.
57. Persson, M. A., R. H. Caothien, D. R. Burton. 1991. Generation of diverse high-affinity human monoclonal antibodies by repertoire cloning. *Proc. Natl. Acad. Sci. USA* 88: 2432–2436.
58. Burton, D. R., C. F. Barbas, M. A. A. Persson, S. Koenig, R. M. Chanock, R. A. Lerner. 1991. A large array of human monoclonal antibodies to type 1 human immunodeficiency virus from combinatorial libraries of asymptomatic seropositive individuals. *Proc. Natl. Acad. Sci. USA* 88: 10134–10137.
59. Davies, E., J. Smith, C. Birkett, J. Manser, D. Anderson-Dear, J. Young. 1995. Selection of specific phage-display antibodies using libraries derived from chicken immunoglobulin genes. *J. Immunol. Meth.* 186: 125–135.
60. Lang, I., C. R. Barbas, R. Schleef. 1996. Recombinant rabbit Fab with binding activity to type-1 plasminogen activator inhibitor derived from a phage-display library against human alpha-granules. *Gene* 172: 295–298.

61. Persic, L., M. Righi, A. Roberts, H. R. Hoogenboom, A. Cattaneo, A. Bradbury. 1997. Targeting vectors for intracellular immunisation. *Gene* 187: 1–8.
62. Chaudhary, V. K., J. K. Batra, M. G. Gallo, M. C. Willingham, D. J. FitzGerald, I. Pastan. 1990. A rapid method of cloning functional variable-region antibody genes in Escherichia coli as single-chain immunotoxins. *Proc. Natl. Acad. Sci. USA* 87: 1066–1070.
63. Smith, G. P. 1985. Filamentous fusion phage: novel expression vectors that display cloned antigens on the virion surface. *Science* 228: 1315–1317.
64. Kang, A. S., C. F. Barbas, K. D. Janda, S. J. Benkovic, R. A. Lerner. 1991. Linkage of recognition and replication functions by assembling combinatorial antibody Fab libraries along phage surfaces. *Proc. Natl. Acad. Sci. USA* 88: 4363–4366.
65. Barbas, C. F., A. S. Kang, R. A. Lerner, S. J. Benkovic. 1991. Assembly of combinatorial antibody libraries on phage surfaces: The gene III site. *Proc. Natl. Acad. Sci. USA*. 88: 7978–7982.
66. Huse, W., T. Stinchcombe, S. Glaser, L. M. Starr, M., K. Hellstrom, I. Hellstrom, D. Yelton. 1992. Application of a filamentous phage pVIII fusion protein system suitable for efficient production, screening, and mutagenesis of F(ab) antibody fragments. *J. Immunol.* 149: 3914–3920.
67. Garrard, L. J., M. Yang, M. P. O'Connel, R. F. Kelley, D. J. Henner. 1991. Fab assembly and enrichment in a monovalent phage display system. *Bio/Technology* 9: 1373–1377.
68. Breitling, F., S. Dubel, T. Seehaus, I. Klewinghaus, M. Little. 1991. A surface expression vector for antibody screening. *Gene* 104: 147–153.
69. Engelhardt, O., R. R. Grabher, G. Himmler, F. Ruker. 1994. Two-step cloning of antibody variable domains in a phage display vector. *Bio/Techniques* 17: 44–46.
70. Dubel, S., F. Breitling, P. Fuchs, M. Braunagel, I. Klewinghaus, M. Little. 1993. A family of vectors for surface display and production of antibodies. *Gene* 128: 97–101.
71. Lah, M., A. Goldstraw, J. White, O. Dolezal, R. Malby, P. Hudson. 1994. Phage surface presentation and secretion of antibody fragments using an adaptable phagemid vector. *Human Antibodies and Hybridomas* 5: 48–56.
72. Chang, C. N., N. F. Landolfi, C. Queen. 1991. Expression of antibody Fab domains on bacteriophage surfaces. *J. Immunology* 147: 3610–3614.
73. Schier, R., R. F. Balint, A. McCall, G. Apell, J. W. Larrick, J. D. Marks. 1996. Identification of functional and structural amino acid residues by parsimonious mutagenesis. *Gene* 169: 147–155.
74. Hochuli, E., W. Bannwarth, H. Dobeli, R. Gentz, D. Stuber. 1988. Genetic approach to facilitate purification of recombinant proteins with a novel metal chelate adsorbent. *Bio/Technology* 6: 1321–1325.
75. Wong, C., S. Chen, P. Amersdorfer, T. Smith, S. Desphande, R. Finneran, R. Sheridan, J. D. Marks. 1997. Molecular characterization of the murine humoral immune response to Botulinum neurotoxin type A binding domain as assessed using phage antibody libraries. *Infect. Immun.* 65(9): 3743–3752.
76. Kettleborough, C., K. Ansell, R. Allen, E. Rosell-Vives, D. Gussow, M. Bendig. 1994. Isolation of tumor cell-specific single-chain Fv from immunized mice using

phage-antibody libraries and the reconstruction of whole antibodies from these antibody fragments. *Eur. J. Immunol.* 24: 952–958.

77. Barbas, C. F., T. A. Collet, W. Amberg, P. Roben, J. M. Binley, D. Hoekstra, D. Cababa, T. M. Jones, A. Williamson, G. R. Pilkington, N. L. Haigwood, E. Cabezas, A. C. Satterthwait, I. Sanz, D. R. Burton. 1993. Molecular profile of an antibody response to HIV-1 as probed by combinatorial libraries. *J. Mol. Biol.* 230: 812–823.
78. Binley, J. M., H. J. Ditzel, C. F. Barbas III, N. Sullivan, J. Sodroski, P. W. H. I. Parren, D. R. Burton. 1996. Human antibody responses to HIV type 1 glycoprotein 41 cloned in phage display libraries suggest three major epitopes are recognized and give evidence for conserved antibody motifs in antigen binding. *AIDS Res. Hum. Retroviruses.* 12: 911–924.
79. Zebedee, S. L., C. F. Barbas, Y.-L. Hom, R. H. Cathien, R. Graff, J. DeGraw, J. Pyatt, R. LaPolla, D. R. Burton, R. A. Lerner, G. B. Thornton. 1992. Human combinatorial antibody libraries to hepatitis B surface antigens. *Proc. Natl. Acad. Sci. USA* 89: 3175–3179.
80. Chan, S., J. Bye, P. Jackson, J. Allain. 1996. Human recombinant antibodies specific for hepatitis C virus core and envelope E2 peptides from an immune phage display library. *J. Gen. Virol.* 10: 2531–2539.
81. Barbas, C., J. Crowe, D. Cababa, T. Jones, S. Zebedee, B. Murphy, R. Chanock, D. Burton. 1992. Human monoclonal Fab fragments derived from a combinatorial library bind to respiratory syncytial virus F glycoprotein and neutralize infectivity. *Proc. Natl. Acad. Sci. USA* 89: 10164–10168.
82. Crowe, J., B. Murphy, R. Chanock, R. Williamson, C. Barbas, D. Burton. 1994. Recombinant human respiratory syncytial virus (RSV) monoclonal antibody Fab is effective therapeutically when introduced directly into the lungs of RSV-infected mice. *Proc. Natl. Acad. Sci. USA* 91: 1386–1390.
83. Reason, D., T. Wagner, A. Lucas. 1997. Human Fab fragments specific for the Haemophilis influenzae b polysaccharide isolated from a bacteriophage combinatorial library use variable region gene combinations and express an idiotype that mirrors in vivo expression. *Infection and Immunity* 65: 261–266.
84. Barbas, S., H. Ditzel, E. Salonen, W. Yang, G. Silverman, D. Burton. 1995. Human autoantibody recognition of DNA. *Proc. Natl. Acad. Sci. USA* 92: 2529–2533.
85. Portolano, S., S. McLachlan, B. Rapoport. 1993. High affinity, thyroid-specific human autoantibodies displayed on the surface of filamentous phage use V genes similar to other autoantibodies. *J. Immunol.* 151: 2839–2851.
86. Graus, Y., M. de Baets, P. Parren, S. Berrih-Aknin, J. Wokke, V. Breda, P. Vriesman, D. Burton. 1997. Human anti-nicotinic acetylcholine receptor recombinant Fab fragments isolated from thymus-derived phage display libraries from myasthenia gravis patients reflect predominant specificities in serum and block the action of pathogenic serum antibodies. *J. Immunol.* 158: 1919–1929.
87. Winter, G., C. Milstein. 1991. Man-made antibodies. *Nature* 349: 293–298.
88. Embleton, M. J., G. Gorochov, P. T. Jones, G. Winter. 1992. In-Cell PCR from mRNA: Amplifying and Linking the Rearranged Immunoglobulin heavy and light chain V-genes within single cells. *Nucleic Acids Research* 20: 3831–3837.
89. Kang, A. S., T. M. Jones, D. R. Burton. 1991. Antibody redesign by chain shuffling

from random combinatorial immunoglobulin libraries. *Proc. Natl. Acad. Sci. USA* 88: 11120–11123.

90. Schier, R., J. M. Bye, G. Apell, A. McCall, G. P. Adams, M. Malmqvist, L. M. Weiner, J. D. Marks. 1996. Isolation of high affinity monomeric human anti-c-erbB-2 single chain Fv using affinity driven selection. *J. Mol. Biol.* 255: 28–43.
91. Griffin, H., W. Ouwehand. 1995. A human monoclonal antibody specific for the leucine-33 (P1A1, HPA-1a) form of platelet glycoprotein IIIa from a V gene phage display library. *Blood* 86: 4430–4436.
92. Ping, T., M. Tornetta, R. Ames, B. Bankosky, S. Griego, C. Silverman, T. Porter. 1996. Isolation of a neutralizing human RSV antibody from a dominant, non-neutralizing immune repertoire by epitope-blocked panning. *J. Immunol.* 157: 772–780.
93. Williamson, R., D. Peretz, N. Smorodinsky, R. Bastidas, H. Serban, I. Mehlhorn, S. DeArmond, S. Prusiner, D. Burton. 1996. Circumventing tolerance to generate autologous monoclonal antibodies to the prion protein. *Proc. Natl. Acad. Sci. USA* 93: 7279–7282.
94. Cai, X., A. Garen. 1995. Anti-melanoma antibodies from melanoma patients immunized with genetically modified autologous tumor cells: selection of specific antibodies from single-chain Fv fusion phage libraries. *Proc. Natl. Acad. Sci. USA* 92: 6537–6541.
95. Bruggeman, Y., A. Boogert, A. van Hock, P. Jones, G. Winter, A. Schots, R. Hilhorst. 1996. Phage antibodies against an unstable hapten: oxygen sensitive reduced flavin. *FEBS. Lett.* 388: 242–244.
96. Nissim, A., H. R. Hoogenboom, I. M. Tomlinson, G. Flynn, C. Midgley, D. Lane, G. Winter. 1994. Antibody fragments from a 'single pot' phage display library as immunochemical reagents. *EMBO J.* 13: 692–698.
97. Marks, J. D., W. H. Ouwehand, J. M. Bye, R. Finnern, B. D. Gorick, D. Voak, S. Thorpe, N. C. Hughes-Jones, G. Winter. 1993. Human antibody fragments specific for blood group antigens from a phage display library. *Bio/Technology* 11: 1145–1149.
98. Hughes-Jones, N., J. M. Bye, J. D. Marks, B. Gorick, W. Ouwehand. 1994. Blood group antibodies. *Brit. J. Hemaetol.* 88: 180–186.
99. de Kruif, J., L. Terstappen, E. Boel, T. Logtenberg. 1995. Rapid selection of cell subpopulation-specific human monoclonal antibodies from a synthetic phage antibody library. *Proc. Natl. Acad. Sci. USA* 92: 3938–3942.
100. Cai, X., A. Garen. 1996. A melanoma-specific VH antibody cloned from a fusion phage library of a vaccinated melanoma patient. *Proc. Natl. Acad. Sci. USA* 93: 6280–6285.
101. Schier, R., J. D. Marks, E. J. Wolf, G. Apell, C. Wong, J. E. McCartney, M. A. Bookman, J. S. Huston, L. L. Houston, L. M. Weiner, G. P. Adams. 1995. In vitro and in vivo characterization of a human anti-c-erbB-2 single-chain Fv isolated from a filamentous phage antibody library. *Immunotechnology* 1: 73–81.
102. Griffiths, A. D., M. Malmqvist, J. D. Marks, J. M. Bye, M. J. Embleton, J. McCafferty, M. Baier, K. P. Holliger, B. D. Gorick, N. C. Hughes-Jones, H. R. Hoogenboom, G. Winter. 1993. Human anti-self antibodies with high specificity from phage display libraries. *EMBO J.* 12: 725–734.

103. Finnern, R., J. M. Bye, K. M. Dolman, M.-M. Zhao, A. Short, M. C. Lockwood, W. H. Ouwehand. 1995. Molecular characteristics of antiself antibody fragments against neutrophil cytoplasmic antigens from human V gene phage display libraries. *Clin. Exp. Immunol.* 102: 566–574.

104. Cohen, I. R., A. Cooke. 1986. Natural autoantibodies might prevent autoimmune-disease. *Immunology Today* 7: 363–364.

105. Perelson, A. S., G. F. Oster. 1979. Theoretical studies of clonal selection: Minimal antibody repertoire size and reliability of self non-self discrimination. *J. Theor. Biol.* 81: 645–670.

106. Perelson, A. S. 1989. Immune network theory. *Immunol. Rev.* 110: 5–36.

107. Hoogenboom, H. R., G. Winter. 1992. Bypassing immunisation: human antibodies from synthetic repertoires of germ line VH-gene segments rearranged *in vitro*. *J. Mol. Biol.* 227: 381–388.

108. Barbas, C., W. Amberg, A. Simoncsits, T. Jones, R. Lerner. 1993. Selection of human anti-hapten antibodies from semisynthetic libraries. *Gene* 137: 57–62.

109. de Kruif, J., E. Boel, T. Logtenberg. 1995. Selection and application of human single chain Fv antibody fragments from a semi-synthetic phage antibody display library with designed CDR3 regions. *J. Mol. Biol.* 248: 97–105.

110. Griffiths, A. D., S. C. Williams, O. Hartley, I. M. Tomlinson, P. Waterhouse, W. L. Crosby, R. E. Kontermann, P. T. Jones, N. M. Low, T. J. Allison, T. D. Prospero, H. R. Hoogenboom, A. Nissim, J. P. L. Cox, J. L. Harrison, M. Zaccolo, E. Gherardi, G. Winter. 1994. Isolation of high affinity human antibodies directly from large synthetic repertoires. *EMBO J.* 13: 3245–3260.

111. Wilson, I. A., R. L. Stanfield. 1993. Antibody-antigen interactions. *Curr. Opinion Struct. Biol.* 3: 113–118.

112. Williams, S. C., J.-P. Frippiat, I. M. Tomlinson, O. Ignatovich, M.-P. Lefranc, G. Winter. 1996. Sequence and evolution of the human germline V_λ repertoire. *J. Mol. Biol.* 264: 220–232.

113. Tomlinson, I. M., G. Walter, J. D. Marks, M. B. Llewelyn, G. Winter. 1992. The repertoire of human germline VH sequences reveals about fifty groups of VH segments with different hypervariable loops. *J. Mol. Biol.* 227: 776–798.

114. Tomlinson, I. M., J. P. Cox, E. Gherardi, A. M. Lesk, C. Chothia. 1995. The structural repertoire of the human V_k domain. *EMBO J.* 14: 4628–4638.

115. Chothia, C., A. M. Lesk. 1987. Canonical structures for the hypervariable regions of immunoglobulins. *J. Mol. Biol.* 196: 901–917.

116. Chothia, C., A. M. Lesk, E. Gherardi, I. M. Tomlinson, G. Walter, J. D. Marks, M. Llewelyn, G. Winter. 1992. Structural repertoire of the human VH segments. *J. Mol. Biol.* 227: 799–817.

117. Novotny, J., R. E. Bruccoleri, F. A. Saul. 1989. On the attribution of binding energy in antigen-antibody complexes McPC 603, D1.3, and HyHEL-5. *Biochemistry* 28: 4735–4749.

118. Kelley, R. F., M. P. O'Connell. 1993. Thermodynamic analysis of an antibody functional epitope. *Biochemistry* 32: 6828–6835.

119. Hawkins, R. E., S. J. Russell, M. Baier, G. Winter. 1993. The contribution of contact and non-contact residues of antibody in the affinity of binding to antigen.

The interaction of mutant D1.3 antibodies with lysozyme. *J. Mol. Biol.* 234: 958–964.

120. Waterhouse, P., A. Griffiths, K. Johnson, G. Winter. 1993. Combinatorial infection and in vivo recombination: a strategy for making large phage antibody repertoires. *Nuc. Acid. Res.* 21: 2265–2266.
121. Sheets, M. D., P. Amersdorfer, R. Finnern, P. Sargent, E. Lindqvist, R. Schier, G. Hemmingsen, C. Wong, J. C. Gerhart, J. D. Marks, 1998. Efficient construction of a large non-immune phage antibody library: The production of panels of high-affinity human single-chain antibodies to protein antigens. *Proc. Natl. Acad. Sci. USA* 95(11): 6157–6162.
122. Abergel, C., J.-M. Claverie. 1991. A strong propensity toward loop formation characterizes the expressed reading frames of the D segments at the Ig H and T cell receptor loci. *Eur. J. Immunol.* 21: 3021–3025.
123. Russell, S. J., R. E. Hawkins, G. Winter. 1993. Retroviral vectors displaying functional antibody fragments. *Nucl. Acids Res.* 21: 1081–1085.
124. Somia, N. V., M. Zoppe, I. M. Verma. 1995. Generation of targeted retroviral vectors by using single-chain variable fragment: an approach to in vivo gene therapy. *Proc. Natl. Acad. Sci. USA* 92: 7570–7574.
125. Chen, S. Y., C. Zani, Y. Khouri, W. A. Marasco. 1995. Design of a genetic immunotoxin to eliminate toxin immunogencity. *Gene Therapy* 2: 116–123.
126. Biocca, S., A. Cattaneo. 1995. Intracellular immunization: antibody targeting to subcellular compartments. *Trends Cell Biol.* 5: 248–252.
127. Marasco, W. A. 1997. Intrabodies: turning the humoral immune system outside in for intracellular immunization. *Gene Therapy* 4: 11–15.
128. Holliger, P., T. Prospero, G. Winter. 1993. 'Diabodies': small bivalent and bispecific antibody fragments. *Proc. Natl. Acad. Sci. USA* 90: 6444–6448.
129. Whitlow, M., D. Filpula, M. L. Rollence, S.-L. Feng, J. F. Wood. 1994. Multivalent Fvs: characterization of single-chain Fv oligomers and preparation of a bispecific Fv. *Protein Eng.* 7: 1017–1026.
130. Deng, S.-J., C. R. MacKenzie, T. Hirama, R. Brousseau, T. L. Lowary, N. M. Young, D. R. Bundle, S. A. Narang. 1995. Basis for selection of improved carbohydrate-binding single-chain antibodies from synthetic gene libraries. *Proc. Natl. Acad. Sci. USA* 92: 4992–4996.
131. Meulemans, E. V., R. Slobbe, P. Wasterval, F. C. Ramaekers, G. J. van Eys. 1994. Selection of phage-displayed antibodies specific for a cytoskeletal antigen by competitive elution with a monoclonal antibody. *J. Mol. Biol.* 224: 353–360.
132. Ward, R., M. Clark, J. Lees, N. Hawkins. 1996. Retrieval of human antibodies from phage-display libraries using enzymatic cleavage. *J. Immunol. Meth.* 189: 73–82.
133. Schwab, C., H. R. Boshard. 1992. Caveats for the use of surface-adsorbed protein antigen to test the specificity of antibodies. *J. Immunol. Meth.* 147: 125–134.
134. Balass, M., E. Morag, E. A. Bayer, S. Fuchs, M. Wilchek, E. Katchalski-Katzir. 1996. Recovery of high-affinity phage from a nitrostreptavidin matrix in phage-display technology. *Anal. Biochem.* 243: 264–269.
135. Morag, E., E. A. Bayer, M. Wilchek. 1996. Reversibility of biotin-binding by selective modification of tyrosine in avidin. *Biochem. J.* 316: 193–199.
136. Duenas, M., C. A. Borrebaeck. 1994. Clonal selection and amplification of phage

displayed antibodies by linking antigen recognition and phage replication. *Bio/ Technology* 12: 999–1002.

137. Duenas, M., A. Malmborg, R. Casalvilla, M. Ohlin, C. Borrebaeck. 1996. Selection of phage displayed antibodies based on kinetic constants. *Mol. Immunol.* 33: 279–285.

138. Krebber, C., S. Spada, D. Desplancq, A. Pluckthun. 1995. Co-selection of cognate antibody-antigen pairs by selectively-infective phages. *FEBS Lett.* 377: 227–231.

139. Sanna, P., R. Williamson, A. De Logu, F. Bloom, D. Burton. 1995. Directed selection of recombinant human monoclonal antibodies to herpes simplex virus glycoproteins from phage display libraries. *Proc. Natl. Acad. Sci. USA* 92: 6439–6443.

140. Ditzel, H. J., J. M. Binley, J. P. Moore, J. Sodroski, N. Sullivan, L. S. Sawyer, R. M. Hendry, W. P. Yang, C. F. Barbas, D. R. Burton. 1995. Neutralizing recombinant human antibodies to a conformational V2- and CD4-binding site-sensitive epitope of HIV-1 gp120 isolated by using an epitope-masking procedure. *J. Immunol.* 154: 893–906.

141. Meulemans, E., L. Nieland, W. Debie, F. Ramaekers, G. van Eys. 1995. Phage displayed antibodies specific for a cytoskeletal antigen. Selection by competitive elution with a monoclonal antibody. *Human Antibodies and Hybridomas* 6: 113–118.

142. Ames, R. S., M. A. Tornetta, C. S. Jones, P. Tsui. 1994. Isolation of neutralizing anti-C5a monoclonal antibodies from a filamentous phage monovalent Fab display library. *J. Immunol.* 152: 4572–4581.

143. Bradbury, A., L. Persic, T. Werge, A. Cattaneo. 1993. From gene to antibody: the use of living columns to select phage antibodies. *Bio/Technology* 11: 1565–1569.

144. Hart, S. L., A. M. Knight, R. P. Harbottle, A. Mistry, H.-D. Hunger, D. F. Cutler, R. Williamson, C. Coutelle. 1994. Cell binding and internalization by filamentous phage displaying a cyclic arg-gly-asp-containing peptide. *J. Biol. Chem.* 269: 12468–12474.

145. Pasqualini, R., E. Ruoslahti. 1996. Organ targeting in vivo using phage display peptide libraries. *Nature* 380: 364–366.

146. Gussow, D., T. Clackson. 1989. Direct clone characterization from plaques and colonies by the polymerase chain reaction. *Nucleic Acids Res.* 17: 4000.

147. Barbas, C. F., D. Hu, N. Dunlop, L. Sawyer, D. Cababa, R. M. Hendry, P. L. Nara, D. R. Burton. 1994. In vitro evolution of a neutralizing human antibody to human immunodeficiency virus type 1 to enhance affinity and broaden strain cross-reactivity. *Proc. Natl. Acad. Sci. USA* 91: 3809–3813.

148. Yang, W.-P., K. Green, S. Pinz-Sweeney, A. T. Briones, D. R. Burton, C. F. Barbas. 1995. CDR walking mutagenesis for the affinity maturation of a potent human anti-HIV-1 antibody into the picomolar range. *J. Mol. Biol.* 254: 392–403.

149. Schier, R., A. McCall, G. P. Adams, K. Marshall, M. Yim, H. Merritt, R. S. Crawford, L. M. Weiner, C. Marks, J. D. Marks. 1996. Isolation of picomolar affinity anti-c-erbB2 single-chain Fv by molecular evolution of the complementarity determining regions in the centre of the antibody combining site. *J. Mol. Biol.* 263: 551–567.

150. Foote, J., C. Milstein. 1991. Kinetic maturation of an immune response. *Nature* 352: 530–532.

151. Low, N., P. Holliger, G. Winter. 1996. Mimicking somatic hypermutation: affinity maturation of antibodies displayed on bacteriophage using a bacterial mutator strain. *J. Mol. Biol.* 260: 359–368.

152. Mian, I. S., A. R. Bradwell, A. J. Olson. 1991. Structure, function and properties of antibody binding sites. *J. Mol. Biol.* 217: 133–151.

153. Yelton, D. E., M. J. Rosok, G. Cruz, W. L. Cosand, J. Bajorath, I. Hellstrom, K. E. Hellstrom, W. D. Huse, S. M. Glaser. 1995. Affinity maturation of the BR96 anti-carcinoma antibody by codon-based mutagenesis. *J. Immunol.* 155: 1994–2004.

154. Wells, J. A. 1990. Additivity of mutational effects in proteins. *Biochemistry* 29: 8509–8517.

155. Balint, R. F., J. W. Larrick. 1993. Antibody engineering by parsimonious mutagenesis. *Gene* 137: 109–118.

156. Schier, R. S., J. D. Marks. 1996. Efficient in vitro selection of phage antibodies using BIAcore guided selections. *Human Antibodies and Hybridomas* 7: 97–105.

157. Figini, M., J. D. Marks, G. Winter, A. D. Griffiths. 1994. In vitro assembly of repertoires of antibody chains on the surface of phage by renaturation. *J. Mol. Biol.* 239: 68–78.

158. Jesper, L., A. Roberts, S. Mahler, G. Winter, H. Hoogenboom. 1994. Guiding the selection of human antibodies from phage display repertoires to a single epitope of an antigen. *Bio/Technology* 12: 899–903.

7

BISPECIFIC FUSION PROTEINS

JOEL GOLDSTEIN, ROBERT F. GRAZIANO, AND MICHAEL W. FANGER
Dartmouth Medical School, Lebanon, NH 03756 and Medarex, Inc., Annandale, NJ 08801

7.1 INTRODUCTION

Bispecific molecules—proteins with dual specificities—have been utilized for a variety of purposes (see Refs. 1–4 for reviews). One of the major uses of bispecific molecules has been to redirect cytotoxic effector cells to destroy tumors or infectious agents expressing a target antigen. Cells of both lymphoid and myeloid lineages are capable of this type of cytotoxicity, but to obtain efficient destruction of a tumor cell or pathogen by a cytotoxic effector cell, a trigger molecule on the surface of that effector cell must be engaged (see Table 7.1).

Cytolytic T lymphocytes (CTL) are potent effector cells. The primary cytotoxic trigger complex on CTL is the T-cell receptor (TCR) complex that consists of a noncovalent association of T_i heterodimers (α/β or λ/δ) with the CD3 molecular complex (see Box 7.1). Although normally antigen-specific and MHC restricted, bispecific molecules can circumvent MHC restriction, and thus redirect the action of virtually all cytolytic T cells to cells expressing the antigen of choice.[5] Although linkage to any TCR moiety probably initiates cytotoxicity, most studies of redirected cytotoxicity have involved binding to the CD3 complex. CD2 is also capable of initiating T-cell mediated cytotoxicity.[6,7] In addition, several other T cell and NK cell markers have been shown to trigger cytotoxicity under certain circumstances (see Table 7.1).

Antibody Fusion Proteins, Edited by Steven M. Chamow and Avi Ashkenazi
ISBN 0471-18358-X

TABLE 7.1 Cytotoxic Trigger Molecules

Molecule	Effector Cell	Reference
TCR/CD3 complex	T cells	5, 84
CD2	T cells, NK cells	6, 7
CD44	T cells, NK cells	85, 86
CD69	T cells, NK cells	87
Mel14	T cells	88
Ly-6.2C	T cells	89
CD16 (FcγRIII)	NK cells, macrophages	19
CD32 (FcγRII)	Monocytes, macrophages, neutrophils, eosinophils	90
CD64 (FcγRI)	Monocytes, macrophages, activated neutrophils	90

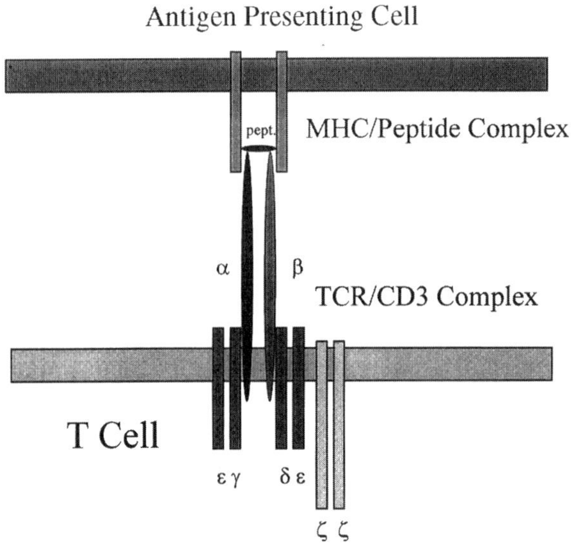

Box 7.1 Structure of the TCR/CD3 Complex.

The T-cell receptor is a multisubunit complex that recognizes processed antigen in the context of the major histocompatibility complex (MHC). The α-β heterodimer recognizes antigen and is associated with three other dimeric structures that make up the CD3 complex, CD3$\gamma\varepsilon$, CD3$\delta\varepsilon$, and CD3$\zeta\zeta$. The δ, ε, and γ subunits each contain one immunoreceptor tyrosine-based activation motif (ITAM), and the ζ subunits contain three ITAMs. The ITAMs are responsible for transmitting signals that activate the T cell. Monoclonal antibodies specific for CD3 subunits can also generate signals that may activate the T cell.

Numerous reports describing the use of bispecific molecules for redirected killing by T cells have been reviewed.[8,9,10] These studies of cytotoxicity have primarily involved bispecific molecules that link the CD3 complex on T cells to a variety of tumor-associated antigens. Potent in vitro cytotoxicity, as well as in vivo nude mouse models, have demonstrated the utility of directing T cells to destroy tumor targets using bispecific molecules. Further studies demonstrated that CD8+ tumor infiltrating lymphocytes (TIL), when targeted with the appropriate bispecific molecules, mediated efficient lysis of tumor targets.[11,12] In addition to direct cytotoxicity, bispecific molecules also trigger the release of cytokines, including TNF-α and IFN-γ, that block tumor growth in vitro and probably in vivo.[13,14]

Bispecific molecules that engage additional cell surface molecules can further augment T-cell cytotoxicity. In particular, two antitumor bispecific antibodies (BsAb) in concert, one directed to a trigger molecule and the other to a molecule involved in activation, can simultaneously activate effector cells and trigger cytotoxicity in the absence of exogenously added cytokines. For example, anti-CD28 is able to activate the lytic potential and proliferative capacities of cytotoxic T cells.[15] Since B7-1 or B7-2, the normal ligands for CD28, are expressed on monocytes, macrophages, and activated B cells, but not on carcinoma cells, the use of anti-CD28 BsAbs to stimulate T cells could be useful for upregulating cytolytic activity. In fact, combinations of anti-CD3 × anti-tumor antigen with anti-CD28 × anti-tumor antigen BsAbs either as Fab × Fab hybrids or whole BsAbs (prepared from hybrid-hybridomas) were effective in both activating and mediating T cell killing.[16] In this regard, bispecific Fab × Fab hybrids are potentially more useful than whole bispecific antibodies since Fab × Fab hybrids would not be removed prematurely by Fc receptor-bearing cells of the reticuloendothelial system.

Human myeloid cells (monocytes, macrophages, eosinophils, and neutrophils) can also be potent tumoricidal cells. The cytotoxic and phagocytic trigger molecules on these cells include the three classes of FcγR: FcγRI (CD64), FcγRII (CD32), and FcγRIII (CD16). FcγRI is a 70 kDa glycoprotein that is highly expressed only on monocytes, macrophages, and cytokine-activated polymorphonuclear cells (PMNs). It binds monomeric human IgG1 and IgG3, and mouse IgG2a and IgG3 with high affinity, and is a potent cytotoxic trigger molecule on all these cells, mediating both antibody dependent cellular cytotoxicity (ADCC) and phagocytosis (see Box 7.2). Two monoclonal antibodies, mAb32 and mAb22, have been described[17,18] that bind to FcγRI and trigger effector cells expressing this receptor. Since these mAbs bind to FcγRI at a location outside the ligand binding site, cytotoxicity mediated by BsAb prepared using these anti-FcγRI antibodies is not blocked by human IgG.[2,19,20] FcγRI has also recently been shown to be expressed by dendritic cells.[21] FcγRII, a 40 kDa glycoprotein that reacts with IgG immune complexes and opsonized particles, is a trigger molecule on monocytes, macrophages, PMNs, and eosinophils and mediates lysis and/or phagocytosis of tumor cell targets. However, FcγRII is also expressed on noncytotoxic cells such as B

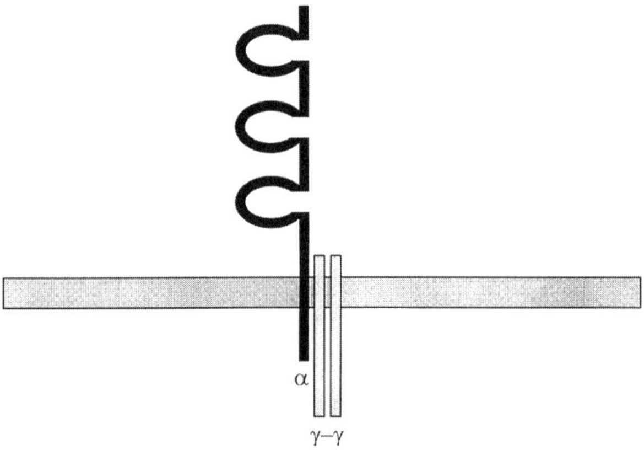

Box 7.2 Structure of CD64 (FcγRI).

CD64 is the high-affinity receptor for the Fc portion of IgG. CD64 recognizes and binds to the Fc portion of human IgG1 and IgG3 with high affinity. CD64 is expressed on the surface of monocytes, macrophages, dendritic cells, and cytokine-activated polymorphonuclear leukocytes. Monoclonal antibodies have been generated that bind to CD64 at a site outside of the Fc binding site, and are able to trigger function of the receptor. The ligand binding α-chain of CD64 has three immunoglobulin-like domains and is noncovalently associated with the FcRγ chain homodimer. Signalling is mediated through the γ-chain, which contains immunoreceptor tyrosine-based activation motifs.

lymphocytes and platelets. FcγRIII is a 50–70 kDa glycoprotein present on macrophages, PMNs, and the natural killer (NK) cell/large granular lymphocyte (LGL) population. This receptor binds immune complexes and opsonized particles, and has two isoforms. FcγRIIIA is expressed on macrophages and LGL/NK cells and is a transmembrane and cytotoxic trigger molecule.[19] FcγRIIIB is found on PMNs linked to the surface via a phosphatidylinositol glycan linkage and does not mediate cytotoxicity of tumor target cells.

Bispecific molecules have been made both by cell fusion and chemical methods (for review see Ref. 22). Nisonoff and Rivers initially generated BsAb by mixing Fab′ fragments of polyclonal rabbit antibodies directed against two different antigens.[23] Upon reoxidation of the interheavy chain disulfide bonds, BsAb were formed. With the advent of monoclonal antibody technology and the development of a variety of specific chemical cross-linking reagents, the development and production of BsAb became more sophisticated. BsAb could be produced biologically by fusing the two hybridoma cell lines that produced the antibodies of interest. However, due to the often random H-H chain and H-L chain associations, only 1/10 of the antibody species produced would have the desired bispecificity.[24] Notwithstanding this limitation, this method has been used to produce BsAb that have subsequently been purified to homogeneity and used in clinical trials.[25] Chemical cross-linkers initially used to produce BsAb generally led to the creation of heterogeneous products due to random cross-linking.[5] However, homobifunctional cross-linkers such as 5, 5'-dithiobis(2-nitrobenzoic acid) (DTNB) and *o*-phenylenedimaleimide (*o*-PDM) have been used to prepare homogeneous, readily purified bispecific molecules, some of which are in clinical trials.[22,26] DTNB and *o*-PDM react with free thiols generated by reduction of the inter-heavy chain disulfide bonds of $F(ab')_2$ fragments. DTNB acts to regenerate disulfide bonds between the two distinct Fabs, whereas *o*-PDM acts to form a thioether bond between the two Fabs.

The preparation of homogeneous Ab fragments by proteolytic digestion can be difficult, and often conditions have to be optimized for each fragment. Recombinant DNA technology has therefore been used to produce defined immunoglobulin fragments such as Fab fragments (50 kDa)[27] and Fv fragments (25 kDa).[28,29] Smaller fragments may be preferable over whole IgG (150 kDa), because they can penetrate into tissue from the vascular system more effectively.[30] Single-chain Ab or single-chain Fv (sc-Fv) in which the 2 V-region domains are tethered by a covalent peptide linker,[31,32] or disulfide stabilized Fv in which the V-regions are disulfide linked with cysteines substituted in the framework regions,[33,34] are two strategies that have been developed to stabilize Fv fragments.

This chapter reviews the various bispecific fusion proteins that have been generated as single chain molecules and the approaches used in their preparation (see also Chapter 5). In general, these molecules can be classified as either Ab x Ab fusion proteins or as Ab × ligand fusion proteins. Many of the molecules that we discuss have been generated with a potential therapeutic role in mind, which also is discussed.

7.2 SINGLE-CHAIN BISPECIFIC ANTIBODIES (sc-BsAb, Ab × Ab fusion proteins)

The fusion of two different sc-Fv fragments separated by a peptide linker, resulting in the generation of a single chain-bispecific antibody (sc-BsAb), was first reported by George et al.,[35] who fused the sc-Fv from an anti-CD3 mAb (OKT3) with the sc-Fv for an anti-DNP mAb (U7.6) and expressed the bispecific molecule in bacteria. The potential for more stable and homogeneous preparations using such constructs offer advantages over chemical conjugation methods. Other genetic bispecific constructs that employ domain–domain associations through dimerization domains [e.g., amphipathic helices that include leucine zippers (see Chapter 5) and helix turn helix motifs,[36,37,38] or preferential pairing of chains with VH and VL from two different Abs (diabodies)[39,40]] employ two or more polypeptides, and consequently are not considered here. This section focuses on construction, expression, purification, and characterization of sc-BsAb; considering first those constructs that have been successfully expressed in bacteria, followed by sections on constructs designed for eukaryotic expression systems. The details for each of the constructs that we discuss are summarized in Tables 7.2 and 7.3, and Figure 7.1 depicts a schematic representation of these molecules.

7.2.1 Bacterial Expression

7.2.1.1 ***Anti-fluorescein × anti-ssDNA.*** Mallender and Voss generated a sc-BsAb specific for fluorescein and for single-stranded DNA.[41] In a construct which they called a bispecific single-chain antibody, the sc-Fv of the anti-fluorescein Ab (4420) was linked to the sc-Fv of the anti-single-stranded DNA Ab (BV04-01) by a flexible linker that was modeled after a secreted fungal cellulase protein. The 24 amino acid linker peptide of *Trichoderma reesi* cellobiohydrolase I (CBH1)[42,43] was previously shown to be an effective interdomain linker in an sc-Fv expressed and secreted by *E. coli.*[44] A "cassette" method was employed where the CBH1 linker was added to individual sc-Fv using polymerase chain reaction (PCR), and then one sc-Fv/linker gene was cloned adjacent to the other sc-Fv/linker gene. The sc-BsAb gene could not be constructed solely by PCR because of the high degree of primary structure identity between the two sc-Fv proteins. The cassette design allows for

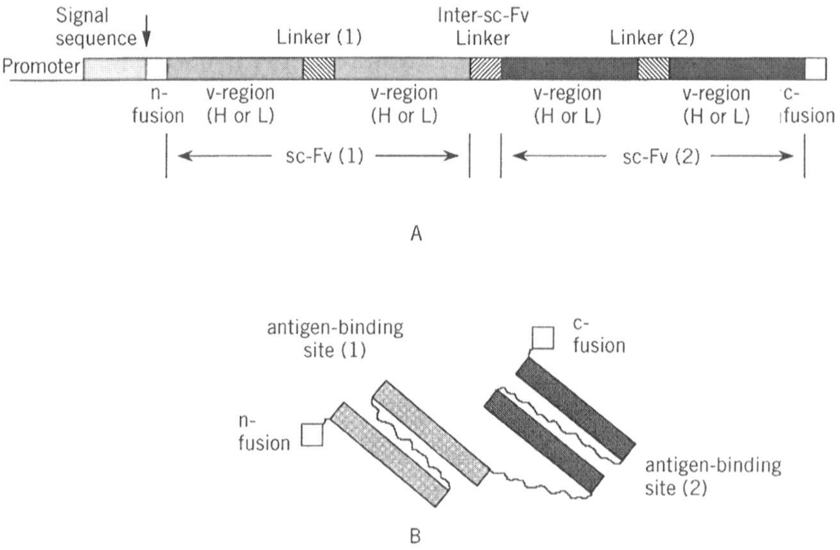

Figure 7.1 Schematic representation of Ab x Ab (sc-BsAb) constructs. *A*. Genetic structure. The arrow points to the signal sequence protease cleavage site. *B*. Peptide structure. n-fusion = amino-terminal fusion tag; c-fusion = carboxy-terminal fusion tag.

modification of the sc-BsAb gene by replacement of sc-Fv genes or linker sequences. The sc-BsAb was expressed in *E. coli* under the control of a hybrid O_L/P_R λ phage promoter and, although fused to the OmpA signal sequence, was isolated from insoluble inclusion bodies and subsequently renatured. Fluorescein-specific, affinity-purified fractions contained sc-BsAb plus smaller active protein fragments, while DNA-specific, affinity-purified fractions contained only sc-BsAb. The sc-BsAb was able to bind simultaneously to both ligands as determined by a solid phase ELISA. Fluorescence resonance energy transfer studies indicated that this sc-BsAb undergoes dynamic structural fluctuations in solution in a manner similar to the two Fab domains of an $F(ab')_2$ fragment, suggesting that the flexibility of the inter-sc-Fv linker acts in a manner analogous to an Ab hinge region.[45] A possible disadvantage of this and similar constructs is the potential immunogenicity of the CBH1 linker when used in a clinical setting. This study, however, provided the proof of principle that sc-BsAb form stable fusion proteins.

7.2.1.2 Anti-T-Cell Receptor (TCR) × anti-fluorescein. Using a genetic construction and *E. coli* expression system similar to the one just described, Gruber et al.[46] created a sc-BsAb consisting of the sc-Fv of an anti-TCR mAb (1B2) fused to an anti-fluorescein sc-Fv (4420). In this construct, the interchain linker was 25 amino acid residues in length and consisted mainly of Asp, Lys, and Ala. In addition, a 10-residue c-myc tag was inserted at the carboxy-terminus to allow detection with any anti-c-myc mAb. The 57 kDa anti-TCR × anti-fluorescein sc-BsAb was isolated from inclusion bodies and affinity purified on a fluorescein column with a yield of 1 mg/liter of bacterial culture. The refolded bispecific protein had an affinity for fluorescein that was nearly identical to that of the 4420 monospecific sc-Fv (despite the presence in the bispecific molecule of anti-TCR variable regions at the amino-terminus and the c-myc peptide at the carboxy-terminus). This was not the case, however, for TCR. Although the sc-BsAb also bound the TCR of the appropriate CTL, it required protein concentrations that were 300-fold higher than the 1B2 Fab fragment. By adding an ion-exchange chromatography step to further fractionate sc-BsAb after affinity purification, Kranz et al. found that a significant amount of the sc-BsAb had reduced binding ability, indicating that a significant portion of affinity purified protein was misfolded.[47] Despite a low yield of properly folded, functional sc-BsAb, the protein was capable of redirecting murine CTL to lyse fluoresceinated human tumor cells, thus demonstrating in principle the potential therapeutic value of sc-BsAb.

7.2.1.3 Anti-BCL1 lymphoma × anti-CD3. A sc-BsAb generated by de Jonge et al.[48,49] redirected T cells to kill BCL1 lymphoma cells in a mouse model. In this construct, DNA encoding an sc-Fv directed against a tumor antigen expressed by the murine BCL1 lymphoma was fused to DNA encoding an anti-mouse CD3 sc-Fv (2C11). A $(Gly_4Ser)_3$ linker was used for both intra-sc-Fv

TABLE 7.2 sc-BsAb Constructs (Bacterial Expression)

	anti-fluorescein x anti-ssDNA	anti-TCR x anti-fluorescein	anti-BCL1 x anti-CD3	anti-lysozyme x anti-lysozyme
Expression	*E. coli*	*E. coli*	*E. coli*	*E. coli*
Promoter	$O_L/P_R\lambda$ phage promoter	$O_L/P_R\lambda$ phage promoter	lac/T5	IPTG inducible promoter
Signal sequence	OmpA	OmpA	PelB	PelB
n-fusion	—	—	—	—
sc-Fv (1)	anti-fluorescein (4-4-20) VL-VH	anti-TCR (1B2) VL-VH	anti-BCL1 VH-VL	anti-lysozyme (D1.3) VH-VL
Linker (1)	212 (14 aa) GSTSGSGKSSEGKG	212 (14 aa) GSTSGSGKSSEGKG	Gly_4Ser (15 aa) GGGGSGGGGSGGGGS	Gly_4Ser (15 aa) GGGGSGGGGSGGGGS
Inter-sc-Fv linker	CBH1 (24 aa) PGGNRGTTRPATS GSSPGPTNSHY	205C′ (25 aa) ASADDAKKDAAKK DDAKKDDAKKDL	Gly_4Ser (15 aa) GGGGSGGGGSGGGGS	(18 aa) GSSSGSDGKASGGSGSGG
sc-Fv (2)	anti-ssDNA (04-01) VL-VH	anti-fluorescein (4-4-20) VL-VH	anti-murine CD3ε chain (2C11) VH-VL	anti-lysozyme (HyHEL-10) VL-VH
Linker (2)	212 (14 aa) GSTSGSGKSSEGKG	205C (25 aa) SSADDAKKDAAKK DDAKKDDAKKDG	Gly_4Ser (15 aa) SGGGSGGGGSGGGGS	Gly_4Ser (15 aa) SGGGSGGGGSGGGGS
c-fusion	—	c-myc	His-6	Cys + c-myc
Reference	41	46	48	51

	anti-TfR x anti-CD3	anti-c-erb-2 x anti-EGF-R x ETA	anti-c-erbB-2 x anti-CD16
Expression	*E. coli*	*E. coli*	*E. coli*
Promoter	T7lac	Tac	lac
Signal sequence	—	OmpA	PelB
n-fusion	—	FLAG epitope + His-6	—
sc-Fv (1)	anti-murine CD3ε chain (2C11) VL-VH	anti-c-erbB-2 (FRP5) VH-VL	anti-c-erbB-2 (C6.5) VH-VL
Linker (1)	Gly_4Ser (15 aa) GGGGSGGGGSGGGGS	Gly_4Ser (15 aa) GGGGSGGGGSGGGGS	Gly_4Ser (15 aa) GGGGSGGGGSGGGGS
Inter-sc-Fv linker	(23 aa) EFAKTTAPSVYPL APVLESSGSG	(> 120 aa) His-6 + ETA domain II (aa 252 – 366)	(16 aa) GSSGGGGSGGGGSGGS
sc-Fv (2)	anti-human transferrin receptor (OKT9) VL-VH	anti-EGF-R (225) VH-VL	anti-human CD16 (NM3E2) VH-VL
Linker (2)	Gly_4Ser (15 aa) GGGGSGGGGSGGGGS	Gly_4Ser (15 aa) GGGGSGGGGSGGGGS	Gly_4Ser (15 aa) GGGGSGGGGSGGGGS
c-fusion	—	His-6 + ETA domains II, Ib, & III	c-myc + His-6
Reference	52	56	59

TABLE 7.3 sc-BsAb Constructs (Eukaryotic Expression)

	anti-L6 x anti-CD3	anti-TfR x anti-CD3	anti-17-1A x anti-CD3	anti-CD64 x anti-CEA
Expression	mammalian transient-COS	mammalian transient-COS-7	mammalian stable-CHO	mammalian stable-NS0
Promoter	CMV	CMV	human elongation factor 1α	CMV
Signal sequence	L6 kappa	Ig light chain	eukaryotic	Ig
n-fusion	—	—	FLAG epitope	—
sc-Fv (1)	anti-L6 VL-VH	anti-murine CD3ε chain (2C11) VL-VH	anti-17-1A (M79) VL-VH	anti-human CD64 (H22) VH-VL
Linker (1)	Gly_4Ser (15 aa) GGGGSGGGGSGGGGS	Gly_4Ser (15 aa) GGGGSGGGGSGGGGS	Gly_4Ser (15 aa) GGGGSGGGGSGGGGS	Gly_4Ser (15 aa) GGGGSGGGGSGGGGS
Inter-sc-Fv linker	helical linker (27 aa) DQSNSEEAKKEEAKK EEAKKSNSLESL	(23 aa) EFAKTTAPSVYPL APVLESSGSG	Gly_4Ser (5 or 15 aa) GGGGS or GGGGSGGGGSGGGGS	(10 aa) SSCSSGGGGS
sc-Fv (2)	anti-human CD3ε (G19-4) VL-VH	anti-human transferrin receptor (OKT9) VL-VH	anti-human CD3 (TR66) VH-VL	anti-human CEA (MFE-23) VH-VL
Linker (2)	Gly_4Ser (15 aa) GGGGSGGGGSGGGGS	Gly_4Ser (15 aa) GGGGSGGGGSGGGGS	(18 aa) VEGGSGGSGGSGGS-GGVD	Gly_4Ser (15 aa) GGGGSGGGGSGGGGS
c-fusion	human IgG1 hinge + Fc	c-myc	His-6	c-myc + His-6
Reference	60	61	62	J. Goldstein, unpublished results

	anti-CD16 x anti-c-erbB-2	anti-TMV x anti-TMV
Expression	insect (Sf9) (baculovirus)	plant (tobacco)
Promoter	polyhedrin promoter	35s promoter of CaMV
Signal sequence	honeybee melittin	+/− eukaryotic signal sequence
n-fusion	—	—
sc-Fv (1)	anti-human CD16 VH-VL	anti-17kDa coat protein of TMV (scFv29) VL-VH
Linker (1)	Gly_4Ser (15 aa) GGGGSGGGGSGGGGS	212 (14 aa) GSTSGSGKSSEGKG
Inter-sc-Fv linker	(16 aa) GSSGGGGSGGGGSGGS	CBH1 (24 aa) PGGNRGTTRPATS GSSPGPTNSHY
sc-Fv (2)	anti-c-erbB-2 (C6.5) VH-VL	anti-intact virion neotopes of TMV (scFv24) VL-VH
Linker (2)	Gly_4Ser (15 aa) GGGGSGGGGSGGGGS	212 (14 aa) GSTSGSGKSSEGKG
c-fusion	c-myc + His-6	His-6 +/− KDEL
Reference	McCall & Weiner, unpublished results	66

linkers and for the inter-sc-Fv linker in this construct. This linker is commonly used in many sc-Fv constructs.[31,50] The construct was under the control of an IPTG inducible lac/T5 promoter in a pQE expression plasmid (Qiagen). Notwithstanding a PelB signal sequence that had previously directed expression of the parental sc-Fvs to the periplasm, sc-BsAb needed to be isolated from insoluble membrane fractions. Changing culture growth conditions, including temperature, medium, and time of induction, did not influence the solubility of the expressed sc-BsAb. The fusion protein was made with a carboxy-terminal His-6 tail, facilitating purification of resolubilized protein by immobilized metal affinity chromatography. After this step, the yield of denatured sc-BsAb was 8 mg/liter. After a second affinity purification step on a BCL1 antigen Sepharose column, the final yield was approximately 100 μg/liter. This purified molecule, with a functional anti-BCL1 sc-Fv, was not further purified by affinity to CD3, and therefore some of the material may not have CD3 binding activity. SDS-PAGE analysis revealed a single protein of 55 kDa after the BCL1 affinity column. A flow cytometry assay was used to show that the purified protein was able to bind to both target antigens simultaneously. Biological activity was tested by showing that this bispecific molecule could induce T-cell proliferation in the presence of mitomycin C-treated BCL1 cells and was able to induce lysis of BCL1 cells by allo-reactive CTL. In addition, immunotherapy with this sc-BsAb resulted in tumor elimination in BCL1-bearing mice.

7.2.1.4 ***Anti-lysozyme × anti-lysozyme.*** To achieve high avidity binding to a polymeric antigen using this technology, one could conceivably fuse two of the same sc-Fvs together with an interdomain linker in order to achieve the monospecificity of an Ab, but on a small single-chain molecule. An alternative approach taken by Neri et al. was to use two different sc-Fvs that bind adjacent and nonoverlapping epitopes simultaneously on a target antigen.[51] In this way, the sc-BsAb was used to harness a "chelate effect"; thus, it was called a chelating recombinant antibody (CRAb). A natural example of the chelate effect is the interaction of pentameric IgM with repeated epitopes of a polymeric Ag, leading to high avidities of synergistic binding. Neri et al. utilized two Abs against different epitopes of hen egg lysozyme (HEL). The sc-Fv fragment of mAb D1.3 was fused to the sc-Fv fragment of mAb HyHEL-10^{TF}, using splice overlap extension to create an 18 amino acid flexible linker consisting mainly of Gly and Ser residues. A carboxy-terminal fusion was created consisting of a heptapeptide terminating with a Cys residue for chemical modification, followed by a c-myc tag for detection and purification. Using an IPTG inducible promoter and PelB signal sequence, fusion protein was isolated from bacterial culture supernatants by affinity purification with an anti-c-myc antibody (9E10). Typical yields after purification ranged from 0.2 to 0.5 mg/liter of culture. This sc-BsAb was shown to have 20- to 100-fold higher affinity for HEL than either of the parental sc-Fv fragments, measured by a variety of parameters. Currently, the design of epitope-specific CRAbs requires the use of computer modeling. A possibly easier random approach could be developed that involves screening of large antibody fragment libraries expressed on the surface of phage particles. sc-Fv fragments could be assembled together in a random combinatorial manner as a repertoire of CRAbs displayed on phage, which could then be screened for high-affinity binding to the antigen.

7.2.1.5 ***Anti-Transferrin Receptor (TFR) × anti-CD3.*** Kurucz et al.[52] linked an sc-Fv specific for mouse CD3ε to an sc-Fv specific for human transferrin receptor (TfR) in a construct they called a single-chain bispecific Fv-dimer ($Bs[scFv]_2$). The authors used sc-Fv derived from mAb 2C11 (anti-mouse CD3ε) and sc-Fv derived from mAb OKT9 (anti-TfR), attached via a 23 amino acid inter-sc-Fv linker derived from an Ab CH1 region followed by a small flexible domain. Expression was under the control of the T7lac promoter/operator in a pET-11 bacterial expression vector with no signal sequence. Induction with IPTG, therefore, led to cytoplasmic production with $Bs[scFv]_2$ sequestered in inclusion bodies. Recovery from these inclusion bodies was enhanced by dissolving the protein in the weak ionic detergent sodium lauroylsarcosine, which allowed correct disulfide formation upon oxidation in air. Ion exchange was used to remove the detergent, and 18% of the protein applied to a size exclusion column was eluted as purified $Bs[scFv]_2$. FITC-labeled $Bs[scFv]_2$ was shown to be capable of binding to both CD3ε- and TfR-expressing cells by flow cytometry. The binding affinity of each arm of the $Bs[scFv]_2$ was similar to that of the parental Fab or sc-Fv, suggesting that most of the purified protein had been correctly folded. Furthermore, the $Bs[scFv]_2$ could redirect either a murine alloreactive CTL line or activated murine splenocytes to mediate lysis of TfR^+ target cells (TfR targeting in a different context is given in Chapter 2).

7.2.1.6 ***Anti-c-erbB-2 × anti-EGF-R × exotoxin A.*** c-erbB-2 and the receptor for epidermal growth factor (EGF-R) are members of the Type-I receptor tyrosine kinase family. This receptor family plays an important role in the development of human malignancies, and its members are overexpressed on a variety of human tumors. The highest incidence of c-erbB-2 overexpression has been found to occur in breast and ovarian tumors.[53] (See Sec. 7.3.1.1. for discussion on EGF-R overexpression in tumors.) A method other than re-directing cytotoxic effector cells for destroying such tumors is to utilize a toxin (see also Chapter 4). Wels et al have used such an approach by fusing the enzymatic domains of *Pseudomonas* exotoxin A (ETA) to sc-Fv specific for c-erbB-2 and for EGF-R.[54,55] Since both of these receptors are often expressed simultaneously on tumor lines, targeting ETA to both receptors could augment tumor killing.[56] This molecule was constructed as a multidomain fusion protein in the following manner from amino to carboxy terminus (see Fig. 7.2A): FLAG epitope; His-6; anti-c-erbB-2 sc-Fv (FRP5); His-6; 114 amino acid ETA domain II (for inter-sc-Fv linker); anti-EGF-R sc-Fv (225); His-6; ETA domains II, Ib, and III. This construct lacks the Ia domain of ETA, which

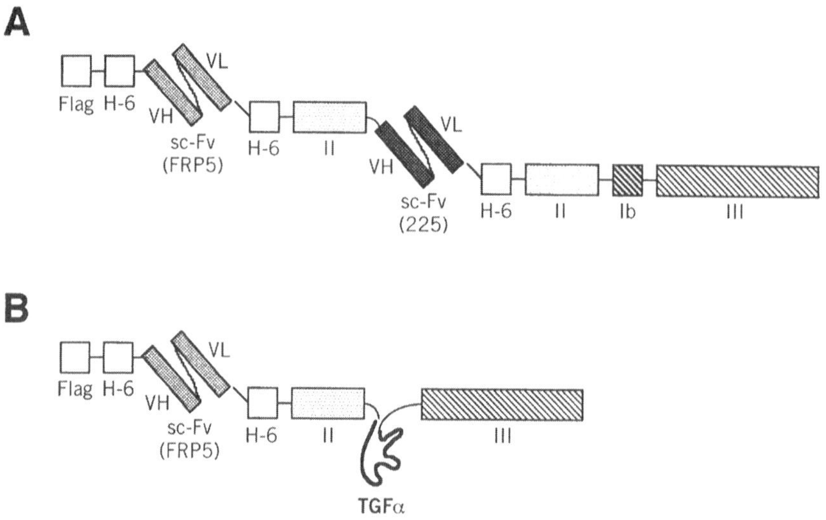

Figure 7.2 Schematic representation of *Pseudomonas* exotoxin A (ETA) fusion proteins. *A*. sc-BsAb-ETA fusion protein.[56] *B*. Ab x ligand-ETA fusion protein.[57] See text for details.

is responsible for cell recognition. Domain II of ETA is the translocation domain, and domains Ib and III mediate ADP ribosylation of the eukaryotic elongation factor 2. The resulting expression plasmid contained an IPTG-inducible tac promoter and an OmpA signal sequence. Nevertheless, sc-BsAb was isolated from total bacterial lysates and purified by immobilized metal affinity chromatography with a yield of 1 mg/liter of culture. SDS-PAGE analysis revealed that approximately 70% of the purified protein was the expected molecular weight of 107 kDa. Immunoprecipitation experiments showed that the fusion protein bound both receptors, and ELISA experiments demonstrated that both arms of the bispecific molecule bound to their respective Ags with affinity constants similar to those of the parental sc-Fv-ETA proteins. Furthermore, the activity of the sc-BsAb-ETA fusion protein was enhanced over the parental sc-Fv-ETA fusions to kill a cell line expressing both target Ags; competition experiments indicated that both sc-Fv domains contributed to the cell-killing activity of the molecule. A similar construct was made by Schmidt and Wels[57] in which the anti-EGF-R sc-Fv was replaced by an EGF-R ligand, TGFα (see Sec. 7.3.1.1; Fig. 7.2B).

7.2.1.7 ***Anti-c-erbB-2*** × ***anti-CD16.*** A human filamentous phage antibody library was used to obtain sc-Fv for human CD16 and for c-erbB-2.[58] A sc-BsAb was then constructed using intra- and inter-sc-Fv linkers consisting of Gly and Ser amino acids.[59] Under the control of a lac promoter, the fusion protein was directed to the *E. coli* periplasmic compartment by the PelB signal sequence. It was expressed at a concentration of 0.3 mg/liter. A 6-His tail facilitated purification by immobilized metal affinity chromatography; by SDS-PAGE a single band of approximately 55 kDa indicated that only monomers were formed. Binding activities of either arm of the bispecific molecule was demonstrated by BIAcore analysis. Subsequently, affinity mutants of this sc-BsAb were reconstructed using the anti-c-erbB-2 sc-Fv to determine the biological effect of increasing the affinity of the tumor-targeting arm of the bispecific molecule. It was found that reconstructed sc-BsAb with higher affinity anti-c-erbB-2 sc-Fv resulted in significantly greater target cell killing.

7.2.2 Mammalian Expression

7.2.2.1 ***Anti-L6*** × ***anti-CD3.*** The first report of expression of sc-BsAb in mammalian cells was by Hayden et al.[60] The sc-Fv of an antitumor associated antigen (L6) was joined through a 27 amino acid helical peptide linker to the sc-Fv of an anti-human CD3ε antibody. The linker sequence motif $(EEAKK)_n$ was chosen because it is particularly hydrophilic but has a neutral net charge. The peptide can form $(i + 3)$ and $(i + 4)$ ionic interactions if helical conformation is adopted. The sc-BsAb sequence was preceded by a DNA cassette encoding the anti-L6 immunoglobulin light chain signal sequence to facilitate secretion and followed by a DNA cassette encoding the Fc portion of human IgG1 that was used as a molecular "tag." (For additional examples of Fc fusions, see Part II). The Fc and hinge domain Cys residues were mutated to Ser to reduce intrachain disulfide bonding, and a point mutation at residue 238 (Pro to Ser) abolished Fc-mediated ADCC and complement killing in standard assays. Replacing the Fc with other molecular tags (Ab C_k domain, FLAG peptide, peptide from the V3 loop of gp120 from HIV) resulted in Ab fragments that bound Ag with varying affinities. Human IgG1 Fc was the best tag for reproducing the binding characteristics of the native Abs. The sc-Fv genes were inserted as interchangeable cDNA cassettes into the modified eukaryotic expression vector containing the cytomegalovirus (CMV) promoter. This cassette construction design facilitates the substitution of various sc-Fv and tag sequences. The construct, which was made in the transient transfection vector pCDM8, was transfected into a COS cell line. The 94 kDa sc-BsAb was then purified from culture supernatants by affinity chromatography on immobilized protein A. While the sc-BsAb was found to be predominantly in the monomeric form, molecules did associate through either the CH3 domain or by nonspecific interactions to form homodimers (approximately 30%) and higher order oligomers (10% or less). The sc-BsAb promoted conjugation of human T cells to L6-positive tumor cells and stimulated T cells to proliferate and mediate killing of L6-positive tumor cells.

7.2.2.2 ***Anti-TfR x anti-CD3.*** The anti-TfR X anti-CD3 sc-BsAb construct described for bacterial expression in Section 7.2.1.5 was also expressed in a mammalian system.[61] Transient expression in COS-7 cells was driven by a strong CMV promoter and secreted via a light chain leader sequence into the culture medium. The rate of secretion was found to be equal to that of the two parental sc-Fvs, indicating that the sc-BsAb was secreted as a properly folded molecule. A c-myc tag was also added to the carboxy-terminus of the protein for detection. Radiolabeled protein was used to determine that the secreted sc-BsAb retained both anti-CD3 and anti-TfR binding activities. This sc-BsAb was shown to be stable after overnight incubation at 37°C as well as after long-term storage at 4°C . Furthermore, COS-7 cell supernatant containing the sc-BsAb was able to efficiently redirect lysis of TfR^+ target cells by CTL generated by a mixed lymphocyte reaction.

7.2.2.3 ***171A x anti-CD3.*** A stable transfection system in CHO cells was used to express a 17-1A sc-Fv fused with a Gly_4Ser linker to an anti-human CD3 sc-Fv.[62,63,64] This bispecific single-chain fusion protein was made for the purpose of redirecting human CTLs to kill colorectal cancer cells by targeting the 171A or EpCAM antigen. To produce the sc-BsFv, expression was driven by the promoter for human elongation factor 1α, and a eukaryotic signal sequence directed secretion of the protein from the cell. In addition, a FLAG epitope was fused to the amino-terminus for detection and a His-6 tail was fused to the carboxy-terminus for purification. Two versions of the Gly_4Ser inter-domain linker were used: 1 Gly_4Ser unit for a total of 5 amino acids, and 3 units for a total of 15 amino acids. An internal ribosome binding site was engineered downstream of the fusion protein coding sequence to allow bicistronic expression with dihydrofolate reductase, which was used as a marker to select for stable transfectants. After selection of transformants, expression was enhanced by exposing transfected dihydrofolate reductase-deficient CHO cells to methotrexate, which leads to gene amplification. A yield of 12–15 mg/liter was obtained after purification on a nickel-nitrilotriacetic acid column. Flow cytometry showed that the fusion protein bound both 17-1A$^+$ and CD3$^+$ cells, and both ELISA and flow cytometry revealed that affinities for these antigens were comparable to the parental sc-Fvs alone.

This sc-BsAb mediated specific lysis of 17-1A$^+$ tumor cells by unstimulated peripheral blood mononuclear cells from healthy donors. No difference in cytotoxic activity was observed between interdomain linker versions of the fusion protein, indicating that variability in linker length is well tolerated for functional sc-BsAb. It is interesting to note that Mack et al. were unable to obtain a functional protein when they attempted bacterial expression of the same sc-BsAb in the periplasm of *E. coli* using the OmpA signal sequence. This *E. coli* compartment may be insufficient for functional expression of this particular fusion protein, which contains more than two immunoglobulin domains on a single polypeptide chain. It should be noted that another sc-BsAb (see Anti-c-erbB-2 X anti-CD16 in Section 7.2.1.7) that was similarly directed to the periplasm apparently did fold properly.[59]

7.2.2.4 ***Anti-CD64 x anti-carcinoembryonic antigen.*** We have recently stably expressed a sc-BsAb in a mammalian system (J. Goldstein, unpublished results). This fusion protein consisted of the sc-Fv of a humanized anti-CD64 mAb (H22) linked by ten amino acids consisting mainly of Gly and Ser to an anti-carcinoembryonic antigen (MFE-23) sc-Fv.[65] A c-myc tag and His-6 tail were also fused to the carboxy-terminus of the sc-BsAb. The construct was made in the mammalian expression vector, pcDNA3 (Invitrogen). Using a CMV promoter and antibody signal sequence, sc-BsAb was expressed and secreted into the culture medium of the transfected murine myeloma line, NS0. Harvested supernatant bound both carcinoembryonic antigen and CD64 in a bispecific ELISA, whereas the parental sc-Fvs had no such dual-binding activity. Affinity purification by immobilized metal affinity chromatography, however, resulted in a significant loss in this bispecific binding activity. Since elution in this mode of chromatography is achieved using mild conditions, we theorize that this sc-BsAb may be unstable in purified form. We are currently attempting to improve the stability of this fusion protein so that its ability to mediate cytotoxicity of carcinoembryonic antigen-positive tumor cells can be investigated.

7.2.3 Insect Cell Expression

7.2.3.1 ***Anti-CD16 x anti-c-erbB-2.*** A baculovirus expression system was recently used to express an sc-BsAb consisting of an sc-Fv specific for CD16, fused with a 16 amino acid linker consisting of Gly and Ser, to an sc-Fv specific for c-erbB-2 (C6.5) (A. McCall and L. Weiner, unpublished results). A c-myc tag and His-6 tail were fused to the carboxy-terminus of the fusion protein for detection and purification. Expression was carried out in Sf9 cells using the vector pMelBAcB (Invitrogen) under the control of a polyhedrin promoter. Although there was a honeybee melittin signal sequence to facilitate secretion, the fusion protein could only be isolated from the cytoplasmic compartment. Yields of at least 100 μg per 2×10^8 cells were achieved. SDS-PAGE analysis revealed that this sc-BsAb formed not only monomers of 55kDa, but also oligomers as evidenced by aggregates of purified protein. BIAcore analysis demonstrated that the fusion protein was able to bind to both Ags.

7.2.4 Plant Expression

7.2.4.1 ***Anti-coat protein x anti-virion.*** Schumann et al.[66] expressed an sc-BsAb in both transgenic tobacco plants and tissue culture cells. This sc-BsAb, recognizing two different epitopes on tobacco mosaic virus (TMV), was used to study structure–function relationships of viral targets in plants by expressing the sc-BsAb in different cell compartments. One sc-Fv was specific for the 17 kDa coat protein monomer, and the other sc-Fv bound to neotopes of intact virions. The two sc-Fvs were fused with the same flexible fungal CBH1 linker as the one used by Mallender and Voss[41] to construct their sc-BsAb in bacteria. Three constructs were made to target the expressed fusion protein to different cellular compartments. The sc-BsAb was expressed either without or with a signal sequence to target the fusion protein to the cytosol or the apoplast of plant cells, respectively. Third, endoplasmic reticulum signal and retention sequences (KDEL) were added to the construct. In all cases, expression was under the control of a 35S promoter from CaMV in a plant expression vector, and due to the presence of a carboxy-terminal His-6 tail, purification was achieved using immobilized metal affinity chromatography. The fusion proteins were successfully localized in the three plant cell compartments, although there was significantly less functionally active sc-BsAb in the cytosol as compared to the secreted versions. The three sc-BsAbs had binding activities for both TMV epitopes in an ELISA assay, and simultaneous binding to both intact and dissociated virions was detected by ELISA and surface plasmon resonance.

7.2.5 Summary

sc-BsAb are small, single-peptide molecules with dual antigen-binding specificities. The small size of these reagents offers the potential for increased tumor penetration. Homogeneous bispecific molecules obtained directly from the fermenter greatly facilitate manufacturing procedures. Although a number of investigators have had success in generating these bispecific molecules, others have run into difficulties. Kipriyanov et al.[67] made an anti-CD3 x anti-CD19 sc-BsAb with a 20 amino acid linker that failed to recognize human CD3 when expressed in *E. coli*. The optimal system for generic production of sc-BsAb has yet to be defined; indeed, the production system may need to be tailored to each individual sc-BsAb. Furthermore, processes used for production of sc-BsAb may or may not apply directly to other types of bispecific molecules that are not sc-BsAb.[36,39,40]

Bacterial and mammalian expression systems both have advantages and drawbacks. The expression of antibody fragments in bacteria provides a means of generating large quantities of sc-BsAb. In contrast to parental sc-Fv, however, sc-BsAb must often be recovered from inclusion bodies and refolded in vitro,[41,46] and the yield of active protein may be low.[47,48] Renaturing active sc-BsAb from the inclusion bodies may be difficult. However, Kurucz et al.[52] were able to increase refolding efficiency by using weak ionic detergents. Also, the protein may be toxic to the cell when expressed behind a signal sequence that targets the protein to the periplasm.[68] Although periplasmic expression for one such sc-BsAb did result in active product,[59] it did not for another one.[62] In addition, although Neri et al.[51] were able to secrete sc-BsAb into the culture medium, other constructs with signal sequences were primarily localized in inclusion bodies[41,46] or membrane fractions.[48] Furthermore, bacteria may not be the best host for expression of sc-BsAb, because the duplication present in four homologous variable region repeats may potentially lead to deletion of a portion of the DNA sequence encoding the sc-BsAb.

For production in mammalian cells, large scale production of sc-BsAb from transiently transfected COS cells[60,61] is unfeasible. The system described by Mack et al.[62] for stable transfection in CHO cells offers the potential for large, functional, homogeneous production of sc-BsAb. Baculovirus and plant cell expression systems[66] also offer potential for sc-BsAb production.

With regard to the design of these molecules, sc-BsAb constructs exhibit a certain degree of flexibility. V-region order has been successfully tried in all four combinations: VL-VH X VL-VH, VH-VL X VH-VL, VL-VH X VH-VL, VH-VL X VL-VH. Both carboxy-terminal and, to a lesser extent, amino-terminal tag fusions have been employed. Moreover, the variation in inter-sc-Fv linker sequences and lengths indicates a general tolerance for different structures for this purpose. Gly-Ser sequences, charged residue (Asp- Lys) sequences, immunoglobulin hinge regions, fungal sequences, ETA translocation domain — all ranging from 5 to at least 120 amino acids — have been successfully used. The type of linker can, however, affect the tendency of these fusion proteins to aggregate.

7.3 BISPECIFIC FUSION PROTEINS (Ab x LIGAND FUSION PROTEINS)

Another method for generating bispecific molecules as fusion proteins is to have one binding specificity derived from immunoglobulin variable regions linked to a non-immunoglobulin ligand with a unique binding specificity. This approach was first demonstrated by Liu et al.[69] who showed that an anti-CD3 mAb that was chemically coupled to an analog of melanocyte-stimulating hormone could target T cells to lyse human melanoma cells. The Ab portion may be a Fab, an sc-Fv, or whole antibody. If an sc-Fv fragment is used in a fusion protein construct, then the resulting molecule is a true single-chain fusion protein. This section focuses on antibody-ligand fusion proteins.

Ligands that have biological function include, but are not limited to, growth factors, cytokines, and hormones. The non-immunoglobulin ligand polypeptide may be fused to either the amino-terminus or carboxy-terminus of the immunoglobulin polypeptide with a linker separating the two domains. Since Ab-ligand fusion proteins are dealt with in more detail in Chapter 2, we will cover these constructs selectively. We focus on a few examples including a TGFα-toxin fusion protein expressed in *E. coli* and an epidermal growth factor (EGF) fusion protein linked to an anti-CD3 mAb expressed in mammalian cells. In addition, we describe our work on anti-CD64 Fab fusions to growth factors such as human EGF and heregulin (HRG) (see Fig. 7.3). In addition, we discuss an anti-CD64 X antigen fusion protein that has been developed to enhance antigen presentation.

7.3.1 Bacterial Expression

7.3.1.1 Anti-c-erbB-2 x TGFα x ETA. This construct made by Schmidt and Wels[57] is similar to the sc-BsAb described in Section 7.2.1.6 for anti-c-erbB-2 x anti-EGF-R X ETA, except that the second sc-Fv (anti-EGF-R) is replaced by TGFα, a ligand for EGF-R. The overexpression of either the normal or mutated form of EGF-R is a characteristic feature of squamous cell carcinomas and is found in a number of human malignancies (for review see Ref. 70). In particular, native EGF-R is overexpressed in almost all head and neck tumors, about one-third of breast and ovarian tumors,[71] and can be overexpressed in cancers of the prostate, kidney, bladder, lung, brain, pancreas, and gastrointestinal system.

The differences between these two toxin fusion proteins (anti-c-erbB-2 × TGFα × ETA vs. anti-c-erbB-2 × anti-EGFR × ETA) can be seen in Fig. 7.2. One potential advantage of the TGFα-containing fusion protein is that its smaller size (73 vs. 107 kDa) may allow better tumor penetration. For details of the construction and expression of this multidomain single-chain fusion protein, see Section 7.2.1.6. The protein was purified to approximately 70% homogeneity using immobilized metal affinity chromatography, resulting in a yield of 1 mg/liter of bacterial culture. Activity of the anti-c-erbB-2 arm of the recombinant protein was assayed by ELISA. Similarly, the ability of the TGF-α arm to bind and activate cells expressing EGF-R or c-erbB-2 was demonstrated. This bispecific molecule was cytotoxic in vitro for tumor cell lines overexpressing either of the target receptors, and it inhibited the growth of established A431 tumor xenografts in nude mice.

7.3.2 Mammalian Expression

7.3.2.1 Anti-CD3 x EGF. Gillies et al.[72] fused human EGF, which like TGF-α is a natural ligand for the EGF-R, to the carboxy-terminus of the CH3 domain of a chimeric anti-CD3 mAb (see Fig. 7.3). The V-regions of mouse anti-CD3 (OKT3) were inserted into a mammalian cell expression vector containing either human genomic Cγ1 or Cγ4. The EGF gene was synthesized from the known protein sequence and attached to CH3 with no linker region. Both H and L chains were under the control of the mouse metallothionine I promoter as separate transcription units on the same vector containing a weakened dihydrofolate reductase selection gene. DNA was transfected stably into a mouse hybridoma cell line (Sp2/0 Ag14) in one step, and the transfected line was screened for resistance to methotrexate. Since the dihydrofolate reductase gene was weakened, cells that were resistant to methotrexate secreted higher levels of the fusion protein. The entire antibody molecule was used in this construct, resulting in a conjugate containing two antigen binding sites at one end and two EGF molecules at the other (see Fig. 7.3). Ab/EGF fusion molecules were purified from culture supernatants by affinity chromatography

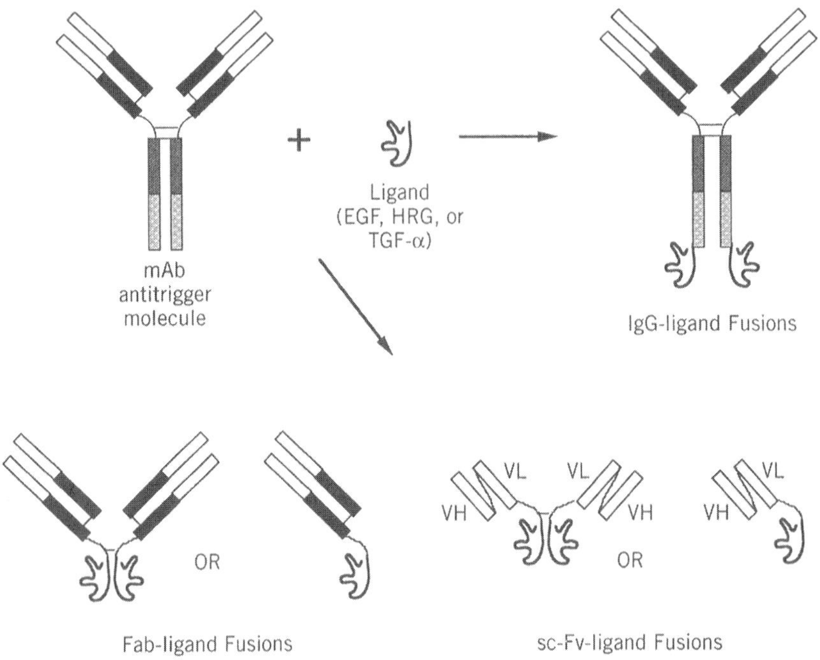

Figure 7.3 Schematic representation of Ab x ligand constructs. Fragments of a monoclonal antibody specific for trigger molecules on cytotoxic immune cells are fused to various ligands, as indicated. In the case of anti-CD3 X EGF (see text; Ref. 72), the whole IgG molecule was used in the fusion protein.

using protein A. The purified fusion protein was able to compete with EGF for its receptor in a competitive binding assay. It was also capable of mediating the lysis of EGF-R-bearing tumor cells by a tumor-infiltrating lymphocyte line and by a CTL line established from peripheral blood.

7.3.2.2 ***Anti-CD64 x EGF.*** We have prepared a different EGF-containing fusion protein using a mAb specific for myeloid cells expressing CD64.[73,74] For this construct, genomic DNA encoding the Fd fragment of a humanized anti-CD64 mAb, H22,[75] which binds CD64 at an epitope distinct from the Fc binding site, was fused at the carboxy-terminus of the hinge region to a cDNA encoding human EGF. This was done by creating unique restriction sites for inserting the ligand gene on the vector, so that any ligand could be inserted as a cassette (see Fig. 7.3 for schematic representation of H22 fusion proteins). EGF was synthesized by polymerase chain reaction using cDNA as template. The natural human IgG1 hinge acted as a linker between the antibody and ligand domains. The resulting H22Fd-EGF expressing vector, which contained the gpt selection marker, was stably transfected into a myeloma cell line. This cell line was previously transfected with a vector containing genomic DNA encoding the H22 kappa light chain.[75] Both H and L chains were under the control of mouse immunoglobulin gene promoter and enhancer.

SDS-PAGE analysis of purified fusion protein demonstrated that the fusion protein was secreted predominantly as H22Fab′-EGF monomer ($\approx$55 kDa), even though a free Cys residue existed in the hinge region of the H22 Fab′ component. Using a novel bispecific flow cytometry binding assay, we demonstrated that H22Fab′-EGF was able to bind simultaneously to soluble CD64- and EGF-R-expressing cells. H22Fab′-EGF inhibited the growth of EGF-R overexpressing tumor cells and mediated dose-dependent cytotoxicity of these cells in the presence of CD64-bearing cytotoxic effector cells. These results suggest that this fusion protein may have therapeutic utility to treat malignancies in which EGF-R is overexpressed.

More recently, we have generated an H22-EGF single-chain bispecific fusion protein (unpublished results). This bispecific molecule is also quite small, with a molecular mass of only 33 kDa. For this construct, the mammalian expression vector pcDNA3 (Invitrogen) was used. Expression was under the control of the CMV promoter, and a neomycin selection marker ensured stable transfection in the presence of G418. Purified fusion protein demonstrated bispecific antigen and receptor binding by flow cytometry. Furthermore, this molecule was able to mediate lysis of EGF-R overexpressing tumor cells by CD64-bearing cytotoxic effector cells. The sc-Fv portion was made in the VL-VH orientation, and the linker region between the sc-Fv and EGF domains consisted of nine amino acids: GSTGGGGSS. Upon purification on a Protein L column, the fusion protein migrated predominantly as a monomer with approximately 10% dimer on SDS-PAGE under both reducing and nonreducing conditions.

7.3.2.3 ***Anti-CD64 x Heregulin.*** Heregulin (HRG) is a ligand for the HER3 (c-erbB-3) and HER4 (c-erbB-4) membrane proteins,[76] which are overexpressed in tumors such as those originating from breast and thyroid tissues.[77,78] Both HER3 and HER4 may form homodimers, heterodimers with each other, or heterodimers with HER2 (c-erbB-2).[76,79] To target HER3- and HER4-positive cells for killing, we have constructed by genetic means a bispecific reagent consisting of the Fab′ fragment of H22 and the EGF domain of HRG-β2.[74] For this fusion protein, construction, expression and purification were accomplished using the same procedures as for H22Fab′-EGF above (Sec. 7.3.2.2). Using flow cytometry it was found that this molecule was able to bind to the HER3/4-expressing tumor cell line, SKBR-3, as well as to CD64-expressing cells. This H22-HRG fusion protein inhibited the growth of SKBR-3 tumor cells and mediated killing of SKBR-3 cells by CD64-bearing cytotoxic effector cells. Furthermore, H22-HRG upregulated ICAM 1 in a dose- and time-dependent fashion in SKBR-3 cells, indicating that the fusion protein mediates cell signalling events. Since ICAM 1 is a ligand for CD11a and CD11b, molecules that are expressed by immune effector cells of both lymphoid and myeloid lineages, H22-HRG may indirectly recruit immune effector cells other than CD64-bearing cells.

7.3.2.4 ***Anti-CD64 x tetanus toxoid peptides.*** Previous studies have shown that the immune response to a particular antigen can be enhanced by using anti-CD64 mAbs to direct the antigen to antigen-presenting cells.[80] In one study, enhanced presentation of antigenic and antagonistic peptides was demonstrated by targeting them to CD64 on human monocytes.[81] Specifically, the T helper epitope of tetanus toxoid, TT830, and the antagonistic peptide for TT830, TT833S, were fused to anti-CD64 Fab′, expressed and purified in the same way as the Fab fusion proteins described in Sections 7.3.2.2 and 7.3.2.3. These CD64-targeted peptides TT830 and TT833S were 1,000- and 100-fold more efficient than the parent peptides for T-cell stimulation and antagonism, respectively, suggesting that such fusion proteins could effectively increase the delivery of antigenic peptides to antigen-presenting cells in vivo. Moreover, the CD64-targeted antagonistic peptide inhibited proliferation of TT830-specific T cells, even when antigen-presenting cells were first pulsed with native peptide, a situation comparable to that which would be encountered in vivo when attempting to ameliorate an autoimmune response. These results suggest that a strategy such as targeted presentation of antagonistic peptides could enable better antigen-specific therapies for autoimmune disease.

7.3.3 Summary

There are a number of potential advantages to using the natural ligand over an anti-receptor antibody in a bispecific reagent. Even if an antibody has been humanized, which is a tedious task in itself, the antibody still may elicit an anti-idiotypic response in the course of human therapy. A natural ligand, on the other hand, is fully human. Furthermore, since a natural ligand may bind to a receptor with higher affinity than an Ab under physiological conditions, a relatively lower amount of the fusion protein may be sufficient to mediate an optimum in vivo response. A ligand may also elicit modulatory properties on target cells. For cancer treatment, this may be advantageous in the event of growth inhibition, but deleterious if the opposite effect occurs. Finally, properly folded antibody X ligand fusion protein may be easier to obtain, since only two immunoglobulin V-regions need to combine in a functional manner, as opposed to four V-regions in an sc-BsAb.

In addition to the anti-CD64 fusion proteins described previously, we are developing other fusion proteins such as H22 X CD4 (for HIV therapy) using the same cassette system. For targeted antigen presentation, we are developing H22 x tumor-associated antigens (for cancer vaccines), and H22 x gp120 (for HIV vaccine; Ref. 82). Moreover, we have successfully constructed and expressed an H22 sc-Fv that was used to generate a true single-chain fusion protein (H22 sc-Fv -EGF), and plan to use this system to generate other fusions.

7.4 CONCLUSIONS AND FUTURE DIRECTIONS

Considerable progress has been made in recent years in the development and characterization of bispecific fusion proteins. These molecules offer several advantages over bispecific molecules made chemically or via hybrid-hybridomas. Since these molecules are made as single or covalently attached polypeptides, losses incurred due to purification and conjugation of two distinct antibodies are eliminated. Furthermore, because bispecific fusion proteins are often small relative to full-length antibodies, tumor penetration may be enhanced and immunogenicity decreased. Whether bispecific fusion proteins made as covalently linked single chain molecules will have advantages over other types of genetically engineered bispecific molecules remains to be determined. Future studies must also address the stability, immunogenicity, and efficacy of these molecules in vivo. In addition, novel methods for developing bispecific fusion proteins have been postulated. For example, Keck and Huston[83] propose engineering a second antigen-binding site within an existing sc-Fv, thus generating what they refer to as a "chimeric bispecific antibody binding site." Procedures for making these types of reagents will continue to evolve as researchers and clinicians continue to develop imaginative uses for them.

ACKNOWLEDGMENTS

We are grateful to A. McCall, L. Weiner, and R. Fischer for sharing their unpublished results.

REFERENCES

1. Fanger, M. W., ed., 1995. *Bispecific Antibodies*, Austin, TX: R. G. Landes Company.
2. Fanger, M. W., P. M. Guyre. 1991. Bispecific antibodies for targeted cytotoxocity. *Trends in Biotech.* 9: 375–380.
3. Fanger, M. W., R. F. Graziano, P. M. Guyre. 1994. Production and use of anti-FcR bispecific antibodies. *Immunomethods* 4: 72–81.
4. van de Winkel, J. G. J., B. Bast, G. C. de Gast. 1997. Immunotherapeutic potential of bispecific antibodies. *Immunol. Today* 18: 562–564.
5. Perez, P., R. W. Hoffman, S. Shaw, J. A. Bluestone, D. M. Segal. 1985. Specific targeting of cytotoxic T cells by anti-T3 linked to anti-target cell antibody. *Nature* 316: 354–356.
6. Scott, C. F. Jr., J. M. Lambert, R. S. Kalish, C. Morimoto, S. F. Schlossman. 1988. Human T cells can be directed to lyse tumor targets through the alternative activation/TII-E rosette receptor pathway. *J. Immunol.* 140: 8–14.
7. van de Griend, R. J., R. L. Bolhuis, G. Stoter, R. C. Roozemond. 1987. Regulation of cytolytic activity in CD3− and CD3+ killer cell clones by monoclonal

antibodies (anti-CD16, anti-CD2, anti-CD3) depends on subclass specificity of target cell IgG-FcR. *J. Immunol.* 138: 3137–3144.

8. Segal, D. M., D. P. Snider. 1989. Targeting and activation of cytotoxic lymphocytes. *Chem. Immunol.* 47: 179–213.
9. Segal, D. M., C. E. Urch, A. J. T. George, C. R. Jost. 1992. Bispecific antibodies in cancer treatment. In *Biologic Therapy of Cancer*, Vol. 2, ed. V. T. DeVita, Jr., S. Hellman, S. A. Rosenberg, Philadelphia: J. B. Lippincott Co., p. 1.
10. Segal, D. M., T. Bakacs, C. R. Jost, I. Kurucz, G. Sconocchia, J. A. Titus. 1995. T cell-targeted cytotoxicity. In *Bispecific Antibodies*, ed. M. W. Fanger, Austin, TX: R. G. Landes Company, pp. 27–42.
11. Gorter, A., K. M. Kruse, P. I. Schrier, G. J. Fleuren, R. J. van de Griend. 1992. Enhancement of the lytic activity of cloned human CD8 tumour-infiltrating lymphocytes by bispecific monoclonal antibodies. *Clin. Exp. Immunol.* 87: 111–116.
12. Reid, I., J. Lundy, J. H. Donohue. 1992. Enhancement of in vitro tumor-infiltrating lymphocyte cytotoxicity by heteroconjugated antibodies. *J. Immunol.* 148: 2630–2635.
13. Qian, J-H., J. Titus, S. M. Andrews, D. Mezzanzanica, M. A. Garrido, J. R. Wunderlich, D. M. Segal. 1991. Human peripheral blood lymphocytes targeted with bispecific antibodies release cytokines that are essential for inhibiting tumor growth. *J. Immunol.* 146: 3250–3256.
14. Segal, D. M., J. H. Qian, S. M. Andrew, J. A. Titus, D. Mezzanzanica, M. A. Garrido, J. R. Wunderlich. 1991. Cytokine release by peripheral blood lymphocytes targeted with bispecific antibodies, and its role in blocking tumor growth. *Ann. N. Y. Acad. Sci.* 636: 288–294.
15. Jung, G., H. J. Muller-Eberhardt. 1988. An in vitro model for tumor immunotherapy with antibody heteroconjugates. *Immunol. Today* 9: 257–260.
16. Jung, G., U. Freimann, Z. Von Marschall, R. A. Reisfeld, W. Wilmanns. 1991. Target cell-induced T cell activation with bi- and trispecific antibody fragments. *Eur. J. Immunol.* 21: 2431–2435.
17. Anderson, C. A., P. M. Guyre, J. C. Whitin, D. H. Ryan, R. J. Looney, M. W. Fanger. 1986. Monoclonal antibodies to Fc receptors for IgG on human mononuclear phagocytes: antibody characterization and induction of superoxide production in a monocyte cell line. *J. Biol. Chem.* 261: 12856–12864.
18. Guyre, P. M., R. F. Graziano, B. A. Vance, P. M. Morganelli, M. W. Fanger. 1989. Monoclonal antibodies that bind to distinct epitopes on FcγRI are able to trigger receptor function. *J. Immunol.* 143: 1650–1655.
19. Fanger, M. W., L. Shen, R. F. Graziano, P. M. Guyre. 1989. Cytotoxicity mediated by human Fc receptors for IgG. *Immunol. Today* 10: 92–99.
20. Shen, L., P. M. Guyre, C. L. Anderson, M. W. Fanger. 1986. Heteroantibody-mediated cytotoxicity: antibody to the high affinity Fc receptor for IgG mediates cytotoxicity by human monocytes that is enhanced by interferon-gamma and is not blocked by human IgG. *J. Immunol.* 137: 3378–3382.
21. Fanger, N. A., K. Wardwell, L. Shen, T. F. Tedder, P. M. Guyre. 1996. Type I (CD64) and type II (CD32) Fcγ receptor-mediated phagocytosis by human blood dendritic cells. *J. Immunol.* 157: 541–548.
22. Graziano, R. F., C. Somasundaram, J. Goldstein. 1995. The production of bispecific antibodies. In *Bispecific Antibodies*, ed. M. W. Fanger, Austin, TX: R. G. Landes Company, pp. 1–26.

23. Nisonoff, A., M. M. Rivers. 1961. Recombination of a mixture of univalent antibody fragments of different specificity. *Arch. Biochem. Biophys.* 93: 460–462.

24. Nolan, O., R. O'Kennedy. 1990. Bifunctional antibodies: concept, production and applications. *Biochim. Biophys. Acta* 1040: 1–11.

25. Weiner, L. M., J. I. Clark, M. Davey, S. Wei, I. Garcia de Palzzo, D. B. Ring, R. K. Alpaugh. 1995. Phase I trial of 2B1, a bispecific monoclonal antibody targeting c-erB-2 and FcγRIII. *Cancer Res.* 55: 4586–4593.

26. Deo, Y. M., R. F. Graziano, R. Repp, J. G. J. van de Winkel. 1997. Clinical significance of IgG Fc receptors and FcγR-directed immunotherapies. *Immunol. Today* 18: 127–135.

27. King, D. J., J. A. Adair, S. Angal, D. C. Low, K. A. Proudfoot, J. C. Lloyd, M. Bodmer, G. T. Yarranton. 1992. Expression, purification and characterization of a mouse:human chimeric antibody and chimeric Fab′ fragment. *Biochem. J.* 281: 317–323.

28. Riechmann, L., J. Foote, G. Winter. 1988. Expression of an antibody Fv fragment in myeloma cells. *J. Mol. Biol.* 203: 825–828.

29. Skerra, A., A. Plückthun. 1988. Assembly of a functional immunoglobulin Fv fragment in *Escherichia coli*. *Science* 240: 1038–1041.

30. Yokota, T., D. E. Milenic, M. Whitlow, J. Schlom. 1992. Rapid tumor penetration of a single-chain Fv and comparison with other immunoglobulin forms. *Cancer Res.* 52: 3402–3408.

31. Bird, R. E., K. D. Hardman, J. W. Jacobson, S. Johnson, B. M. Kaufman, S-M. Lee, T. Lee, S. H. Pope, G. S. Riordan, M. Whitlow. 1988. Single-chain antigen-binding proteins. *Science* 242: 423–426.

32. Huston., J. S., D. Levinson, M. Mudgett-Hunter, M-S. Tai, J. Novotny, M. N. Margolies, R. Ridge, R. E. Bruccoleri, E. Haber, R. Crea, H. Oppermann. 1988. Protein engineering of antibody binding sites: recovery of specific activity in an anti-digoxin singe-chain Fv analogue produced in *Escherichia coli*. *Proc. Natl. Acad. Sci. USA* 85: 5879–5883.

33. Reiter, Y., I. Pastan. 1996. Antibody engineering of recombinant Fv immunotoxins for improved targeting of cancer: disulfide-stabilized Fv immunotoxins. *Clin. Cancer Res.* 2: 245–252.

34. Reiter, Y., U. Brinkmann, B. Lee, I. Pastan. 1996. Engineering antibody Fv fragments for cancer detection and therapy: disulfide-stabilized Fv fragments. *Nature Biotechnol.* 14: 1239–1245.

35. George, A. J. T., S. M. Andrew, P. Perez, P. J. Nicholls, J. S. Huston, D. M. Segal. 1991. Production of a bispecific antibody by linkage of two recombinant single chain Fv molecules. *J. Cell. Biochem. Supp.* 15E: 127.

36. Kostelny, S. A., M. S. Cole, J. Y. Tso. 1992. Formation of a bispecific antibody by the use of leucine zippers. *J. Immunol.* 148: 1547–1553.

37. Pack, P., A. Plückthun. 1992. Miniantibodies: Use of amphiphatic helices to produce functional, flexibly linked dimeric F_V fragments with high avidity in *Escherichia coli*. *Biochemistry* 31: 1579–1584.

38. Pack, P., M. Kujau, V. Schroeckh, U. Knupfer, R. Wenderoth, D. Riesenberg, A. Plückthun. 1993. Improved bivalent miniantibodies, with identical avidity as whole

antibodies, produced by high cell density fermentation of *Escherichia coli*. *Bio/Technology* 11: 1271–1277.

39. Holliger, P., T. Prospero, G. Winter. 1993. "Diabodies": Small bivalent and bispecific antibody fragments. *Proc. Natl. Acad. Sci. USA* 90: 6444–6448.
40. Holliger, P., G. Winter. 1997. Diabodies: small bispecific antibody fragments. *Cancer Immunol. Immunother.* 45: 128–130.
41. Mallender, W. D., E. W. Voss. 1994. Construction, expression, and activity of a bivalent bispecific single-chain antibody. *J. Biol. Chem.* 269: 199–206.
42. Schmuck, M., I. Pilz, M. Hayn, H. Esterbauer. 1986. Investigation of cellobiohydrolase from *Trichoderma reesei* by small angle x-ray scattering. *Biotechnol. Lett.* 8: 397–402.
43. Abuja, P. M., M. Schmuck, I. Pilz, P. Tomme, M. Claeyssens, H. Esterbauer. 1988. Structural and functional domains of cellobiohydrolase I from *Trichoderma reesei*. *Eur. Biophys. J.* 15: 339–342.
44. Takkinen, K., M. L. Lankkanen, D. Sizmann, K. Alfthan, T. Immonen, L. Vanne, M. Kaartinen, J. K. C. Knowles, T. T. Teeri. 1991. An active single-chain antibody containing a cellulase linker domain is secreted by *Escherichia coli*. *Protein Eng.* 4: 837–841.
45. Mallender, W. D., S. T. Ferreira, E. W. Voss, T. Coelho-Sampaio. 1994. Inter-active site distance and solution dynamics of a bivalent-bispecific single-chain antibody molecule. *Biochemistry* 33: 10100–10108.
46. Gruber, M., B. A. Schodin, E. R. Wilson, D. M. Kranz. 1994. Efficient tumor cell lysis mediated by a bispecific single chain antibody expressed in *Escherichia coli*. *J. Immunol.* 52: 5368–5374.
47. Kranz, D. M., M. Gruber, E. R. Wilson. 1995. Properties of bispecific single chain antibodies expressed in *Escherichia coli*. *J. Hematotherapy* 4: 403–408.
48. de Jonge, J., J. Brissink, C. Heirman, C. Demanet, O. Leo, M. Moser, K. Thielemens. 1995. Production and characterization of bispecific single-chain antibody fragments. *Mol. Immunol.* 32: 1405–1412.
49. de Jonge, J., C. Heirman, M. De Veerman, S. Van Meirvenne, C. Demanet, J. Brissinck, K. Thielemens. 1997. Bispecific antibody treatment of murine B cell lymphoma. *Cancer Immunol. Immunother.* 45: 162–165.
50. Huston, J. S., M. Mudgett-Hunter, M-S. Tai, J. McCartney, F. Warren, E. Haber, H. Oppermann. 1991. Protein engineering of single-chain Fv analogs and fusion proteins. *Meth. Enzymol.* 203: 46–88.
51. Neri, D., M. Momo, T. Prospero, G. Winter. 1995. High-affinity antigen binding by chelating recombinant antibodies (CRAbs). *J. Mol. Biol.* 246: 367–373.
52. Kurucz, I., J. A. Titus, C. R. Jost, C. M. Jacobus, D. M. Segal. 1995. Retargeting of CTL by an efficiently refolded bispecific single-chain Fv dimer produced in bacteria. *J. Immunol.* 154: 4576–4582.
53. Slamon, D. J., W. Godolphin, L. A. Jones, J. A. Holt, S. G. Wong, D. E. Keith, W. J. Levin, S. G. Stuart, J. Udove, A. Ullrich, M. F. Press. 1989. Studies of the HER2/*neu* proto-oncogene in human breast and ovarian cancer. *Science* 244: 707–712.
54. Wels, W., I. M. Harwerth, M. Mueller, B. Groner, N. E. Hynes. 1992. Selective inhibition of tumor cell growth by a recombinant single-chain antibody-toxin specific for the erbB-2 receptor. *Cancer Res.* 52: 6310–6317.

55. Wels, W., R. Beerli, P. Hellman, M. Schmidt, B. M. Marte, E. S. Kornilova, A. Hekele, J. Mendelsohn, B. Groner, N. E. Hynes. 1995. EGF receptor and $p185^{erbB-2}$-specific single-chain antibody toxins differ in their cell-killing activity on tumor cells expressing both receptor proteins. *Int. J. Cancer* 60: 137–144.

56. Schmidt, M., N. E. Hynes, B. Groner, W. Wels. 1996. A bivalent single-chain antibody-toxin specific for erbB-2 and the EGF receptor. *Int. J. Cancer* 65: 538–546.

57. Schmidt, M., W. Wels. 1996. Targeted inhibition of tumour cell growth by a bispecific single-chain toxin containing an antibody domain and TGFα. *Br. J. Cancer* 74: 853–862.

58. Schier, R., J. D. Marks, E. J. Wolf, G. Apell. C. Wong, J. E. McCartney, M. A. Bookman, J. S. Huston, L. L. Houston, L. M. Weiner, G. P. Adams. 1995. In vitro and in vivo characterization of a human anti-c-erbB-2 single-chain Fv isolated from a filamentous phage antibody library. *Immunotechnology* 173–181.

59. McCall, A. M., A. R. Amoroso, L. Zhang, R. Schier, J. D. Marks, L. M. Weiner. 1997. Improving the killing efficiency of bispecific agents through the construction of bispecific $scFv_2$ specific for HER2/*neu* and human FcγRIII. 8th annual IBC conference on Antibody Engineering, Coronado, CA. International Business Communications, Southborough, MA.

60. Hayden, M. S., P. S. Linsley, M. A. Gayle, J. Bajorath, W. A. Brady, N. A. Norris, H. Fell, J. A. Ledbetter, L. K. Gilliland. 1994. Single chain mono- and bispecific antibody derivatives with novel biological properties and anti-tumor activity from a COS cell transient expression system. *Therapeutic Immunol.* 1: 3–15.

61. Jost, C. R., J. A. Titus, I. Kurucz, D. M. Segal. 1996. A single-chain bispecific Fv_2 molecule produced in mammalian cells redirects lysis by activated CTL. *Molecular Immunol.* 33: 211–219.

62. Mack, M., G. Riethmüller, P. Kufer. 1995. A small bispecific antibody construct expressed as a functional single-chain molecule with high tumor cell cytotoxicity. *Proc. Natl. Acad. Sci. USA* 92: 7021–7025.

63. Mack, M., R. Gruber, S. Schmidt, G. Riethmüller, P. Kufer. 1997. Biologic properties of a bispecific single-chain antibody directed against 17-1A (EpCAM) and CD3. *J. Immunol.* 158: 3965–3970.

64. Kufer, P., M. Mack, R. Gruber, R. Lutterbüse, F. Zettl, G. Riethmüller. 1997. Construction and biological activity of a recombinant bispecific single-chain antibody designed for therapy of minimal residual colorectal cancer. *Cancer Immunol. Imunother.* 45: 193–197.

65. Verhaar, M. J., K. A. Chester, P. A. Keep, L. Robson, R. B. Pedley, J. A. Boden, R. E. Hawkins, R. H. J. Begent. 1995. A single chain Fv derived from a filamentous phage library has distinct tumour targeting advantages over one derived from a hybridoma. *Int. J. Cancer* 61: 497–501.

66. Schumann, D., S. Schillberg, S. Zimmermann, Y-C. Liao, M. Sack, R. Fischer. 1996. Expression and characterization of bispecific scFv-fragments produced in transgenic plants. 7th annual IBC conference on Antibody Engineering, Coronado, CA. International Business Communications, Southborough, MA.

67. Kipriyanov, S. M., G. Moldenhauer, G. Strauss, M. Little. 1997. Bispecific diabody for lysis of malignant human B cells. 8th annual IBC conference on Antibody

Engineering, Coronado, CA. International Business Communications, Southborough, MA.

68. Darveau, R. P., J. E. Somerville, H. P. Fell. 1992. Expression of antibody fragments in *Escherichia coli*. *J. Clin. Immunoassay* 15: 25–29.
69. Liu, M. A., S. R. Nussbaum, H. N. Eisen. 1988. Hormone conjugated with antibody to CD3 mediates cytotoxic T cell lysis of human melanoma cells. *Science* 239: 395–398.
70. Modjtahedi, H., C. Dean. 1994. The receptor for EGF and its ligands: expression, prognostic value and target for therapy in cancer (Review). *Int. J. Oncol.* 4: 277–296.
71. Owens, O. J., C. Stewart, I. Browne, R. E. Leake. 1991. Epidermal growth factor receptors (EFGR) in human ovarian cancer. *Brit. J. Cancer* 64: 907–910.
72. Gillies, S. D., J. S. Wesolowski, K.-L. Lo. 1991. Targeting human cytotoxic T lymphocytes to kill heterologous epidermal growth factor receptor-bearing cells. *J. Immunol.* 146: 1067–1071.
73. Goldstein, J., R. F. Graziano, K. Sundarapandiyan, C. Somasundaram, Y. M. Deo. 1997. Cytolytic and cytostatic properties of an anti-human FcγRI (CD64) x epidermal growth factor bispecific fusion protein. *J. Immunol.* 158: 872–879.
74. Graziano, R. F., J. Goldstein, K. Sundarapandiyan, C. Somasundaram, T. Keler, Y. M. Deo. 1997. Targeting tumor cell destruction with CD64-directed bispecific fusion proteins. *Cancer Immunol. Immunother.* 45: 124–127.
75. Graziano, R. F., P. R. Tempest, P. White, T. Keler, Y. Deo, H. Ghebremariam, K. Coleman, L. C. Pfefferkorn, M. W. Fanger, P. M. Guyre. 1995. Construction and characterization of a humanized anti-γ-Ig receptor type I (FcγRI) monoclonal antibody. *J. Immunol.* 155: 4996–5002.
76. Carraway, K. L., L. C. Cantley. 1994. A new acquaintance for erbB3 and erbB4: a role for receptor heterodimerization in growth signaling. *Cell* 78: 5–8.
77. Lemoine, N. R., D. M. Barnes, D. P. Hollywood, C. M. Hughs, P. Smith, E. Dublin, S. A. Prigent, W. J. Gullick, H. C. Hurst. 1992. Expression of the erbB3 gene product in breast cancer. *Br. J. Cancer* 66: 1116–1121.
78. Faksvåg Haugen, D. R., L. A. Akslen, J. E. Varhaug, J. R. Lillehaug. 1996. Expression of c-erbB-3 and c-erbB-4 proteins in papillary thyroid carcinomas. *Cancer Res.* 56: 1184–1188.
79. Sliwkowski, M. X., G. Schaefer, R. W. Akita, J. A. Lofgren, V. D. Fitzpatrick, A. Nuijens, B. M. Fendly, R. A. Cerione, R. L. Vandlen, K. L. Carraway III. 1994. Coexpression of erbB-2 and erbB-3 proteins reconstitutes a high-affinity receptor for heregulin. *J. Biol. Chem.* 269: 14661–14665.
80. Gosselin, E. J., K. Wardwell, D. R. Gosselin, N. Alter, J. L. Fisher, P. M. Guyre. 1992. Enhanced antigen presentation using human Fcγ receptor (monocyte/macrophage)-specific immunogens. *J. Immunol.* 149: 3477–3481.
81. Liu, C., J. Goldstein, R. F. Graziano, J. He, J. K. O'Shea, Y. Deo, P. M. Guyre. 1996. FcγRI-targeted fusion proteins result in efficient presentation by human monocytes of antigenic and antagonistic T cell epitopes. *J. Clin. Invest.* 98: 2001–2007.
82. Guyre, P. M., R. F. Graziano, J. Goldstein, P. K. Wallace, P. M. Morganelli, K. Wardwell, A. L. Howell. 1997. Increased potency of Fc-receptor-targeted antigens. *Cancer Immunol. Immunother.* 45: 146–148.

83. Keck, P. C., J. S. Huston. 1996. Symmetry of Fv architecture is conducive to grafting a second antibody binding site in the Fv region. *Biophys. J.* 71: 2002–2011.

84. Staerz, U. D., O. Kanagawa, M. J. Bevan. 1985. Hybrid antibodies can target sites for attack by T cells. *Nature* 314: 628–631.

85. Galandrini, R., N. Albi, G. Tripodi, D. Zarcone, A. Terenzi, A. Moretta, C. E. Grossi, A. Verlandi. 1993. Antibodies to CD44 trigger effector functions of human T cell clones. *J. Immunol.* 150: 4225–4235.

86. Sconocchia, G., J. A. Titus, D. M. Segal. 1994. CD44 is a cytotoxic trigger molecule in human peripheral blood NK cells. *J. Immunol.* 153: 5473–5479.

87. Moretta, A., A. Poggi, D. Pende, G. Tripodi, A. M. Orengo, N. Pella, R. Augugliaro, C. Bottino, E. Ciccone, L. Moretta. 1991. CD69-mediated pathway of lymphocyte activation: Anti-CD69 monoclonal antibodies trigger the cytolytic activity of different lymphoid effector cells with the exception of cytolytic T lymphocytes expressing T cell receptor α/β. *J. Exp. Med.* 174: 1393–1398.

88. Seth, A., L. Gote, M. Nagarkatti, P. S. Nagarkatti. 1991. T-cell receptor-independent activation of cytolytic activity of cytotoxic T lymphocytes mediated through CD44 and gp90MEL14. *Proc. Natl. Acad. Sci. USA* 88: 7877–7881.

89. Leo, O., M. Foo, D. M. Segal, E. Shevach, J. A. Bluestone. 1987. Activation of murine T lymphocytes with monoclonal antibodies: Detection on Lyt2+ cells of an antigen not associated with the T cell receptor complex but involved in T cell activation. *J. Immunol.* 139: 1214–1222.

90. Wallace, P. K., F. H. Valone, M. W. Fanger. 1995. Myeloid cell-targeted cytotoxicity of tumor cells. In *Bispecific Antibodies*, ed. M. W. Fanger, Austin, TX: R. G. Landes Company, pp. 43–76.

PART II

Fc FUSION PROTEINS

8

IMMUNOGLOBULIN FUSION PROTEINS

ALEJANDRO ARUFFO

Bristol Squibb Pharmaceutical Research Institute, Princeton, NJ 08543

8.1 INTRODUCTION

Rapid advances in molecular biology, protein expression, and protein purification techniques have allowed the preparation of novel chimeric polypetides with multiple functional domains. Among the most common of these are immunoglobulin (Ig) fusion proteins. These proteins consist of the constant regions of immunoglobulin, typically mouse or human, fused to an unrelated protein or protein fragment. Ig fusion proteins have become valuable laboratory reagents for the study of protein structure and function in vitro and in vivo and in some instances might have applications in a clinical setting. In this chapter the following topics are covered: (1) design of chimeric genes encoding Ig fusion proteins, (2) expression, purification, and characterization of the Ig fusion proteins, and (3) applications of Ig fusion proteins as laboratory tools and in the clinic.

8.2 Ig FUSION PROTEINS: DESIGN OF THE CHIMERIC GENES ENCODING Ig FUSION PROTEINS

The conceptual root of Ig fusion proteins comes from the application of molecular biology techniques to the manipulation of immunoglobulin genes[1–3] and the finding that many cell surface receptors contain extracellular

Antibody Fusion Proteins, Edited by Steven M. Chamow and Avi Ashkenazi
ISBN 0471-18358-X

domains that are homologous to the domains found in the immunoglobulin proteins. It was reasoned that it should be possible to generate chimeric genes in which the DNA fragments (genomic or cDNA) encoding the variable regions of an immunoglobulin heavy chain could be replaced with a DNA fragment encoding the Ig-like domains of other cell surface proteins. Among the first such DNA constructs were those encoding CD4-Ig fusion proteins.[4–6] In these constructs the DNA encoding the Ig heavy chain variable region was replaced with a DNA fragment encoding the extracellular domain of CD4 which is composed of four Ig-like domains. When placed in a mammalian expression vector, these chimeric genes directed the expression of CD4-Ig. It was proposed that these CD4-Ig fusion proteins could be used to slow down the progression of HIV infection by binding to virions via interactions with gp120 leading to virus neutralization. While these CD4-Ig fusion proteins have not been effective in the clinic, they have established that it is possible to replace the N-terminal Ig domain corresponding to the antigen-combining (variable) region of the antibody with an Ig-like domain from an unrelated protein and generate an Ig fusion protein in which both the receptor and Ig domains retain their biological activity. In recent years, a wide range of Ig fusion proteins have been made including Ig fusions of both type I and II cell surface proteins, with and without Ig-like domains, and Ig fusions of cytoplasmic proteins (for a review see Refs. 7 and 8). Thus, Ig fusion protein technology can be broadly applied to generate novel laboratory tools to study protein structure and function, as well as potentially novel drugs.

8.2.1 Construction of the Chimeric Gene Encoding the Ig Fusion Protein

In most instances the gene encoding the fusion protein will contain the cDNA or cDNA fragment encoding the polypetide of interest fused in frame with either a cDNA fragment or a genomic DNA fragment encoding the hinge (H), and CH2 and CH3 domains of an immunoglobulin heavy chain. The location of the DNA fragments encoding these two regions of the fusion protein relative to one another is dependent on the characteristics of the protein under study and is discussed in a following section. While this is the most common format for the preparation of chimeric genes encoding Ig fusion proteins, it should be noted that constructs containing CH2 and CH3 or only CH3 can also be generated.

Theoretically, these chimeric genes direct the expression of covalent Ig fusion protein homodimers. The covalent nature of the homodimer is driven by the presence of the cysteine (Cys) residues that form the immunoglobulin interchain disulfide bonds in the H domain of the Ig constant region. If the fusion protein is made with an Ig domain containing a mutant H domain in which the Cys residues have been replaced by serine (Ser) residues, the Ig fusion protein is likely to be expressed as a noncovalent homodimer. Practically, it has been observed that many of these fusion proteins form higher aggregates. For this reason, care should be taken to characterize the aggregation of the fusion protein.

8.2.2 Species, Isotype, and Functional Properties of the Ig Domain

The first consideration when designing the construction of the chimeric gene encoding an Ig fusion protein is its intended use(s). The use dictates the species and isotype of the Ig domain used in the construction of the fusion protein; it also determines whether the Ig domain should be wild-type or mutated in such a way that its own functional properties (complement and Fc-receptor binding) are altered. The choice of Ig domain also impacts the valance of the fusion protein and the type of Ig binding reagents available for purifying and manipulating the fusion protein (Box 8.1). These issues are perhaps best illustrated with a few examples:

8.2.2.1 Identification of Tissues Expressing the Ligand for a Novel Human Cell Surface Receptor. In this example, the Ig fusion protein is used to carry out immunohistology on a number of human tissues in an effort to identify tissues and cells expressing the receptor's ligand(s). In this case, the Ig domain should be derived from mouse IgG. This will ensure the fusion protein can be easily purified on a Protein A or G column and that the anti-IgG reagents used in the immunohistology study will not cross-react with tissues containing human B cells expressing Ig. Conversely, if the immunohistology is done using mouse tissues, the Ig domain for the fusion protein should be derived from human IgG. If binding is detected using the fusion proteins, care should be taken to demonstrate that this binding is not mediated by the Ig domain. This can be accomplished by preblocking Fc-receptors or using an Ig fusion protein containing an Ig domain with an isotype or point mutant that no longer binds Fc-receptors. Examples of this application include the identification of cells expressing the CD6, CD27, CD30, CD44, and CD62E ligands.[9–13]

8.2.2.2 Blocking Receptor-Ligand Interaction in vivo by Injecting Soluble Receptor. In this example, the physiologic role of the interaction between a mouse cell surface receptor and its ligand is studied in vivo by interrupting it with a receptor Ig fusion protein. In this case, the Ig domain should be derived from mouse IgG. The mouse IgG domain allows the facile purification of the fusion protein and is not antigenic when injected into mice. If a human IgG domain were used in this instance, the fusion protein would be very antigenic when injected into mice. Consequently, multiple dosing of a human IgG chimera would rapidly result in the generation of antihuman IgG antibodies that would dramatically reduce the half-life of the fusion protein in vivo and limit its efficacy. If the fusion protein is to be used in vivo, the isotype of the fusion protein should also be taken into consideration. Is it an isotype that can fix complement leading to the depletion of the cells expressing the receptor ligand? Is it an isotype that can bind Fc receptor resulting in unintended Fc

Box 8.1 Properties of the Human Immunoglobulin Fc Region.

There are five types of human immunoglobulin Fc regions with different effector and pharmacokinetic properties: IgG, IgA, IgM, IgD, and IgE. IgG is a monomeric molecule (single heterodimer) and is the most abundant immunoglobulin in serum (75–85% of serum Ig). IgG has the longest half-life in serum of any Ig (23 days) and a synthesis rate of 33 mg/kg/day. Unlike other Ig, IgG is efficiently recirculated following binding to FcRn (Fc receptor identified on neonatal intestine). There are four IgG subclasses G1, G2, G3, and G4, each of which have different effector functions. G1, G2, and G3 can bind C1q and fix complement while G4 can not. Even though G3 is able to bind C1q more efficiently than G1, G1 is more effective at mediating complement-directed cell lysis. G2 fixes complement very inefficiently. The C1q binding site in IgG is located at the carboxy terminal region of the CH2 domain. All IgG subclasses are capable of binding to Fc receptors (CD16, CD32, CD64) with G1 and G3 being more effective than G2 and G4. The FcR binding region of IgG is formed by residues located in both the hinge and the carboxy terminal region of the CH2 domain. Interactions between IgG and FcRn also allow the transfer of this immunoglobulin across the neonatal intestine and the placenta. IgA can exist both in a monomeric and dimeric form held together by a J-chain. IgA is the second most abundant Ig in serum (7–15%). IgA is the most heavily synthesized Ig (65 mg/kg/day) but is rapidly turned over in the serum where it has a half-life of 6 days. It is the most abundant Ig in secretions (tears, milk, etc.). IgA secretion is mediated by binding to pIgR, the polymeric Ig receptor that mediates the transport of polymeric Ig (IgA and IgM) across mucosal epithelium. IgA has three effector functions. It binds to an IgA specific receptor on macrophages and eosinophils, which drives phagocytosis and degranulation, respectively. It can also fix complement via an unknown alternative pathway. IgM is expressed as either a pentamer or a hexamer, both of which are held together by a J-chain. IgM is secreted during a primary immune response and represents 5–10% of serum Ig. IgM is synthesized at the rate of 7 mg/kg/day and has a serum half-life of 5 days. IgM binds weakly to C1q via a binding site located in its CH3 domain. Due to the multimeric nature of IgM, it can efficiently fix complement at low antigen densities, with the hexameric form being more efficient than the pentameric form. IgM can also bind C3b and mediate phagocytosis by binding to C3b receptors on macrophages. Like IgA, IgM binds to the pIgR and is transported across mucosal epithelium and is found in secretions. IgD is a monomeric molecule that represents 0.3% of serum Ig. It has a half-life of 3 days in serum and is synthesized at the rate of 0.4 mg/kg/day. To date, no effector functions have been ascribed to its Fc domain. IgE is a monomeric Ig and is the least abundant serum Ig (0.02%). It has a serum half-life of 2.5 days and it is synthesized at the rate of 0.016 mg/kg/day. IgE binds to two Fc receptors, the high-affinity FcεR1 and the low-affinity CD23. Binding of IgE to FcεR1 on basophils drives degranulation and results in the release of proinflammatory agents. Binding of IgE to CD23 triggers a down regulatory feedback loop that results in decreased IgE production.

receptor-mediated activities? These are important considerations which, if not taken into account, could lead to the incorrect interpretation of the in vivo effects of the fusion protein. For an extensive review of the functional properties of the human and murine Ig Fc region as well as detailed information on which residues are involved in Fc receptor and complement binding please see Refs. 14 and 15. It should be noted that the functional activity of the Ig domain of the fusion can be addressed by changing isotypes and/or using Ig domains with point mutations that abolish function. Examples of this application include the use of CTLA4-Ig[16] to examine the effect of blocking the interaction between CD28 and CTLA4 and their ligands B7-1 and B7-2, the use of CD40-Ig to study the interaction between CD40 and its ligand (CD154) in the immune response,[17] the use of CD44-Ig to examine the effect of blocking interaction between CD44 and hyaluronan on tumor cell growth,[18] and the use of L-selectin-Ig to study the role of this receptor in leukocyte recruitment to sites of inflammation[19] among others.

8.2.2.3 Identification of Ligand Domains and Residues Responsible for Receptor Binding. In this example, an Ig fusion of the receptor and an Ig fusion of the ligand are used to determine the ligand domains, and residues within the binding domain, which are responsible for ligand binding in vitro. For these studies a human IgG domain could be appended to the receptor and immobilized on plastic using an antihuman IgG antibody. The binding of wild-type or mutant (truncated or point mutant) mouse Ig fusion proteins of the ligand to the immobilized receptor could then be monitored by ELISA using antimouse IgG reagents that do not bind to human IgG. Conversely, these experiments could be done with a receptor mouse IgG and a ligand human IgG fusion protein. Examples of this application include the identification of the ALCAM domains and residues responsible for CD6 binding,[20] the identification of ICAM-3 domains involved in LFA-1 binding,[21] and the identification CD31 domains involved in receptor binding,[22] among others.

These three examples illustrate that while Ig fusion proteins are very useful tools, they have inherent limitations, which vary depending on the functions of the Ig domain used to prepare the fusion protein. For this reason, careful consideration should be given to the choice of Ig domain. Because these limitations have such a profound effect on the use of such fusion proteins, investigators who use this technology will usually generate more than one fusion protein, each containing a different Ig domain.

8.2.3 Placement of the Ig Domain in the Fusion Protein

Perhaps the greatest concern for an investigator using an Ig fusion protein is that the inclusion of the Ig domains would perturb the structure and function of the protein domain(s) under study. If this were to happen, observations regarding the structure and function of the protein being studied that were made with the fusion protein may not be physiologically relevant. For this reason, it is important to consider carefully how the protein or protein fragment to be studied and the Ig domains will be joined. If the three-dimensional structure of the protein to be studied is known or can be predicted with certainty, this information should be used to generate the fusion protein. The Ig domain should be placed at either the N- or C-terminus of the protein depending on which end of the protein is free and does not participate directly in determining its three-dimensional structure. Consideration should also be given to the location of the functional regions of the protein. The Ig domain should be added to the protein in the location least likely to affect the function of the native protein. However, in most instances nothing is known about the three-dimensional structure of the protein being studied or the location of its functional domains. Indeed, in many instances the Ig fusion protein is the tool used to explore the function of a novel protein. In these instances the following guidelines should be used.

8.2.3.1 Ig Fusions with the Extracellular Domain of Cell Surface Proteins. If the goal is to study the extracellular domain of either a type I or type II membrane protein, the DNA sequence encoding the Ig domain should replace those sequences that encode the transmembrane and cytoplasmic domain of the cell surface protein. In this class of proteins the cytoplasmic membrane is a sharp boundary separating structural elements on the inside and outside of the cell. Thus, placing an unrelated protein domain in this location will, most likely, not perturb the structure of the extracellular domain of the protein being studied. In the case of type I and II membrane proteins, the cDNA or genomic fragment encoding the extracellular domain of the protein of interest should be fused to a cDNA or genomic fragment encoding the constant domain of Ig in place of the sequences encoding the transmembrane and cytoplasmic domain of the surface protein. Care should be taken in the design of the nucleotide sequence at the junction of the extracellular domain and the Ig domain to ensure that the sequence encoding the Ig domain will be in frame with the sequences encoding the extracellular domain of the protein of interest. In the case of Ig fusion proteins with type I membrane proteins, the N-terminal secretory signal sequence is contained within the nucleotide sequence encoding the cell surface protein of interest. Unless there is a specific reason to use a signal sequence derived from another protein (poor protein expression, etc.), the natural N-terminal secretory signal sequence should be used. Doing so will ensure that the N-terminal secretory signal sequence is clipped from the mature protein in the same location as in the native cell surface protein, ensuring that the N-terminal amino acid sequence of the cell surface receptor and its secreted Ig fusion counterpart are identical (Fig. 8.1). For type II membrane proteins, an exogenously derived N-terminal secretory signal sequence must be added to the sequences encoding the constant region of the Ig domain which will in turn be located 5′ of the sequences encoding the extracellular domain of the surface protein (Fig. 8.1).[23]

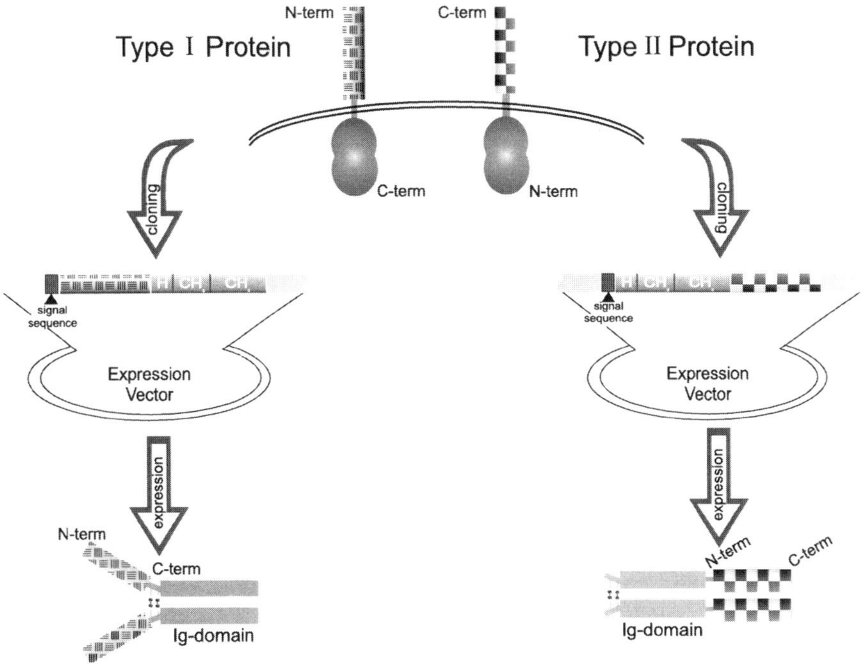

Figure 8.1 Construction of chimeric genes encoding Ig fusion proteins. Schematic representation of the chimeric genes encoding Ig fusion proteins of either a Type I or Type II cell surface protein. In both instances, the Ig constant domains are fused to the cDNA fragment encoding the extracellular domain of the protein to be studied, in place of the DNA encoding the transmembrane and extracellular domain of the cell surface protein.

If the Ig fusion protein is to be used to study a fragment of the extracellular domain of a cell surface protein, the domain organization of the extracellular region of the protein should be taken into account when designing the fusion protein. The sequences encoding the Ig region should be put in place of the sequences encoding one or more extracellular domains. Preserving the integrity of domains should minimize the possibility that inclusion of the Ig constant regions will affect the structure and function of the polypeptide fused to it. If the sequences encoding the extracellular domain fragment to be studied do not contain an N-terminal secretory signal sequence, one should be added.

8.2.3.2 Ig Fusions of Cytoplasmic Proteins. In principle, the Ig domain can be added to either the N- or C-terminus of the cytoplasmic protein to be studied;[24] however, placement of the Ig domain sequences should be dictated by which end of the protein is free. This will minimize the possibility that the Ig domain will perturb the structure and function of the cytoplasmic protein being studied. In those cases in which three-dimensional structure information is available for the protein of interest (or a closely related protein), this information should be used in the design of the fusion protein. If there is no information on the structure of the protein, it might be prudent to experiment with both N- and C-terminus Ig fusion proteins. As is the case for extracellular proteins, any information on the domain organization of the cytoplasmic proteins should be taken into account when designing the construction of the chimeric gene encoding a fusion protein containing only a fragment(s) of a cytoplasmic domain protein. If desired, an N-terminal secretory signal sequence can be added to the cytoplasmic-Ig fusion protein to direct its secretion, allowing its purification from the supernatant of transfected cells.

8.2.3.3 Ig Fusions of Secreted Proteins. In most cases, secreted proteins have an N-terminal secretory signal sequence that is cleaved from the mature protein.[25] In order to leave the N-terminus of the secreted Ig fusion protein intact, the Ig domain is placed at the C-terminus of the fusion protein. From a structural standpoint, however, this might not be the most favorable place for the Ig domain. Thus, as in the case of the cell surface and cytoplasmic proteins, any available information on the three-dimensional structure of the secreted protein should be used in the design of the chimeric protein.

While Ig fusion protein technology has been used successfully to study the structure and function of many proteins (>70 different proteins to date), it is not a panacea. In many instances it is not possible to append Ig domains onto a polypeptide without affecting its structure and function. Each case is unique and investigators using this technology should carefully characterize the Ig fusion proteins that they prepare to ensure that their activity is as close to physiological as possible. In this regard, if the investigator finds that a given Ig fusion construct is unusable due to a perturbation of the structure and/or function of the protein under study, changing the location of the Ig domain (for example by adding a linker between the polypetide being studied and the Ig domain) will sometimes ameliorate this problem.

8.2.4 Cleavable Ig Fusion Proteins

For some applications it is convenient to have an Ig domain for purification and other purposes (histology, ADCC, etc.), however, the Ig domain can also interfere with additional experiments planned for that protein. This has led to the construction of Ig fusion proteins that have an engineered protease cleavage site at the junction between the polypeptide encoding the protein of interest and the Ig domain (Fig. 8.2). Two types of cleavable Ig fusion proteins have been described: one in which the cleavage site is located at the Ig-domain junction,[26,27] and the other in which the proteinase site is located after the hinge domain of the Ig. Proteinase digestion of the former leads to the separation of Ig domain and the polypeptide to which it has been fused. In the latter case, if the hinge sequence of the Ig domain being used has wild-type

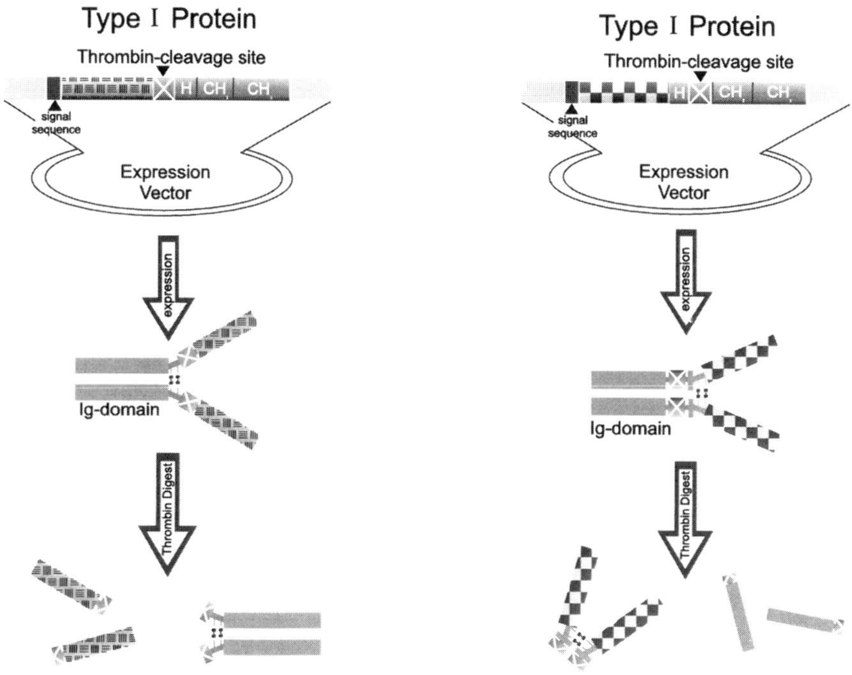

Figure 8.2 Construction of chimeric genes containing cleavable Ig domains. Schematic representation of Ig fusion proteins containing throbin cleavage sites upstream or downstream of the hinge (H) domain. In the former case, treatment of the fusion protein with thrombin results in the release of the monomer of the polypeptide fused to the Ig constant region. In the latter, thrombin treatment results in the release of a dimer of the polypeptide fused to the Ig constant region.

sequence, proteinase cleavage results in a covalent dimer of the protein of interest that no longer has the functional activities contained within the CH2 and CH3 domains of the Ig constant domain. Examples of protease recognition sites engineered at the junction of Ig fusion proteins domains include thrombin[26] and Genenase I.[27] It should be noted that the hinge region of human IgG1 contains the sequence E-P-K, which is similar to a thrombin cleavage site in actin.[28] In some Ig fusion proteins this nonclassical thrombin site is susceptible to enzyme digestion,[29] allowing the separation of the IgG1 domains from the polypeptide to which it has been fused.

8.2.5 Additional Considerations

While the first concern with the use of the Ig fusion protein to study protein structure and function is the effect the Ig domain will have on the polypeptide being studied, consideration should also be given to the impact that the addition of a foreign polypeptide will have on the Ig domain structure and function. To prevent potentially unwanted domain interactions in the chimeric proteins, small spacers (typically Ser-Gly rich polypeptides) can be included at the junction between the Ig and the polypeptide fused to it. However, careful characterization of the Ig domain of the fusion protein should be undertaken prior to its use.

8.2.6 Summary of Design Considerations

The intended use of the Ig fusion protein is the most important issue to be considered when designing the chimeric gene encoding the Ig fusion protein. Use will dictate the species, isotype, and functional properties of the Ig domain to be used in the construction. This choice affects the type of reagents available to study and manipulate the fusion protein, as well as the overall activity of the protein. Then consideration should be given to the placement of the Ig domain in the fusion protein. The Ig domain should be placed in the location where it is least likely to perturb the structure and function of the protein being studied. When an Ig fusion of a type I cell surface protein or a secreted protein is designed, it is preferable to use the endogenous N-terminal signal sequence to ensure that the fusion protein will have N-terminal sequence identical to that of the native protein. Just as the presence of the Ig domain can perturb the structure and function of the polypeptide to which it is fused, it is possible that the additional sequence might perturb the structure and function of the Ig domain of the fusion protein. Thus, it is important that this domain be fully characterized to ensure that the fusion protein has the expected activities (binding Protein A, complement, Fc receptors, etc.)

8.3 Ig FUSION PROTEINS: EXPRESSION, PURIFICATION, AND CHARACTERIZATION

This section reviews the methods used to produce small quantities of proteins for research use. Details of the techniques required for large-scale expression are discussed in Chapter 10.

8.3.1 Expression of the Ig Fusion Protein

The chimeric gene encoding the Ig fusion protein is typically subcloned into a mammalian expression vector designed for either transient and/or stable expression in mammalian cell lines (Fig. 8.3). Regardless of the final method of protein production, it is always prudent to test the construct by transient expression in an easily transfected cell line such as the SV40 transformed African green monkey kidney cell line (COS). The goal of this test is to ensure that the expression vector directs the expression of an Ig fusion protein of the expected molecular weight. Also, if the fusion protein is expected to be secreted,

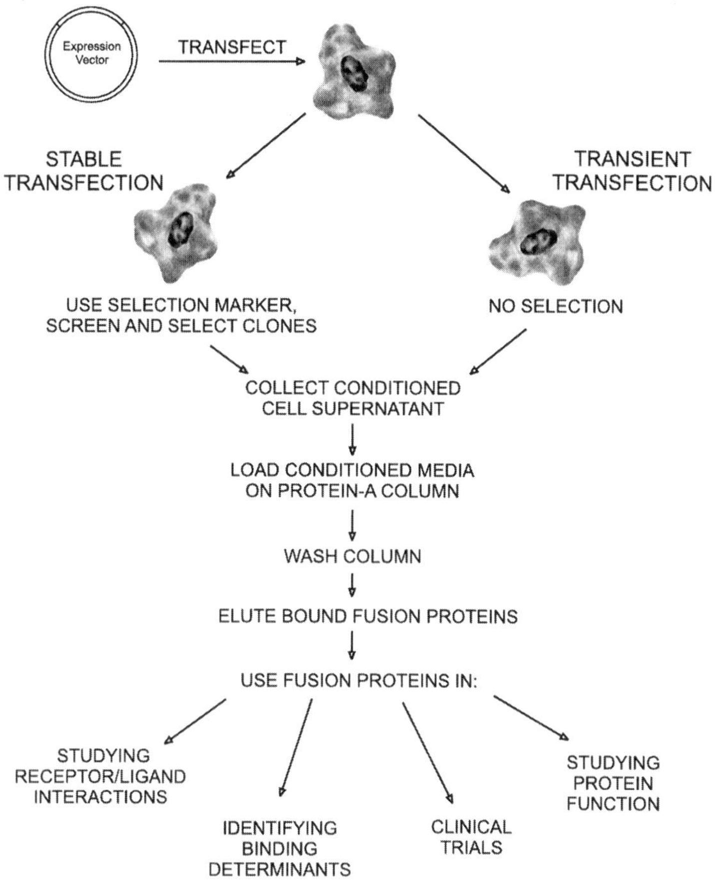

Figure 8.3 Production, purification, and use of the Ig fusion protein. Flow diagram showing how the plasmid encoding the Ig fusion protein directs either the stable or transient expression of the fusion protein. The fusion protein is purified from the spent supernatant of these cells and used as indicated.

a sample of spent tissue culture media can be tested for the presence of protein with reagents directed to the Ig moiety.[30]

In a typical experiment, 48 h following the transfection of cells with the mammalian expression vector, the cells are fixed, permeabilized, and the intracellular expression of the Ig fusion protein examined by immunofluoresence using fluorescein conjugated anti-Ig antibodies. If the fusion protein cannot be detected intracellularly, the most likely explanation is that there is a problem with the chimeric gene, the sequence of which should be checked to ensure that the gene fragments that make up the chimeric gene are in frame and correctly located between the promoter and sequences required for mRNA polyadenylation. Less likely, but possible, is that the fusion protein is made at very low levels, and/or is very unstable. While the first problem is easy to fix, the second is difficult. The researcher can begin to investigate if the problem has to do with gene expression or protein stability by determining if the transfected cells produce mRNAs encoding the fusion protein. If they do, then changes in the design of the gene encoding the chimeric protein can be made to enhance expression and/or stability. Potential changes would include changing the location of the junction between the polypeptide being studied and the Ig domain and/or changing the N-terminal signal sequence. Unfortunately, there are no guidelines that can be systematically applied to address this problem, leaving the investigator no alternative but to change the design of the chimeric gene and hope that these changes lead to a more stable protein product.

If the Ig fusion protein is expected to be secreted, a sample of the tissue culture media from the transfected cells can be examined 48 to 72 h posttransfection for the presence of the Ig domain. This can be done by metabolically labeling the protein, purifying the fusion protein from the media by adsorption and elution from an Ig affinity matrix (Protein A, anti-IgM sepharose, etc.), and analysis of the purified protein by SDS-PAGE and autoradiography. Alternatively, the unlabeled protein can be detected by Western blotting or ELISA following affinity matrix purification.

8.3.2 Purification of the Ig Fusion Proteins

Most of the Ig fusion proteins to date have been made with the constant domains of IgG. This has allowed the purification of these proteins by adsorption and elution from a Protein A or Protein G affinity matrix. However, this technique results in the purification of all IgG-containing proteins found in the spent supernatant of the transfected cells. Thus, if bovine serum were added to the tissue culture media, Protein A or Protein G purification would result in the isolation of both the Ig fusion protein and the bovine serum-derived Ig. If this isolation is a concern, the transfected cells should be placed in serum-free media. The method of elution from the affinity matrix should also be considered. Low or high pH buffers are often used to elute the fusion protein from the affinity matrix. However, these protocols can lead to an irreversible denaturation of the binding and/or active site of the non-Ig portion of the fusion protein, rendering the fusion protein unusable for most applications. Thus if at all possible, mild elution conditions should be employed when eluting the fusion protein. For example, imidazole-containing buffers can often be used with good results.[31] However, a variety of purification protocols may need to be evaluated for each protein. Because of the possibility that the purification protocol might affect the activity of the non-Ig portion of the fusion protein, the investigator should examine the intended use of the fusion protein and avoid protein purification if not needed. For many applications it is not necessary to purify the fusion protein.

8.3.3 Characterization of the Ig Fusion Proteins

It is prudent that the characteristics of the fusion protein relative to the native protein from which it is derived be examined prior to use. These include function (ligand binding if a receptor, enzymatic activity, etc.), structure (ability to be recognized by mAb and in particular non-blotting mAb which recognize conformational epitopes), and its posttranslational modification (glycosylation, phosphorylation, etc.). In many instances there is little or no information on the function of the native protein, indeed in many cases Ig fusion proteins have been used as a tool to investigate the function of proteins. Also, the number of reagents available to characterize the fusion protein is frequently limited. Yet, it is important that as many aspects as possible of the structure and posttranslational processing of the fusion protein be examined prior to use. It should be noted that, at least in some cases, it is possible to manipulate the posttranslational modification of the fusion protein by transfecting the DNA encoding it into a different cell line and/or by cotransfecting plasmids encoding enzymes involved in posttranslational modification of proteins into the COS cells.

Likewise, the structure and function and posttranslational modification of the Ig domain of the fusion protein should be examined. This would include testing the ability of this domain to interact with Protein A or G, complement, Fc-receptors, and anti-Ig monoclonal and polyclonal reagents. While robust, the structure and function of the Ig domain can also be perturbed by presence of the adjacent polypeptide.

Consideration should also be given to the aggregation state of the fusion protein. For example, although it would be expected that Ig fusion proteins with IgG domains would be homodimers, many of these proteins form higher aggregates (tetramers, hexamers, etc.). If the aggregation state of the fusion protein will impact its use or the interpretation of the data obtained with it, it is important that it be examined directly before using the fusion protein. For example, if the fusion protein will be used in vivo, its aggregation state will affect its pharmacokinetics with higher aggregates having a shorter half-life than a homodimer.

8.3.4 Summary of Ig Fusion Protein Expression, Purification, and Characterization

While Ig fusion proteins can be made in a number of different expression systems, the chimeric gene can easily be tested by transient expression in a suitable mammalian cell line such as COS. The Ig domain can be used as a tag polypeptide both to examine the level of expression and for purification. The investigator should be aware that sometimes the gene encoding the chimeric protein encodes the desired protein, however this protein is made but not secreted. In this instance, changing the N-terminal signal sequence and/or the location of the junction between the two proteins may result in secretion of the protein. The fusion protein should be purified with reagents directed to the Ig domain. Typically Protein A or G are used, however antibody columns can be used. Regardless of the affinity matrix used to purify the fusion protein, whenever possible mild elution conditions should be used to avoid denaturation of the fusion proteins. Once the protein has been prepared and purified, it is important that as many aspects of its structure and function be examined. Also, as the protein will be used over time, it is prudent to test its stability. Often Ig fusion proteins will aggregate over time affecting both their in vitro and in vivo properties.

8.4 Ig FUSION PROTEINS: APPLICATIONS

The use of the Ig fusion protein technology to study protein structure and function has been very successful and growing steadily since this technology was introduced almost 10 years ago. However, while there was early excitement about the possible application of this technique to develop novel drugs, it has only been recently that these fusion proteins are showing promise as potential drugs in the clinic. This section reviews the use of the Ig fusion protein as a laboratory tool and in the clinic.

8.4.1 Ig Fusion Proteins as Laboratory Tools

To date, the fusions of >70 different proteins with Ig constant domains have been reported in the literature. In most of these cases, the Ig fusion proteins have been used to identify novel receptor ligand interactions, map the binding determinants of proteins, and study the functional consequences of a receptor ligand interaction.

8.4.1.1 Use of Ig Fusion Proteins to Identify Novel Receptor Ligand interactions. Ig fusion protein technology has probably had its most significant scientific impact in this application. Ig fusion proteins of cell surface receptors or secreted proteins have been used to identify tissues or cells expressing molecules that can bind to the non-Ig moiety of the fusion protein, leading in many cases to its molecular characterization. The number of cases in which this technology has been successfully applied continues to grow steadily.[7,8] The initial application of this technology for this purpose was in the identification of tissue distribution and molecular identification of the ligands for the cell adhesion molecules known as the selectins[12] and CD44.[9] In these cases, Ig fusion proteins of the extracellular domains of E-selectin, L-selectin and P-selectin and CD44 were used to identify cells and tissues that express the ligands. These studies were then extended to identify their carbohydrate ligands as well as proteins that are decorated with these carbohydrates. These same fusion proteins were also used to determine which domains of the selectin proteins were required for ligand binding, demonstrating that this technology could also provide important information about the activity of different domains of a protein (see Section 8.4.1.2). Other examples of the use of Ig fusion proteins to identify novel receptor ligand pairs include the identification of the Htk, REK7, CD22, and Sp,[32–35] among others.

These early studies also showed that Ig fusion protein technology can result in unique laboratory tools. As was mentioned above, CD44-Ig was used to demonstrate that CD44 is a receptor for the glycosaminoglycan hyaluronan. Hyaluronan, due to its conserved structure across species, is nonimmunogenic.[9] Thus antihyaluronan antibodies of good quality, which can be used to study the tissue distribution of this very important molecule, have not been available. CD44-Ig was the first antibody-like reagent that became available to study hyaluronan, and it led to the finding that CD44 is a hyaluronan receptor.

Another example of a fusion protein that was initially used in a ligand identification study but that also was found to be a reagent with unique properties is CD6. CD6-Ig was used to identify ALCAM as a CD6 ligand. Repeated attempts to obtain mouse antihuman ALCAM mAb that blocked CD6-ALCAM were unsuccessful. When the region of ALCAM responsible for CD6 binding was mapped and its amino acid sequence compared in human and mouse ALCAM it was found to be identical. This provided a molecular explanation of why it was not possible to generate mouse antihuman ALCAM antibodies that bound to the region of ALCAM responsible for CD6 binding. Therefore CD6-Ig is the only anti-ALCAM reagent currently available for blocking CD6-ALCAM binding.[13]

8.4.1.2 Use of Ig Fusion Proteins to Identify Binding Determinants. Ig fusion protein technology has also been used successfully to map regions as well as individual residues in a protein involved in binding. The utility of the Ig-fusion proteins to map binding determinants was recognized early on and was first applied to mapping the domain requirements for the ligand binding for E-selectin.[12] In these studies, Ig fusion proteins of E-selectin and L-selectin and chimeric E-/L-selectin proteins containing different E-selectin and L-selectin domain combinations were used to show that the ligand binding specificity of the selectins is mediated predominantly by the amino terminal C-type lectin domain and the adjacent EGF-like domain is needed for the selectin ligand binding. The use of Ig fusion proteins to examine the activity of the different protein domains is now well established. Other examples of the use of this technology to map functional domains include studies on CD31[22] and human natriuretic peptide receptor,[36,37] among others.

Ig fusion proteins containing single point mutations have also been used to map the location of residues involved in binding. For example, Ig fusion proteins of E- and P-selectin containing single point mutations in the lectin domain of the selectins were used to identify residues that are key to the interaction between the selectins and their carbohydrate ligands. Similarly, Ig fusion proteins of CD22, gp39, CD40, and Fas, among others, were used to identify residues that are key to the interaction between these receptors and their ligands.[38–42]

8.4.1.3 Use of Ig Fusion Proteins to Study Protein Function. Ig fusion proteins have been used to study protein function. Examples include the use of CD80 and CD86 fusion proteins to characterize the signals sent by the two counter receptors for these proteins, CD28 and CTLA4, respectively.[43] Similarly, an Ig fusion protein of VCAM-1 was used to show that the interaction between VCAM-1 and its receptor VLA-4 results in the trigger leukocyte activation.[44] The use of Ig fusion proteins to study protein function has now become a widely implemented technique that has been applied to study the function of Heregulin, and Trk A, B, and C, among others.[25,26]

8.4.1.4 Heterodimeric Ig Fusion Proteins. The ability of Ig fusion proteins to assemble in the cell to form a homodimer can be exploited to generate heterodimeric Ig fusion proteins. In this case plasmids encoding two different Ig fusion proteins can be cotransfected into a mammalian cell line. These cells then direct the expression of three different proteins: the two homodimeric Ig fusion proteins encoded by each of the chimeric genes and the heterodimer in which the two arms of the fusion protein are encoded by the two different transfected chimeric genes. These heterodimeric fusion proteins can be used to study the effects of cross-linking the molecules that bind to each of the two proteins fused to the Ig domains. Examples of this type of heterodimeric Ig fusion protein include LFA-3/VCAM-1-Ig and LFA-3/ICAM-1-Ig.[46] Similar techniques can be used to generate heterodimers between an Ig fusion protein and an immunoglobulin heavy and light chain. An example of such a bispecific molecule is the anti-CD3/CD4-Ig fusion protein. In this heterodimeric construct, one arm of this is derived from the heavy and light chain of a humanized anti-CD3 mAb. The other arm is derived from a CD4-Ig. This heterodimeric fusion protein was designed to redirect $CD8^+$ CTLs to kill the remaining human immunodeficiency virus (HIV) infected cells in an HIV-infected patient.[47]

8.4.2 Ig Fusion Proteins as Therapeutics

The use of Ig fusion proteins in a clinical setting has yet to be realized; however, there are indications that in the near future some Ig fusion proteins will become pharmaceutical agents (see, for example, Chapter 10, Table 10.1). This expectation comes from the very promising results obtained in human clinical trials of the tumor necrosis factor receptor (TNFR) and CTLA4-Ig. Recently it has been reported that an Ig fusion of the p75 subunit of the TNFR has shown significant benefit in the treatment of rheumatoid arthritis,[48] while an Ig fusion of the p55 subunit of the TNFR is being tested in patients with severe sepsis and septic shock[49] for a detailed discussion of pSSTNFR-IgG, see Chapter 9). Also, an Ig fusion of CTLA4 has shown efficacy in alleviating clinical symptoms in psoriasis.[50,51] These preliminary findings suggest that in the coming years, sufficient data will be available from a number of different clinical trials to determine if Ig fusion proteins will become commercially viable drugs. In addition to these clinical trials, some Ig fusion proteins are showing significant promise in preclinical animal models. For example, LFA3-Ig has been shown to prolong cardiac allograft survival in primates.[52]

8.4.3 Summary of Ig Fusion Protein Application

The use of Ig fusion protein technology to study protein structure and function is now well established and easily accessible to most laboratories. Fusion proteins have been used successfully to identify new receptor/ligand interactions, to determine protein function(s), and domains or individual responsible for these functions. While Ig fusion proteins have yet to reach the clinic as novel therapeutic agents, promising results with at least two such proteins suggest that in the near future, Ig fusion proteins will likely become useful drugs.

8.5 CONCLUSION

Immunoglobulin fusion protein technology has proved to be a powerful tool to study the structure and function of proteins. While it is not the only technology presently available to tag proteins functionally, it is one of the most robust and accessible. A number of factors have led to the increased use of this technology. These include: (1) the very large number of research tools developed for use with antibodies that can be used with Ig fusion proteins, (2) the fact that the structural and functional characteristics of immunoglobulin constant regions are very well understood, (3) the robust nature of the Ig domain structure that is not easily perturbed by the presence of other structural motiffs, and (4) the favorable in vivo properties (long half-life, etc.) of immunoglobulins retained by the Ig fusion proteins. While this technology has had an important impact in the laboratory, its clinical impact has yet to be demonstrated. However, the results of clinical and preclinical trials with four Ig fusion proteins, TNFR-p55, TNFR-p75, LFA3, and CTLA4, suggest that in the near future Ig fusion technology might yield its first drug.

ACKNOWLEDGMENTS

I thank Barb Thorne for providing key information on the interaction between Ig-Fc domain and Ig receptors, Edith Wolff for critical review of this manuscript, and Debby Baxter and Diane Horner for help in its preparation.

REFERENCES

1. Morrison, S. L., M. J. Johnson, L. A. Herzenberg, V. T. Oi. 1984. Chimeric human antibody molecules: mouse antigen-binding domains with human constant region domains. *Proc. Natl. Acad. Sci. USA* 81: 6851–6855.
2. Gascoigne, N. R. J., C. C. Goodnow, K. L. Dudzik, V. T. Oi, M. M. Davis. 1987. Secretion of a chmeric T-cell receptor-immunoglobulin protein. *Proc. Natl. Acad. Sci. USA* 84: 2936–2940.
3. Riechmann, L., M. Clark, H. Waldmann, G. Winter. 1988. Reshaping human antibodies for therapy. *Nature* 332: 323–327.
4. Capon, D. J., S. M. Chamow, J. Mordenti, S. A. Marsters, T. Gregory, H. Mitsuya, R. A. Byrn, C. Lucas, F. M. Wurm, J. E. Groopman, S. Broder, D. H. Smith. 1989. Designing CD4 immunoadhesins for AIDS therapy. *Nature* 337: 525–531.
5. Traunecker, A., J. Schneider, H. Kiefer, K. Karjalainen. 1989. Highly efficient neutralization of HIV with recombinant CD4-immunoglobulin molecules. *Nature* 339: 86–70.
6. Byrn, R. A., J. Mordenti, C. Lucas, D. Smith, S. A. Marsters, J. S. Johnson, P. Cossum, S. M. Chamow, F. M. Wurm, T. Gregory, J. E. Groopman, D. J. Capon. 1990. Biological properties of a CD4 immunoadhesin. *Nature* 344: 667–670.
7. Ashkenazi, A., D. J. Capon, R. H. R. Ward. 1993. Immunoadhesins. *Intern. Rev. Immunol.* 10: 219–227.
8. Chamow, S. M., A. Ashkenazi. 1996. Immunoadhesins: principles and applications. *Trends in Biotechnol.* 14: 52–60.
9. Aruffo, A., I. Stamenkovic, M. Melnick, C. B. Underhill, B. Seed. 1990. CD44 is the principal cell surface receptor for hyaluronate. *Cell* 61: 1303–1313.
10. Goodwin, R. G., M. R. Alderson, C. A. Smith, R. J. Armitage, T. VandenBos, R. Jerzy, T. W. Tough, M. A. Schoenborn, T. Davis-Smith, K. Hennen, et al. 1993. Molecular and biological characterization of a ligand for CD27 defines a new family of cytokines with homology to tumor necrosis factor. *Cell* 73: 447–456.
11. Smith, C. A., H. J. Gruss, T. Davis, D. S. Anderson, T. Farrah, E. Baker, G. R. Sutherland, C. I. Brannan, N. G. Copeland, N. A. Jenkins, et al. 1993. CD30 antigen, a marker for Hodgkin's lymphoma, is a receptor whose ligand defines an emerging family of cytokines with homology to TNF. *Cell* 73: 1349–1360.
12. Walz, G., A. Aruffo, W. Kolanus, M. Bevilacqua, B. Seed. 1990. Recognition by ELAM-1 of the sialyl-le^x determinant of myeloid and tumor cells. *Science* 250: 1132–1135.
13. Bowen, M. A., D. D. Patel, X. Li, B. Modrell, A. R. Malacko, W. C. Wang, H. Marquardt, M. Neubauer, J. M. Pesando, U. Francke, et al. 1995. Cloning, mapping, and characterization of activated leukocyte-cell adhesion molecule (AL-CAM), a CD6 ligand. *J. Exp. Med.* 181: 2213–2220.
14. Ward, E. S., V. Ghetie. 1995. The effector functions of immunoglobulins: implications for therapy. *Therapeu. Immunol.* 2: 77–94.
15. Hogarth, P. M., F. L. Ierino, M. D. Hulett. 1994. Characterization of FcR Ig-binding sites and epitope mapping. *Immunomethods* 4: 17–24.
16. Linsley, P. S., W. Brady, L. Grosmaire, A. Aruffo, N. K. Damle, J. A. Ledbetter. 1991. Binding of the B cell activation antigen to B7 to CD28 costimulates T cell proliferation and Interleukin-2 mRNA accumulation. *J. Exp. Med.* 173: 721–730.

17. Hollenbaugh, D., L. S. Grosmaire, C. D. Kullas, N. J. Chalupny, S. Braesch-Andersen, R. J. Noelle, I. Stamenkovic, J. A. Ledbetter, A. Aruffo. 1992. The human T cell antigen gp39, a member of the TNF gene family, is a ligand for the CD40 receptor: expression of a soluble form of gp39 with B cell c-stimulatory activity. *EMBO J.* 11: 4313–4321.

18. Bartolazzi, A., R. Peach, A. Aruffo, I. Stamenkovic. 1994. Interaction between CD44 and hyaluronate is directly implicated in the regulation of tumor development. *J. Exp. Med.* 180: 53–66.

19. Watson, S. R., Y. Imai, C. Fennie, J. S. Geoffroy, S. D. Rosen, L. A. Lasky. 1990. A homing receptor-IgG chimera as a probe for adhesive ligands of lymph node high endothelial venules. *J. Cell Biol.* 110: 2221–2229.

20. Whitney, G. S., G. C. Starling, M. A. Bowen, B. Modrell, A. W. Siadak, A. Aruffo. 1995. The membrane proximal scavenger receptor cysteine-rich domain of CD6 contains the activated leukocyte cell adhesion molecule binding site. *J. Biol. Chem.* 270: 18187–18190.

21. Holness, C. L., P. A. Bates, A. J. Little, C. D. Buckley, A. McDowall, D. Bossy, H. Hogg, D. L. Simmons. 1995. Analysis of the binding site on intracellular adhesion molecule 3 for the leukocyte integrin lymphocyte function associated antigen 1. *J. Biol. Chem.* 270: 877–884.

22. Fawcett, J., C. Buckley, C. L. Holness, I. N. Bird, J. H. Spragg, J. Saunders, A. Harris, D. L. Simmons. 1995. Mapping the homotypic binding sites in CD31 and the role of CD31 adhesion in the formation of interendothelial cell contacts. *J. Cell. Biol.* 128: 1229–1241.

23. Daniels, B. F., M. C. Nakamura, S. D. Rosen, W. M. Yokoyama, W. E. Seaman. 1994. Ly-49A, a receptor for H-2D^{d}, has a functional carbohydrate recognition domain. *Immunity* 1: 785–792.

24. Kolanus, W., W. Nagel, B. Schiller, L. Zeitlmann, S. Godar, H. Stockinger, B. Seed. 1996. $\alpha L\beta 2$ integrin/LFA-1 binding to ICAM-1 induced by cytohesin-1 a cytoplasmic regulatory molecule. *Cell* 86: 233–242.

25. Culouscou, J. M., G. W. Carlton, A. Aruffo. 1995. HER4 receptor activation and phosphorylation of shc proteins by recombinant heregulin-Fc fusion proteins. *J. Biol. Chem.* 270: 12857–12863.

26. Hollenbaugh, D., J. Douthwright, V. McDonald, A. Aruffo. 1995. Cleavable CD40Ig fusion proteins and the binding of gp39. *J. Immunol. Methods* 188: 1–7.

27. Beck, J. T., S. A. Marsters, R. J. Harris, P. Carter, A. Ashkenazi, S. M. Chamow. 1994. Generation of soluble interleukin-1 receptor from an immunoadhesion by specific cleavage. *Mol. Immunol.* 17: 1335–1344.

28. Muszbek, L., J. A. Gladner, K. Laki. 1975. The fragmentation of actin by thrombin. Isolation and characterization of the split products. *Arch. Biochem. Biophys.* 167: 99–103.

29. Linsley, P. S., S. G. Nadler, J. Bajorath, R. Peach, H. T. Leung, J. Rogers, J. Bradshaw, M. Stebbins, G. Leytze, W. Brady, A. Malacko, H. Marquandt, S-Y. Shaw. 1995. Binding stoichiometry of the cytotoxic T lymphocyte-associated molecule-4 (CTLA-4). *J. Biol. Chem.* 270: 15417–15424.

30. Hollenbaugh, D., A. Aruffo. 1992. Construction of fusion proteins with immunoglobulin. *Curr. Protocols Immunol.* 10: 19.1–19.11.

31. Kiener, P. K., P. Davis, B. Rankin, A. Aruffo, D. Hollenbaugh. 1995. Ligation of CD40 with soluble gp39 induces pro-inflammatory responses in human monocytes. *J. Immunol.* 155: 4917–4925.

32. Bennett, B. D., F. C. Zeigler, Q. Gu, B. Fendly, A. D. Goddard, N. Gillett, W. Matthews. 1995. Molecular cloning of a ligand for the EPH-related receptor protein-tyrosine kinase Htk. *Proc. Natl. Acad. Sci. USA* 92: 1866–1870.

33. Winslow, J. W., P. Moran, J. Valverde, A. Shih, J. Q. Yuan, S. C. Wong, P. S. Tsai, A. Goddard, W. J. Henzel, F. Hefti, K. D. Beck, I. W. Caras. 1995. Cloning of AL-1, a ligand for an Eph-related tyrosine kinase receptor involved in axon bundle formation. *Neuron* 14: 973–981.

34. Stamenkovic, I., D. Sgroi, A. Aruffo, M. S. Sy, T. Anderson. 1991. The B lymphocyte adhesion molecule CD22 interacts with the leukocyte common antigen CD45RO and a2-6 sialyltransferase, CD75, on B cells. *Cell* 66: 1133–1144.

35. Gebe, J. A., P. A. Kiener, H. Z. Ring, L. Xu, U. Francke, A. Aruffo. 1997. Molecular cloning, mapping to human chromosome 1 q21-q23 and cell binding characteristics of Spα, a new member of the scavenger receptor cysteine-rich (SRCR) family of proteins. *J. Biol. Chem.* 272: 6151–6158.

36. Stults, J. T., L. O'Connell, C. Garcia, S. Wong, A. M. Engel, D. L. Garbers, D. G. Lowe. 1994. The disulfide linkages and glycosylation sites of the human natriuretic peptide receptor-C homodimer. *Biochem.* 33: 11372–11381.

37. Bennett, B. D., G. L. Bennett, R. V. Vitangcol, J. R. S. Jewett, J. Burnier, W. Henzel, D. G. Lowe. 1991. Extracellular domain-IgG fusion proteins for three human natriuretic peptide receptors. *J. Biol. Chem.* 266: 23060–23067.

38. van der Merwe, P. A., P. R. Crocker, M. Vinson, A. N. Barclay, R. Schauer, S. Kelm. 1996. Localization of the putative sialic acid-binding site on the immunoglobulin superfamily cell-surface molecule CD22. *J. Biol. Chem.* 271: 9273–9280.

39. Vinson, M., P. A. van der Merwe, S. Kelm, A. May, E. Y. Jones, P. R. Crocker. 1996. Characterization of the sialic acid-binding site in sialoadhesin by site-directed mutagenesis. *J. Biol. Chem.* 271: 9267–9272.

40. Bajorath, J., J. S. Marken, N. J. Chalupny, T. L. Spoon, A. W. Siadak, M. Gordon, R. J. Noelle, D. Hollenbaugh, A. Aruffo. 1995. Analysis of gp39/CD40 interactions using molecular models and site-directed mutagenesis. Biochem. 34: 9884–9892.

41. Bajorath, J., N. J. Chalupny, J. S. Marken, A. W. Siadak, J. Skonier, M. Gordon, D. Hollenbaugh, R. J. Noelle, H. D. Ochs, A. Aruffo. 1995. Identification of residues on CD40 and its ligand which are critical for the receptor-ligand interaction. *Biochem.* 34: 1833–1844.

42. Starling, G. C., J. Bajorath, J. Emswiler, J. A. Ledbetter, A. Aruffo, P. A. Kiener. 1997. Identification of amino acid residues important for ligand binding to Fas. *J. Exp. Med.* 185: 1487–1492.

43. Chen, C., A. Gault, L. Shen, N. Nabavi. 1994. Molecular cloning and expression of early T cell costimulatory molecule-1 and its characterization as B7–2 molecule. *J. Immunol.* 152: 4929–4936.

44. Damle, N., A. Aruffo. 1991. Vascular cell adhesion molecule-1 (VCAM-1) induces T-cell antigen receptor-dependent activation of CD4+ T lymphocytes. *Proc. Natl. Acad. Sci. USA* 88: 6403–6407.

45. D. L. Shelton, J. Sutherland, J. Gripp, T. Camerato, M. P. Armanini, H. S. Phillips,

K. Carroll, S. D. Spencer, A. D. Levinson. 1995. Human trks: molecular cloning, tissue distribution, and expression of extracellular domain immunoadhesins. *J. Neurosci.* 15: 477–491.

46. Dietsch, M. T., P. Y. Chan, S. B. Kanner, L. K. Gilliland, J. A. Ledbetter, P. S. Linsley, A. Aruffo. 1994. Coengagement of CD2 with LFA-1 or VLA-4 by bispecific ligand fusion proteins primes T cells to respond more effectively to T cell receptor-dependent signals. *J. Leukocyte Biol.* 56: 444–452.
47. Chamow, S. M., D. Z. Zhang, X. Y. Tan, S. M. Mhatre, S. A. Marsters, D. H. Peers, R. A. Byrn, A. Ashkenazi, R. P. Junghans. 1994. A humanized, bispecific immunoadhesin-antibody that retargets $CD3^+$ effectors to kill HIV-1 infected cells. *J. Immunol.* 153: 4268–4280.
48. Moreland, L. W., S. W. Baumgartner, M. H. Schiff, E. A. Tindall, R. M. Fleischmann, A. L. Weaver, R. E. Ettlinger, S. Cohen, W. J. Koopman, K. Mohler, M. B. Widmer, C. M. Blosch. 1997. Treatment of rheumatoid arthritis with a recombinant human tumor necrosis factor receptor (p75)-Fc fusion protein. *New England. J. Med.* 337: 141–147.
49. Abraham, E., M. Glauser, T. Butler, J. Garbino, D. Gelmont, P. Laterre, K. Kudsk, H. A. Bruining, C. Otto, E. Tobin, C. Zwingelstein, W. Lesslauer, A. Leighton. 1997. p55 tumor necrosis factor receptor fusion protein in the treatment of patients with severe sepsis and septic shock. *JAMA* 277: 1531–1538.
50. Krueger, J. G., E. Hayes, M. Brown, S. Kang, M. G. Lebwohl, C. A. Guzzo, B. V. Jegasothy, M. T. Goldfarb, D. J. Hecker, R. M. Mann, J. R. Abrams. 1997. Blockade of T-cell costimulation with CTLA4Ig (BMS-188667) reverses pathologic inflammation and keratinocyte activation in psoriatic plaques. *J. Invest. Dermatol.* 108: 555.
51. Lebwohl, M., S. Kang, C. Guzzo, B. Jegasothy, M. Goldfarb, R. Mann, B. Goffe, A. Menter, N. Lowe, G. Krueger, M. Brown, R. Weiner. 1997. CTLA4Ig (BMS-188667)-mediated blockade of T cell costimulation in patients with psoriasis vulgaris. *J. Invest. Dermatol.* 108: 570.
52. Kaplon R. J., P. S. Hochman, R. E. Michler, P. A. Kwiatkowski, N. M. Edwards, C. L. Berger, H. Xu, W. Meier, B. P. Wallner, P. Chisholm, C. C. Marboe. 1996. Short course single agent therapy with an LFA-3-IgG1 fusion protein prolongs primate cardiac allograft survival. *Transplantation.* 61: 356–363.

9

TNF RECEPTOR IgG FUSION PROTEIN: PRINCIPLES, DESIGN, AND ACTIVITIES

WERNER LESSLAUER
F. Hoffmann-La Roche, Ltd.
Basel, Switzerland

9.1 INTRODUCTION

9.1.1 TNFα Activities

9.1.1.1 A Century of TNFα Research: From Tumors and Cachexia to Infection and Sepsis. The early history of the discovery of TNFα starts in the late 1800s with observations, still intriguing today, that intercurrent severe bacterial infections in cancer patients sometimes elicited what at that time was considered curative antitumor effects. Taking up these observations, Coley inoculated bacterial suspensions in terminally ill cancer patients as a therapeutic intervention, and reported antitumor responses in 1893.[1,2] In later decades, similar protocols in animal studies were found to result in tumor necrosis and thought to reflect the earlier clinical antitumor effects. They led to the discovery that the antitumor effect was indirect, caused by an endogenous factor induced in the plasma of animals treated with bacterial products such as lipopolysaccharide (LPS), resulting in a hemorrhagic necrosis of transplanted tumors; this activity therefore was called *tumor necrosis factor*.[3,4]

Antibody Fusion Proteins, Edited by Steven M. Chamow and Avi Ashkenazi
ISBN 0471-18358-X

This line of research, leading to purification and molecular cloning of TNFα (reviewed in Refs. 5 and 6) seemed to place this factor into the context of natural antitumor defense. A large number of clinical studies of TNFα in single or combination cancer therapy were conducted once recombinant TNFα became available (reviewed in Refs. 7 and 8). The results of these studies were generally disappointing for reasons that are not yet entirely clear. Dose-limiting toxicity may have been one reason, and the more recent finding of antitumor effects of high dose TNFα combined with chemotherapy in isolation perfusion may support this view.[9] However, antitumor effects are not the main biological role of TNFα, but they may result from a combination of its many different activities such as cytotoxicity in some tumor cells, and probably more importantly from activities on the tumor neo-vasculature.

A parallel, yet distinct line of research, focusing on the wasting of rabbits chronically infected with *Trypanosoma brucei*, opened a window on TNFα activity in a different context. It resulted in the discovery of a peptide, cachectin, with profound metabolic activities.[5] Neither the historical background nor initial comparative studies pointed to a relation between TNFα and cachectin, but once amino acid sequence information was available, it became clear that cachectin and TNFα were identical.[10] This widened the functional scope of TNFα and triggered the search for further activities. It was soon realized that TNFα is an essential mediator in the state of hemodynamic shock induced by LPS. TNFα is induced by LPS, and the appearance of systemic TNFα parallels host responses in septic disease.[11–15] Furthermore, exogenous TNFα induced shock and tissue injury similar to sepsis.[16–20] Conversely, inhibition of TNFα in many different animal studies protected from lethal outcome of LPS and bacterial challenge.[21–28] The realization that an endogenous factor may have beneficial as well as extremely potent toxic activities in host response to infection opened a new conceptual dimension; even today the factors dissecting beneficial and detrimental roles of TNFα remain ill understood.

9.1.1.2 Concerted Activity in Host Defense. Once recombinant TNFα became available, many different activities in a wide range of cellular systems were discovered.[5,6] TNFα quite generally activates inflammatory cells, has growth factor activity, induces major histocompatibility gene expression in lymphoid cells, and induces tissue factor and adhesion molecules in endothelium, to name only some activities; virtually all nucleated cells have receptors for and respond to TNFα.[5,6] It also emerged that while activated monocytes may be the largest source of TNFα, many other cells also are competent to produce TNFα.

This seemingly incomprehensible diversity of activities may be understood as concerted activity in immune–inflammatory intercellular communication as part of integrated tissue or host responses to infection and injury. A role of TNFα in infection with intracellular pathogens was indicated by the discoveries that neutralisation of TNFα in vivo interfered with defense, whereas administration of TNFα enhanced resistance against acute infection with *Listeria monocytogenes* and other microorganisms.[29–35] Studies of *L. monocytogenes* infection in TNFα receptor deficient mice extended these findings.[36,37] However, TNFα-dependent defense mechanisms also play an essential role in circumscribed local pathologies as shown by the fact that neutralization of TNFα with anti-TNFα antibodies or soluble TNFα receptor constructs interfered with granuloma formation in *Mycobacterium bovis* infection, and increased parasite load and lesion size in parasitic infections.[38–42] An even broader role of TNFα in tissue injury is suggested by an enhanced neuronal and microglial response to ischemic and excitotoxic brain injury in TNFα receptor deficient mice.[43]

TNFα is not a lone effector in immune–inflammatory reactions; it operates in a complex cytokine network. Its role as a proinflammatory mediator broadly overlaps with that of interleukin 1. Furthermore, TNFα belongs to a gene family of peptide ligands that have evolved in parallel to the TNFα receptor gene family.[44,45] Among these, lymphotoxin α (LTα), (also called TNFβ) is a close relative of TNFα; it is relevant as a proinflammatory cytokine and binds the same receptors as TNFα. LTα is encoded by a gene immediately adjacent to the TNFα gene, but displays distinct transcriptional regulation; it has a classical signal sequence and is expressed as a secreted protein, whereas TNFα, as most other members of the TNFα ligand family, is expressed as a type II transmembrane protein; soluble TNFα is generated by proteolytic processing of membrane TNFα by a specific enzyme, TACE.[46–48]

LTα and TNFα, as most members of the TNFα ligand family, are active as trimeric molecules, suggesting three receptor binding sites and ligand-induced microclustering as a likely receptor activation mechanism. TNFα in membrane bound form has effector function and therefore presumably is folded and packed as a trimer at the cell surface.[46,49,50] The role of membrane TNFα has not yet been fully elucidated, but it appears to be the more important ligand for one of the two TNFα receptors, TNFR75, possibly due to different off-rate kinetics.[51] LTα, in keeping with the general pattern of the whole ligand family, also exists as membrane bound ligand when complexed to the integral membrane protein LTβ in the form of LT$\alpha\beta$ heterotrimers. These bind to another member of the TNFα receptor family, LTβ receptor, but LT$\alpha_2\beta$ exhibits also one TNF receptor binding site.[52,53]

9.1.1.3 Disregulated Expression Leading to Toxicity and Tissue Destruction in Chronic Inflammation. TNFα has a role in the maintenance and progression of inflammation and tissue destruction in chronic inflammatory diseases with autoimmune background, such as rheumatoid arthritis (RA). TNFα was found to stimulate collagenase and prostaglandin E2 production in synovial and articular cartilage cells.[54] It stimulated cartilage proteoglycan degradation and inhibited its resynthesis,[55–58] stimulated osteoclast bone resorption, and inhibited bone formation in vitro in tissue culture studies.[59] Furthermore, TNFα and other cytokines were detected in synovial fluid and tissue of RA patients.[60–66] In studies of collagen-induced arthritis in mice (CIA), which reflects aspects of RA, positive effects of TNFα neutralizing treatment supported the view that TNFα has a causal role in the arthritic disease process.[67–71] Furthermore, mice carrying TNFα transgenic constructs leading to disregulated TNFα expression developed arthritis with histologic similarities to RA that responded to anti-TNFα treatment;[72–74] however, consistent with the complex physiology of the cytokine network, the obviously TNFα-driven disease was also controlled by targeting interleukin 1.[73]

TNFα is thought also to have a role in multiple sclerosis (MS). It might be envisioned to act as a mediator in the unknown etiology of MS, which is thought to be driven by an autoimmune–inflammatory reaction to myelin components of the central nervous system. Early focal lesions show perivenular mononuclear cell infiltrates in white matter and edema, presumably involving a disturbance of the blood-brain barrier; the infiltrating cells appear to mediate the selective destruction of myelin sheaths, initially leaving the axons intact. In further disease progression, astrocyte proliferation leads to a gliotic scar with axonal sparing. There is broad, if circumstantial, evidence that TNFα may act as a mediator and effector molecule in all these processes. TNFα-producing cells, mostly reactive fibrous astrocytes, are found in and at the edge of active MS lesions.[75,76] Furthermore, in situ hybridization studies showed high transcript levels of TNFα and other cytokines, particularly in lesions with strong demyelination.[77] In tissue explant cultures, TNFα was found to exert a direct cytotoxic effect on oligodendrocytes and a contrasting mitogenic activity on astrocytes.[78,79] In some clinical studies, TNFα concentrations in serum and cerebrospinal fluid (CSF) of MS patients were found to be elevated and correlated with disease progression; while other studies did not confirm these results, the consensus developed that the MS disease process and an increased presence of TNFα at disease relevant sites correlated.[80–84]

In assessing the significance of these findings, it should be noted that the mere presence of TNFα at critical sites of tissue destruction in chronic inflammatory states does not allow distinction between disease-promoting or protective activity, or default repair activity of TNFα without causal relationship to the disease process. TNFα may exert biphasic concentration-dependent activity, beneficial at one level and flipping to toxicity at another. Furthermore, the role of TNFα is not necessarily the same at all stages and all sites of a disease process. Interestingly, in relapsing/remitting MS patients, studies of cytokine transcription indicate that TNFα mRNA in blood mononuclear cells may be upregulated well in advance of clinical exacerbations,[85] suggesting a broader systemic involvement in MS, and opposing activities of TNFα in separate compartments.

Animal studies have significantly shaped the views on the autoimmune origin and the role of cytokines in human MS. Experimental allergic encephalomyelitis (EAE), a disease thought to reflect various histopathologic and clinical aspects of MS, is a T-cell–mediated autoimmune disease[86] (for reviews see Refs 87–89). EAE can be elicited in rodents and nonhuman primates by immunization with myelin components, or by adoptive transfer of syngeneic T cells specific for such autoantigens into a susceptible host animal. EAE is a complex disease under multifactorial control, but an important pathogenetic role of TNFα is suggested by a number of findings. In transfer EAE, encephalitogenicity strongly depends on the TNFα and LTα production capability of T-cell clones, all recognizing the same antigenic peptide, and on the presence of cells competent to produce proinflammatory cytokines.[90,91] The role of TNFα in the initiation and the effector mechanisms of EAE is further demonstrated by the protective and curative efficacy of treatment with anti-TNFα monoclonal antibodies or soluble TNFα receptor constructs.[92–99] Furthermore, disregulated expression of transgenic membrane-TNFα driven by an astrocyte-specific promoter can trigger a spontaneous neurologic disease in mice with typical characteristics of human inflammatory neurodegenerative disease; other TNFα transgenic mice specifically expressing soluble TNFα in the CNS also develop inflammatory demyelinating disease, which responds to preventive systemic treatment with anti-TNFα antibody.[100,101]

9.1.1.4 The Dual Face of TNFα. The function of TNFα in acute systemic and chronic local perturbations of immune-inflammatory mediator homeostasis can flip from beneficial defense activity to severe toxicity and tissue injury. That such a system could survive in evolution is puzzling. It might be argued that TNFα and the proinflammatory cytokine system as a whole in the balance of all integrated activities provide a selective advantage to the individual despite their risks. The potential toxicity might be the price to be paid for potency. Alternatively, in selection pressures acting on groups, the ultimately lethal toxicity for individuals with distinct sensitivities might provide an advantage for the population as a whole. When developing therapeutic strategies targeting the TNFα system, it is imperative to understand whether TNFα is a player in a disease, whether its role is good or bad, and how it evolves during the disease process and in different body compartments with sensitivities defined by diverse genetic background and individual preexisting conditions. The integrated effect of these factors remains poorly understood, especially since the functional status of the cytokine network notoriously escapes direct measurement and can only be diagnosed indirectly in the clinical setting. All these considerations call for a cautious approach to anti-TNFα therapeutic strategies.

9.2 TNFα RECEPTORS

9.2.1 Discovery

The molecular cloning of TNFα paved the way to a search for its receptors. TNFα-binding studies at the surface of intact cells revealed a single class of binding sites with Kd values in the picomolar range and numbers ranging from a few hundred to a few thousand receptors per cell; TNFα and LTα compete for the same cellular binding sites;[102] (reviewed in Ref. 103). Chemical cross-linking of radiolabeled TNFα at the cell surface revealed various bands of TNFα-binding proteins by SDS-PAGE and autoradiography suggesting a complex receptor structure,[104] (also review of earlier literature). An important finding was that TNFα appeared to have a low level of species specificity,[105] a view to be refined by later discoveries, e.g., that human TNFα selects one of the two mouse receptors.[106] A major advance was the generation of monoclonal antibodies for TNFα receptors.[107] Interestingly, two sets of antibodies were obtained separated by distinct staining and TNFα binding inhibition patterns on two panels of human cell lines, providing firm evidence for the existence of two independent TNFα receptors.[107] The usual initial confusion about nomenclature now has been resolved; they are termed 55kDa TNFα receptor (TNFR55, CD120a) and 75kDa TNFα receptor (TNFR75, CD120b).

The molecular cloning of the TNFα receptors for several years proved to be an elusive goal, but around 1989/90 several groups within a short time period succeeded in isolating TNFα receptor cDNA by three different approaches (reviewed in Ref. 103). In one approach, the receptor proteins were purified, and partial amino acid sequences were determined and used to isolate the cDNA of human TNFR55[108] and TNFR75.[109] An elegant expression cloning approach using radiolabeled TNFα as a probe shortly before had already identified TNFR75 cDNA.[110] Previously, three groups had discovered a TNFα inhibitory activity associated with a partially purified peptide fraction in human urine.[111–113] The first partial amino acid sequences of these inhibitors showed that two distinct peptides represented soluble extracellular domains of the TNFα receptors, termed sTNFR55 (TNF-BPI) and sTNFR75 (TNF-BPII).[114–116] The relationship between TNFR55 and TNF-BPI was confirmed by immunologic cross-reactivity.[117] Antibodies raised to TNF-BPI elicited a functional cell response similar to that of an agonistic anti-TNFR55 antibody or TNFα.[107,118,119] The third approach to the molecular cloning started from the relationship between TNFR55 and TNF-BPI, which had been guessed at an early time. Amino acid sequence information of TNF-BPI was used to generate probes to identify the full length receptor cDNA.[120–122] Within a few months several other groups also reported the isolation of human and rodent TNFα receptor cDNA (reviewed in 103).

9.2.2 Structure and Function

The predicted amino acid sequences of human TNFR55 and TNFR75 revealed molecules of 426 and 439 amino acids, the much larger difference in apparent molecular mass presumably due to differential N- and O-linked glycosylation. Each receptor has a single membrane-spanning domain. The extracellular domains of 182 (TNFR55) and 235 (TNFR75) amino acid residues share only about 28% identity, roughly equivalent to the sequence similarity each of them shares with the other members of the TNFα receptor superfamily such as Fas antigen, CD27, CD30, or CD40. This is remarkable given the selectivity of

TNFR55 and TNFR75 for both TNFα and LTα, which among themselves share but 31% identity consistent with significantly distinct local structures despite a highly conserved molecular core and overall shape.[123–125] When considering the selective specificities of the other members of the receptor and ligand families, a great variety and plasticity of folding and packing of peptides despite a basic sequence similarity emerges.[44] The extracellular sequences of both receptors upon closer inspection can be divided into four conserved linear subdomains, each defined by a six-cysteine consensus sequence motif repeated four times. The later structural analysis has shown that at least the first three domains also corresponded to conserved structural folding domains defined by conserved disulfide connectivity.[126–128] This cysteine consensus sequence motif, typically repeated four times, defines the whole TNFα receptor family.[44] In contrast to the extracellular domains, the intracellular domain sequences of TNFR55 and TNFR75 are entirely unrelated, suggesting different modes of signalling and function.[109]

To visualize the ligand receptor interaction in atomic detail, a protein crystallographic analysis of complexes of recombinant soluble TNFR55 and LTα was performed.[126] An elegant molecular complex of threefold symmetry with three receptors binding to one LTα trimer was revealed. The tightly packed pyramid-shaped LTα molecule with its three identical subunits is centrally placed. The receptors have an elongated structure formed by a nearly end-to-end alignment of the cysteine repeat motif domains and bind in shallow groves running from base to tip between two adjacent subunits of the LTα trimer. The ligand–receptor interaction is mainly established by contacts between the second and third receptor subdomains and residues of both adjacent LTα subunits. The orientation of this receptor ligand complex relative to the full-length receptor at the cell membrane is defined by this structure, since the C-terminal ends of the receptor extracellular domains are located toward the narrower tip of the LTα trimer.[126]

There is much evidence that receptor clustering is the TNFα signal-generating mechanism at the cell surface. Certain anti-TNFα receptor antibodies elicit cellular responses indistinguishable from those of TNFα, whereas Fab fragments of the same antibodies are inactive, but regain agonistic activity upon cross-linking.[118,119,129–131] A very elegant confirmation of receptor activation by ligand-induced clustering was provided by the dominant negative effect on TNFα responses exerted by signaling defective mutants of human TNFR55 transfected in mouse cells; TNFα nonresponsiveness increased in parallel with the expression level of the truncated human gene, but was overcome by agonistic antimouse TNFR55 antibodies selectively recruiting mouse TNFR55, whereas TNFα indiscriminately clustered mouse and nonfunctional human TNFR55.[132,133]

Practically all nucleated cells express both TNFα receptors simultaneously, albeit at varying relative amounts, and each receptor independently is fully able to bind TNFα and LTα. It was a considerable puzzle that the receptors segregate into homogeneous complexes of either TNFR55/TNFα or TNFR75/

TNFα.[131] The later discovery of families of intracellular proteins selectively associating with TNFR55 or TNFR75 through protein interaction domains (beyond the scope of the present review) has shed new light on this previously poorly understood phenomenon; it emphasizes the functional distinctness of the two receptors. A large number of studies have been undertaken to dissect the TNFR55- and TNFR75-mediated TNFα activities; in general, most activities were asigned to TNFR55, TNFR75 being thought to have more of an accessory or modulating activity, but there is also evidence for independent signaling capacity of TNFR75.[129–131,134]

9.3 ANTI-TNFα STRATEGIES

9.3.1 Sites of Intervention in the TNFα System

The pathways of TNFα physiology start with a signal eliciting transcriptional and translational activation of the TNFα gene.[135–137] The 26 kDa membrane TNFα is expressed and processed, and activates receptors and various signal transmission cascades. Along these pathways many sites might become targets of anti-TNFα therapeutic intervention. Among these, the stress kinase pathways controling TNFα expression,[137] and TACE[47,48,138–140] have been the focus of recent investigations. However, the first approach to interfere with TNFα in animal studies, leading into clinical development, was to target TNFα with antibodies that bind and neutralize it.[21,22,70,94,141–164] Once the TNFα receptors were cloned, recombinant soluble TNFα receptor constructs derived from either TNFR55- or TNFR75-extracellular domains were developed and tested as therapeutic anti-TNFα reagents.[23–25,27,70,92,93,96–98,164–169]

9.3.2 Recombinant TNFα Receptor Constructs as Anti-TNFα Agents

The activity of the natural TNFα-binding proteins, TNF-BP I and TNF-BP II, to compete with the cellular receptors for TNFα binding formed the basis of the biological assays leading to their discovery.[112–114,170] It was therefore natural to view the role of the TNF-BP as counterregulatory, TNFα inhibitory elements. They occur as normal constituents of plasma in relatively high concentrations (about 3–5 ng/ml) which might be understood as excess TNFα neutralization capacity, especially since their concentrations rise considerably in conditions accompanied by elevated TNFα.[171] However, closer inspection of the binding properties revealed the potential for a dual function between neutralization and stabilization of TNFα structure, the latter leading to carrier activity, especially pronounced for TNF-BP II.[131,170,172]

The first in vivo studies of recombinant sTNFR55 in nonhuman primates quickly revealed that this relatively small peptide after a single intravenous dose was rapidly eliminated from circulation.[26] Soluble TNFR55, purified from CHO cell expression, had to be continuously administered using implanted osmotic minipumps in order to achieve a sufficient in vivo exposure and a pharmacodynamic effect in CIA.[70,173] Thus, the first goal in the design of a TNFα-neutralising agent derived from natural TNF-BP was to achieve a more prolonged pharmacokinetic half-life by increasing molecular size.

Another aspect in the design of a TNFα receptor construct is provided by the stereochemical mode of receptor–ligand interaction. The early views on TNFα receptor activation, with the trimeric TNFα molecule clustering three receptors (Fig. 9.1), suggested an interest to generate a dimeric receptor construct, combining two extracelluar TNFα receptor moieties, which would simultaneously bind to one TNFα trimer, leading to cooperative stabilization of binding when compared to monomeric sTNFR. Furthermore, with two receptor sites on the TNFα molecule blocked, any clustering of cellular TNFα receptors would be efficiently prevented. Several groups have generated such dimeric soluble TNFα receptor constructs by the recombinant expression of a fusion gene constructed by the ligation of human sTNFR55 or sTNFR75 cDNA to human immunoglobulin heavy chain fragments, a dimerization function being provided by disulfide bonds of the hinge region.[23–25,165,174,175] (Box 9.1). The protein crystallographic structure of the human LTα/sTNFR55 complex allows further discussion of this design strategy. The human sTNFR55 sequence contains 182 amino acid residues. The main ligand interaction contacts occur in the second and third of the four six-cysteine-repeat motifs of

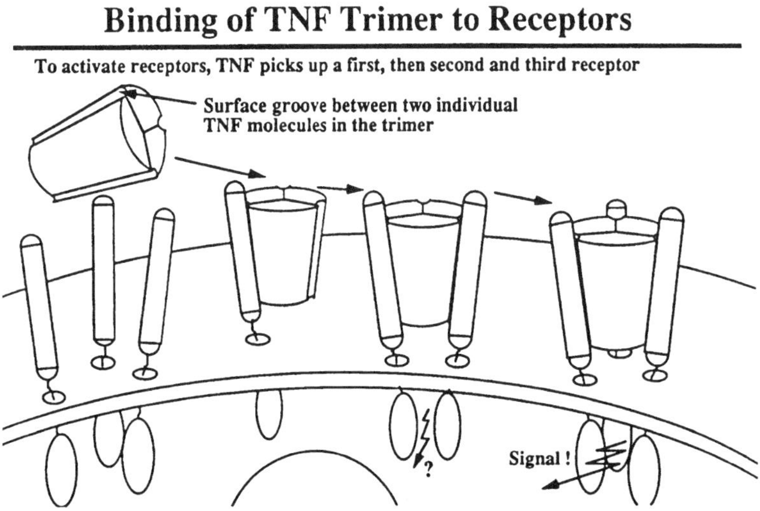

Figure 9.1 Schematic view of TNF receptor activation. The trimeric TNF molecule presents three receptor binding sites. When approaching the cell membrane, TNF binds to three receptor molecules, inducing a microclustering of the receptors, which the cell interprets as a TNF signal. While other activation mechanisms have been discussed, this model represents the most widely accepted mode.

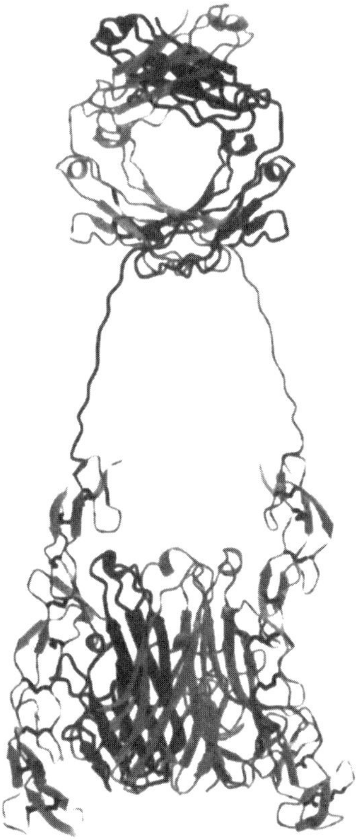

Box 9.1 Strucure of TNFR55-IgG.

The TNF molecule is built from three identical subunits with three receptor binding sites. Receptor activation results when a TNF molecule binds two and three receptors leading to a change in receptor configuration in the membrane microenvironment. The two TNF receptor moieties of TNFR-IgG by design occupy two sites of a TNF molecule, and thus neutralize TNF more effectively than monomeric soluble receptors. The slower off-rate in TNF binding by TNFRR55-derived fusion proteins, when compared to TNFR75-derived constructs, at the same dose results in a more pronounced neutralizing effect. Anti-TNF antibodies due to molecular size and mode of binding tend to cross-link TNF, leading to larger molecular aggregates. (composite of ref. 126 and Ig data base structures, courtesy of C. Broger)

sTNFR55 corresponding to amino acid residues of around 54 to 137; beyond residuc 150, the TNFR55 structure has not been resolved in the complex,[126] but from the general aspect of the structure these residues are very unlikely to be in ligand contact. The axes of the three elongated sTNFR55 at their C-terminal end in the complex are very near to each other. Therefore, it can be easily envisioned that there is sufficient structural flexibility between the hinge disulfide bonds and the ligand interaction domains 2 and 3 of sTNFR55 to allow the two receptor moieties of one TNFR55-IgG molecule to bind to the two receptor sites of the same TNFα trimer (Fig. 9.2)

In one representative study, such chimeric TNFR55-IgG3 and TNFR75-IgG3 constructs were expressed in myeloma cell culture and purified by protein G affinity chromatography.[23] By SDS-PAGE, the purified proteins under nonreducing conditions ran as homogeneous bands of about 135 kDa and 150 kDA apparent molecular masses. Under reducing conditions, they separated into homogeneous subunit bands of about 70 kDa and 80 kDa, consistent with a structure of a heavy chain analogue dimerized by the hinge disulfide bonds, with V_H and C_{H1} domains replaced by sTNFR55 or sTNFR75.[23] Later studies showed that the same constructs were also correctly expressed and processed in CHO cells.[172,173] IgG1 and IgG3 sequences were used in the fusions, and it was established that the fusion partner had no effect on TNFα binding. Other groups in parallel had generated TNF receptor extracellular domain-Ig heavy chain fusion proteins similar in the general principles but distinct in details.[23–25,165,174,175]

The proposed mode of binding between TNFα and TNFR-IgG fusion proteins is supported by substantial indirect evidence. Natural TNFR55 and TNFR75 at the cell surface, and recombinant sTNFR55 and sTNFR75, bind TNFα in equilibrium with Kd values of 350–500 pM and 60–100 pM, respect-

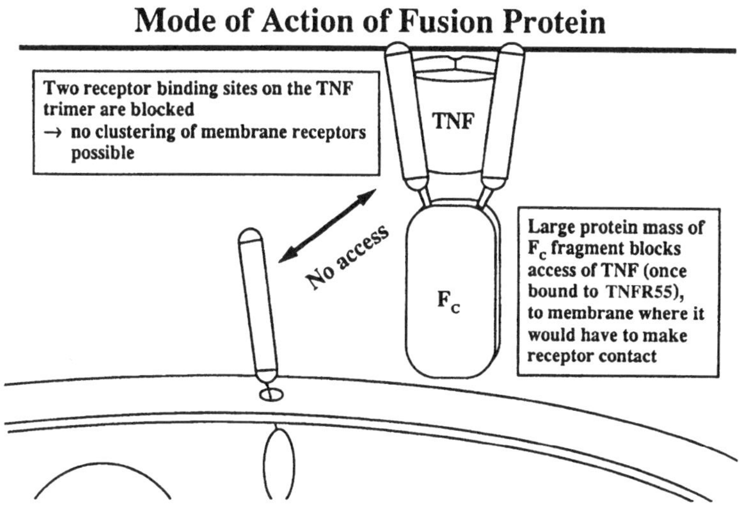

Figure 9.2 TNFR55-IgG–TNF binding mode. The TNFR55-IgG molecule contains two extracellular domains of the TNF receptor fused to an IgG1 heavy chain Fc fragment dimer. The hinge disulfide bonds provide the dimerization function. The two receptor moieties around the fusion site have sufficient flexibility to bind to the same TNF trimer, providing cooperative stabilization of binding.

ively.[104,108,110,120,173,176] In contrast, TNFα binds TNFR55-IgG with about five-fold higher affinity, consistent with the proposed binding mode.[24,172,173] More importantly, the protective activity of TNFR55-IgG against TNFα cytotoxicity is at least 50-fold higher than that of the monomeric soluble receptors.[24,25,173] Studies of the inhibition of cellular TNFα binding using recombinant sTNFR75 as compared to TNFR75-IgG similarly revealed an about 100-fold higher specific activity of the dimeric receptor construct.[27] The analysis of the molecular size of the complex of TNFα and TNFR55-IgG by size exclusion chromatography revealed a major species compatible with a 1:1 TNFR-IgG and THFα molar complex; a second band compatible with a 2:1 TNFα:TNFR-IgG molar complex also occurred, but there was no evidence for larger complexes such as those seen for TNFα and anti-TNFα antibodies, which may form large networks.[177]

The equilibrium TNFα binding affinity of TNFR55 and TNFR75 is quite similar, but there is a pronounced difference in the on/off kinetic rate of TNFα binding; this fact led to the proposition that TNFR75 at low TNFα concentration has an antennalike function, enhancing TNFR55 activation by "ligand passing."[131,172] The TNFα association and dissociation rates at the cell surface are much higher with TNFR75, the dissociation of TNFR55 and TNFR75 corresponding to half-lives of $t_{1/2} > 3\,h$ and $t_{1/2} \simeq 10\,min$, respectively.[131] Similarly distinct binding kinetics were determined with recombinant sTNFR55 and sTNFR75 in solid phase binding assays with half-lives of TNFα exchange of about 0.5 h and 4 min, respectively.[172] These inherent chemical properties of the receptors were preserved in receptor–immunoglobulin fusion proteins, but the differences were even more pronounced; irrespective of whether IgG1 or IgG3 sequences were used for fusion partner, TNFR55-IgG had a half-life of TNFα-exchange of 7–8 h, whereas TNFR75-IgG3 had an exchange half-life of 5–10 min.[172]

It was of considerable interest to compare the TNFα neutralizing capacity and therapeutic effect of TNFR-IgG and anti-TNFα antibodies. Such comparative studies revealed substantial differences in the specific activities of the various antibodies and receptor constructs as expected from their distinct equilibrium affinity and kinetic binding properties. The TNFR-IgG constructs, however, consistently showed higher specific TNFα-neutralizing activity.[25,177–179] One representative anti-TNFα antibody, with Kd values and binding kinetics comparable with TNFR55-IgG, required 50- to 100-fold higher concentrations to achieve half-maximal inhibition in cytotoxicity assays.[177] The differences in equilibrium affinity and binding kinetics of these two reagents do not explain the differences in potency; rather, the stereochemical modes of the interaction of TNFα with TNFR-IgG and anti-TNFα antibody must be considered. It is excluded that the 1:1 molar TNFα:TNFR55-IgG complex, with two receptor moieties of TNFR55-IgG binding one TNFα trimer, can be mimicked by an antibody: for reasons of molecular size and structure, the two antigen-combining sites of the antibody are unlikely to be able to access their symmetry-related epitopes on one TNFα trimer. Instead they bind two TNFα molecules, leading to cross-linking and ultimately to larger TNFα: antibody networks which even may have partial agonist activities.

9.3.3 Effects of TNFR-IgG Treatment in Animal Studies

9.3.3.1 TNFR-IgG Treatment in LPS and Bacterial Challenge. The first studies of the effects of TNFR55-IgG treatment were carried out in mice challenged with LPS,[23,24] where anti-TNFα antibody treatment previously had been shown to provide protection.[21,180] Quite low intravenous doses of TNFR55-IgG in the range of 5–20 μg per mouse were found to fully protect from intraperitoneal administration of a lethal dose of *E. coli* LPS combined with D-galactosamine sensitization, or of intravenous *S. abortus* LPS.[23,24] In these studies the animals were pretreated with TNFR55-IgG to neutralize the early, transient TNFα response, peaking 1–2 h after LPS challenge. More detailed information about the plasma TNFα responses were reported in a study of TNFR75-IgG, where mice had received a partially lethal intravenous dose of *E. coli* LPS;[27] TNFR75-IgG treatment provided protection, but in contrast to TNFR55-IgG, higher doses of 30–100 μg TNFR75-IgG per mouse were required to achieve a 80–90% survival rate.[27] In the same study, treatment with monomeric recombinant sTNFR75 had no effect on survival, even at doses as high as 260 μg per mouse. Interestingly, as a consequence of sTNFR75 treatment, a dramatic increase in plasma TNFα concentration was found 2 h after LPS administration, at a time when plasma TNFα concentrations in placebo treated animals started to decline, suggesting a TNFα carrier activity of the soluble receptor.[27] Analyses of TNFα activity in the plasma of TNFR75-IgG treated mice using cellular assays revealed a further peculiar phenomenon; sera of mice treated with LPS and high dose TNFR75-IgG 2 h after LPS challenge contained little TNFα activity that upon dilution titrated in regular fashion, but sera of mice treated at lower TNFR75-IgG doses contained intermediate active TNFα concentrations that furthermore did not titrate out, suggesting incomplete neutralization and carrier activity in the assay.[27]

To interpret these distinct outcomes of TNFR55-IgG and TNFR75-IgG treatments, the seemingly minor variations in the treatment protocols, the strains of mice, and type of LPS used are recognized to play a role, but the distinct TNFα binding properties of the two types of fusion proteins must also be considered as a determining element. A comparative study of TNFR55-IgG and TNFR75-IgG treatments in LPS/D-galactosamine lethality had shown a pronounced difference in protective efficacy.[172] Further studies compared TNFR55-IgG and TNFR75-IgG in lethal live *E. coli* bacteremia and elucidated mechanisms of in vivo TNFα neutralization.[28] Treatment with TNFR75-IgG at doses as high as 250 μg per mouse here provided only minor protection.[28] Surprisingly, the plasma of these mice for up to 24 h after TNFR75-IgG and

E. coli administration contained persistent TNFα activities, whereas the expected transient TNFα response was seen in the placebo-treated animals, where 5 h after *E. coli* administration TNFα activity had returned to baseline. The persistence of TNFα activity in circulation can be readily understood from the distinct in vivo pharmacokinetic properties of TNFα and TNFR75-IgG, considering the rapid elimination of free TNFα and the TNFα pool bound to long-lived TNFR75-IgG, carrying TNFα along. The lack of protection by TNFR75-IgG treatment resulted from the persistently elevated TNFα activity due to the presence of TNFR75-IgG in plasma, since posttreatment with a neutralizing anti-TNFα antibody 4 h after *E. coli* administration, when in placebo-treated animals TNFα had returned practically to baseline, significantly reduced plasma TNFα in the TNFR75-IgG–treated animals, and protected from the lethal insult.[28] As expected, anti-TNFα antibody treatment at the 4 h time point in the absence of TNFR75-IgG pretreatment, had no protective effect.[28] In conclusion, the TNFR75-IgG treatment had some positive effect, presumably by blunting the TNFα peak response, but also contributed to the lethal toxicity by its carrier activity leading to persistently elevated active TNFα plasma concentrations. In contrast, treatment with TNFR55-IgG even at lower dose in parallel experiments fully neutralized plasma TNFα activity at all times and protected from death.[28] In interpreting these data, the substantial difference in TNFα binding kinetics between TNFR55-IgG and TNFR75-IgG must be considered. The fast exchange rate of TNFα and TNFR75-IgG, relative to the production and elimination fluxes of free TNFα likely had caused the persistent buildup of plasma TNFα activity. In contrast, the slow exchange rate of TNFα and TNFR55-IgG had resulted in an overall dominant TNFα neutralizing activity.

The predictive value of the effects of anti-cytokine intervention in animal models of endotoxemia or bacteremia for human disease is often debated. The contrast between generally high protective efficacy after intravenous bacterial challenge and much poorer efficacy after intraperitoneal challenge, in which sepsis develops more gradually from peritonitis, may cast doubt on the validity of intravenous challenge studies.[34,141,142] However, such animal studies should not focus merely on survival; rather, their value lies in the detailed analysis of organ pathophysiologies and the distinct properties of the different anti-TNFα strategies. In view of the caution mandated by the results of anti-TNFα antibody therapy in peritonitis after cecal ligation and puncture,[34,141,142] TNFR55-IgG was investigated in a generalized peritonitis, developing from intraperitoneal inoculation of live *E. coli* carefully titrated against antibiotic treatment.[172] Pretreatment with TNFR55-IgG at significantly higher doses than those providing full protection from lethal LPS toxicity postponed clinical disease and death by 1 to 2 days, but only marginally improved ultimate survival rates. Interestingly, mice were found to be fully protected from intraperitoneal *S. aureus*-induced lethal toxicity by treatment with quite low doses of TNFR55-IgG if the bacteria were heat inactivated before inoculation, demonstrating that the reduced efficacy of TNFR55-IgG is due to the developing peritonitis and ensuing bacteremia. However, it is noted that live bacterial counts in this experimental animal disease reach values of $>10^6$ cfu/ml plasma,[172] whereas in contrast it is often difficult or impossible to detect causative bacteria in human sepsis.

A study in lethal *E. coli* bacteremia in nonhuman primates, *Papio anubis* (baboon), provided further insight into the effects of TNFR55-IgG treatment on the developing pathologies and organ dysfunctions.[181] TNFR55-IgG at low doses of 0.2 mg/kg body weight fully protected the animals against the lethal toxicity of the challenge. The bacterial challenge elicited a pattern of TNFα and other cytokine releases, of metabolic, hematologic, coagulation, and organ dysfunctions that are similarly affected in the human systemic inflammatory response syndrome and sepsis, suggesting parallel pathogenetic mechanisms irrespective of the initial trigger in baboon and man.[181] TNFR55-IgG treatment showed a protective effect throughout the whole spectrum of the pathologies in all the various organ systems of the baboon, providing a strong rationale for investigating this anti-TNFα strategy in man.[164,181]

9.3.3.2 TNFR-IgG Treatment of Experimental Chronic Inflammatory Disease in Animals. From its early discovery, TNFα was thought to be involved in the pathogenesis of RA. TNFα was found universally to be present in arthritic joints, accompanied by upregulated expression of TNFα receptors.[54–66,182] Studies of CIA in rodents supported this view. CIA develops after immunization with native type II collagen in adjuvant. The disease is mediated by autoreactive T lymphocytes; specific H-2 alleles, T-cell receptor-β genes, and other unknown elements in the genetic background confer susceptibility.[183–185] TNFα could well be imagined to have a role in this disease context; indeed, treatment of CIA with anti-TNFα antibodies delayed the development of arthritis and reduced clinical signs and histopathological severity of disease when started within the first two weeks after immunization.[68,69] A treatment effect was still seen when started as soon as the first clinical signs were observed, but later treatment was ineffective.[68] The critical role of TNFα in CIA in the late immunization phase, shortly before or concomittantly with clinical disease onset, was further documented by treatment with recombinant sTNFR55 continuously administered by osmotic pump, or with TNFR55-IgG in single or repeated dosing.[70,71] The effect of TNFR75-IgG treatment in CIA was studied using essentially similar protocols.[67] Consistent with the other studies, early treatment before disease onset reduced the incidence and severity of CIA. To demonstrate a therapeutic effect, animals with initial signs of arthritis were randomized into treatment and control groups; the TNFR75-IgG–treated animals progressed to a significantly less severe disease than control mice.[67]

In a parallel development, clinical studies had provided evidence suggesting a role of TNFα in MS. TNFα was found to be expressed in and around MS lesions,[75,76,78,79] and TNFα activity in CSF as well as TNFα mRNA expression in blood mononuclear cells appeared to correlate with disease activity.[80–85] This was complemented by animal studies showing that TNFα had a pathogenetic role in EAE,[90,91,186] and that anti-TNFα antibody treatment in EAE had a therapeutic effect.[94,95] These findings provided a rationale for studying TNFα receptor constructs in EAE.[92,93,96–99] The TNFR55-IgG and TNFR75-IgG compounds first used in EAE studies were constructed by the fusion of sTNFR55 or truncated sTNFR75 to a partial J sequence followed by all three constant regions of human IgG1 heavy chain, itself associated with a truncated κ light chain.[92] These constructs and in parallel the TN3.12.19 anti-TNFα antibody administered intraperitoneally and intracranially were used to treat chronic relapsing EAE in Biozzi AB/H mice. Both these TNFR55-IgG and TNFR75-IgG proteins in preventive treatment induced a dose related decrease in disease severity, and a therapeutic effect was observed when treatment started with the onset of clinical signs; TN3.12.19 treatment in general was similarly effective, but 10–100-fold higher doses were required.[92] Clinical signs of active EAE in Lewis rats were ameliorated significantly by the same TNFR55-IgG construct, when the treatment was initiated immediately prior to the onset of clinical disease.[98] Interestingly, no significant difference between TNFR55-IgG- and placebo-treated animals was observed regarding the numbers of $CD4^+$ T cells and macrophages infiltrating the CNS. This indicates a dissociation between CNS cell recruitment, tissue damage, and clinical disease. A more detailed study of CNS-associated leukocytes in passively induced EAE in PVG strain rats allowed the identification of the transferred T cell blasts by genetic marker;[98] it was found that TNFR55-IgG treatment did not impede the early influx of specific T cells, but the early nonspecific inflammatory response was substantially blunted. In the further development of the disease, TNFR55-IgG when compared to controls mitigated the enhancement of influx of MBP-specific T cells thought to represent self-recruitment secondary to tissue damage. The recruitment of nonspecific T cells and of non-T inflammatory cells remained higher in the control animals in the later disease stages. Intriguingly, TNFR55-IgG generally slowed the efflux of inflammatory cells from the CNS irrespective of the preventive and therapeutic effects.[98] A study of experimental autoimmune uveoretinitis extended these findings;[99] approximately equal numbers of $CD4^+$ $\alpha\beta TCR^+$ lymphocytes were found to be present in retinae irrespective of placebo or TNFR55-IgG treatment. However, TNFR55-IgG treatment reduced the number of activated macrophages and granulocytes in the brain, which correlated with substantially reduced retinal target organ damage. This result demonstrated that a tissue that is specifically targeted in an autoimmune process can be protected by interference with TNFα activity despite influx of activated T lymphocytes.[99]

In a further study, a different type of recombinant dimeric sTNFR55 construct, combining two sTNFR55 by attaching a bifunctional 20 kDa polyethylene glycol reagent to a substituted cysteine residue was used to treat active EAE in Lewis rats,[97] resulting in a delay in time of onset and a reduction in severity and duration of the disease depending on dose and dosing schedule.

Interestingly, everyday dosing beginning at immunization was found ineffective, whereas less frequent dosing starting at a later time, but before onset of clinical signs, resulted in beneficial effects.[97] The lack of efficacy observed with prolonged everyday dosing may be explained at least in part by the formation of neutralizing antibodies to the receptor construct. Histopathologic examination revealed that the treatment inhibited the perivascular infiltration of inflammatory cells within the CNS.[97]

The consistently positive anti-TNFα treatment effects in EAE were further confirmed in Lewis rats using a recombinant TNFR55-IgG fusion protein of sTNFR55 and the hinge and downstream constant sequence of the human IgG1 heavy chain.[93] Transfer EAE was induced by intraperitoneal injection of T lymphocytes specific for myelin basic protein (MBP) or MBP peptides. Single doses of this TNFR55-IgG up to the time when clinical signs became apparent resulted in less pronounced weight loss and reduced clinical disease. Rats successfully treated and exposed to a second transfer EAE challenge were sensitive to the renewed challenge, but again responded to TNFR55-IgG treatment irrespective of the first EAE, indicating that the treatment effect in the first disease was not persistent.[93] Furthermore, active EAE was induced by immunization with MBP, or MBP peptide, in adjuvant, characterized by weight loss and severe clinical scores, which became apparent about 9 days post immunization, followed by a recovery phase around day 16. TNFR55-IgG treatment mitigated the weight loss and substantially reduced the clinical signs. Single dose treatment late in the immunization period up to the time of first clinical signs was most effective. Histopathologic examination revealed that TNFR55-IgG treatment did not inhibit but rather delayed cellular infiltration in the CNS.[93] An inflammatory infiltrate in nontreated rats was first seen at day 10, but almost no cell infiltrate could be detected at this time in TNFR55-IgG treated rats, consistent with the findings in a previous report.[97] However, in later stages of the disease, by day 15, the numbers of infiltrating T cells and macrophages were comparable in both TNFR55-IgG treated and control animals.[93] Given the beneficial effect of TNFR55-IgG treatment on clinical signs, this documents a dissociation between cell infiltration and tissue damage and malfunction; it is likely that TNFR55-IgG treatment here interferes with the pathology at a level distinct from cell recruitment, presumably at some terminal effector mechanism. In a further TNFR55-IgG study, a Lewis rat two-stage experimental autoimmune panencephalitis and demyelinating disease, induced by the transfer of S100β specific T cells and subsequent challenge with anti-myelin oligodendrocyte glycoprotein antibodies was investigated.[187] TNFR55-IgG treatment completely prevented clinical disease,[93,187] but in contrast to active EAE, TNFR55-IgG treatment in this condition resulted in a clear inhibition of cellular infiltration into brain tissue.[93]

A number of studies were also performed in allergic inflammation in rats and demonstrated positive TNFR55-IgG treatment effects.[188,189] It may be concluded that generally beneficial treatment effects were seen in the experimental chronic inflammatory diseases studied, but there are clear differences among the various anti-TNFα compounds and studies. One reason for these differences may be the diverse genetic background of the strains of mice and rats used. A further point relates to the distinct pharmacokinetic profiles and TNFα binding properties of the different TNFα receptor constructs. There is no a priori reason why constructs sharing the presence of extracellular TNFα receptor domains, but differing with regards to the specific type of receptor, the presence or absence of distinct immunoglobulin sequence elements, and the expression system used for production, with its distinct posttranslational modifications, all should have the same activities. A further, often forgotten element to consider is that these TNFα receptor constructs in general use human sequences, the low degree of species specificity of TNFα allowing their use in many species, but they induce a strong immune response in animals with neutralizing antibody titers rising rapidly after 5–7 days, leading to efficient elimination of the compounds. Attempts to circumvent this complication have been made either by using high enough doses of the constructs to induce tolerance, or by generating mice expressing human TNFR55-IgG as an endogenous transgenic molecule,[41,42,71]

9.3.4 Clinical Results

The first clinical trial with a recombinant TNFα receptor construct investigated TNFR75-IgG1 in a randomized controlled multicenter study in septic shock with three doses in the range of 0.15 to 1.5 mg/kg body weight.[190] The recruitment criteria of patients included, within the preceding 24 h, a systemic inflammatory response syndrome (SIRS), and hypotension defined by either a systolic pressure $\leqslant$90 mm Hg, a mean arterial pressure $\leqslant$65 mm Hg, or a sustained decrease in systolic pressure $\geqslant$40 mmHg, despite adequate fluid resuscitation, or the need for vasopressors in excess of 5 μg kg^{-1} min^{-1} dopamine. The primary endpoint was all-cause 28 day mortality. Against all expectations from preclinical studies[21–24,27,144,145,181,191] and from clinical sepsis trials with anti-TNFα antibodies,[158,160] an intention-to-treat analysis showed a dose–response relation between TNFR75-IgG treatment and increased mortality.[190] There was a trend to increased rates of mortality with increasing doses of TNFR75-IgG among patients with Gram-positive organisms, whereas patients with Gram-negative or polymicrobial infections had no negative effect. When rewiewing this unexpected result, three key elements should be kept in mind: the compound used (TNFR75-IgG1), the dosing regimen (about 10 to 100 mg for a 70 kg patient), and the targeted patient population (strong emphasis on hypotension).

It may be appropriate first to compare these results with a larger phase III study of a mouse antihuman TNFα monoclonal antibody in sepsis with or without shock.[158,160] The patients recruited here had to show evidence of acute infection, SIRS, and signs of inadequate organ perfusion; all patients had to show an acute change from baseline condition within less than 12 h before enrollment. Single doses in the range from 210 mg to 1.050 g antibody per 70 kg patient were investigated. Shock was defined as a sustained decrease in blood pressure similar to the TNFR75-IgG1 trial, but the decrease had to be present for at least 30 min and nonresponsive to fluid challenge; shock was considered to persist as long as vasopressor administration was required to maintain the systolic pressure $\geqslant 90\,\text{mmHg}$. An interim analysis of one study revealed no evidence of efficacy and a nonsignificant increased mortality among the patients receiving the antibody in the non-shock group; enrollment of these patients was therefore discontinued. In the final analysis, antibody treatment did not reduce 28 day all-cause mortality, although a nonsignificant improvement in survival was seen in the low dose group. Notably, there was no evidence for a detrimental effect that could be compared to that seen in the TNFR75-IgG1 study.[190] Furthermore, in the prospectively defined group of patients with shock, favorable effects with regard to reductions in time to reversal of shock and to development of organ failure, and delay in the onset of organ failure were observed, supporting the view that TNFα is an important mediator in sepsis.

Human sepsis is a highly complex and heterogeneous condition, influenced by diverse preexisting conditions and genetic backgrounds determining the patients' individual sensitivities to microorganisms and SIRS. Recruitment focusing on hypotension, persistent hypotension, or organ dysfunction will select distinct patient populations. All this excludes a direct comparison of the TNFR75-IgG and anti-TNFα antibody studies, but the contrast in the outcome in the two studies, both targeting TNFα, needs further discussion. The anti-TNFα antibody was administered at substantially higher doses, but the negative outcome with TNFR75-IgG was regularly dose related and thus cannot be attributed to underdosing. Furthermore, the specific activities of receptor constructs and antibodies are different. It may even be argued that the approximately 10–100-fold higher doses of the anti-TNFα antibody might reflect the distinct specific activities of receptor constructs and antibodies (see Section 9.3.2). With regard to bacteriology, in contrast to the TNFR75-IgG trial, there were no significant differences in pure Gram-positive, Gram-negative, mixed bacterial, or nonbacterial infections in the anti-TNFα antibody trial.[160] There are overlaps, but also substantial differences in the definitions of the patient groups recruited; it is now recognized that the targeting of the proper patient group is one of the most important factors in sepsis studies, but the different recruitment criteria here do not sufficiently explain the different outcome.

One intriguing peculiarity of TNFR75-IgG is the fast exchange rate of TNFα binding, which had resulted in persistent active TNFα plasma concentrations in animal studies;[27,28] (see section 9.3.3.1). The potential contribution to toxicity of this phenomenon has been analyzed using the available clinical data of the TNFR75-IgG trial.[190] TNFα was detected at baseline in only 4 % of those patients. Free TNFα is rapidly cleared from plasma, yet in the presence of TNFR75-IgG, TNFα is bound and retained in the plasma according to the long half-life of TNFR75-IgG. Bound TNFα was measured by ELISA, and as expected, accumulating total TNFα plasma concentrations were detected in 40% of the patients after TNFR75-IgG administration. To elucidate whether this was paralleled by a buildup of TNFα activity, the plasma IL-6 concentrations were measured as a surrogate marker for proinflammatory cytokine activity; however, elevated IL-6 concentrations were found only in few patients and there was no correlation with the TNFR75-IgG dose.[190]

The sequelae of persistent accumulation of plasma TNFα due to the presence of TNFR75-IgG were further investigated in human volunteers treated with intravenous injection of endotoxin.[192] Intriguingly, the TNFα accumulation was found to be TNFR75-IgG dose related despite the fact that TNFR75-IgG was in large excess of expected free TNFα. However, TNFα cytotoxic activity of the plasma was neutralized, and excess TNFα neutralising activity was found in 24-h plasma samples. Surprisingly, the secondary release of IL-6 and other mediators showed an abnormal dose response with greater suppression effects at lower TNFR75-IgG dose; at high TNFR75-IgG dose, comparable to doses in the clinical trial, IL-6 and MIP-1a responses to endotoxin were as high or slightly higher than in the absence of TNFR75-IgG. Furthermore, fever, stress hormone, and acute phase responses were prolongated or enhanced by TNFR75-IgG treatment, suggesting that the presence of TNFR75-IgG modifies and contributes to the host response to endotoxin.[192] These findings raise a number of questions. Are these phenomena due to some secondary effect of TNFR75-IgG, independent of its TNFα binding activity? Did TNFR75-IgG induce a shift in compartmentalization of TNFα between plasma and tissue? Can TNFα ligand passing related to the fast exchange rate of TNFR75-IgG be rigorously excluded? There are secondary functions in IgG fusion proteins, such as for example the Fc moiety that may be seen by complement or Fcγ receptors; TNFR75-IgG was reported not to activate complement in vitro, but Fcγ receptor cross-linking could not be ruled out. Straightforward answers to these questions are lacking, but a reasoned judgment still can be formed. Given the need for new sepsis therapies, the most important practical question is whether the enhanced endotoxin response in volunteers explains the negative outcome of the clinical TNFR75-IgG trial, and whether it resulted from targeting TNFα in general; there is now substantial evidence supporting the view that the negative outcome of the TNFR75-IgG sepsis phase II trial relates to that specific trial rather than demonstrates the futility of anti-TNFα strategies in sepsis in general. It is noteworthy that this outcome need not be predictive for the effects of TNFR75-IgG treatment in other disease indications (see following).

More encouraging results emerged from a later clinical study, in this case of TNFR55-IgG.[164] The randomized controlled multicenter phase II study of 498 patients was larger than the 141-patient TNFR75-IgG trial. The patients were prospectively stratified into refractory septic shock and severe sepsis/early septic shock groups. The inclusion criteria were objective signs of infection, SIRS, and evidence of at least two instances of inadequate organ function. Inadequate organ functions included refractory hypotension manifested by a similar drop in blood pressure as in previous trials, nonresponding to fluid challenge and requiring vasopressor treatment of $>5\,\mu g\, kg^{-1}\, min^{-1}$ dopamine for $\geqslant 2$ h. Patients with refractory hypotension and one additional inadequate organ function were randomized into the refractory shock group; patients in the severe sepsis group had to fulfill criteria of at least two inadequate organ functions with the exception of refractory hypotension. Signs of organ system dysfunction had to be present within a 12 h period prior to TNFR55-IgG administration. Doses of approximately 3 and 6 mg TNFR55-IgG per 70 kg patient were administered. The 28 day all-cause mortality in the prospectively stratified severe sepsis patients was 23%, 37%, and 36% for the high dose, low dose, and placebo groups, respectively, representing a 36% reduction in mortality among patients treated with the high TNFR55-IgG dose compared to placebo.[164] The trend to improved survival was present in pure Gram-positive as well as Gram-negative infections and was independent of baseline IL-6 levels. In refractory shock patients, no significant increase in survival was observed, although there was a small reduction in mortality in the low TNFR55-IgG dose group; the analysis of all patients revealed a statistically nonsignificant trend toward reduced 28 day all-cause mortality in the TNFR55-IgG treatment group compared to placebo patients.[164] In view of past experiences with anti-cytokine therapeutic trials in sepsis, these promising latter results must be viewed with caution. However, the contradictory nature of these various trial outcomes might be related to the distinct properties of the various compounds and doses, and the differences in patient selection criteria. Only further studies will definitively reveal the true merit or failure of anti-TNFα strategies in sepsis. A pivotal phase III study of TNFR55-IgG in severe sepsis/early septic shock patients is underway and may provide further insight.

Remarkably, in contrast to its detrimental effects in sepsis, TNFR75-IgG has shown positive treatment effects in RA patients.[166,193] In a small toxicity and dose finding trial in refractory RA, TNFR75-IgG doses were escalated in four steps from a 4 mg/m^2 intravenous loading dose (day 1) and 2 mg/m^2 twice weekly maintenance dose (starting at day 4 and ending day 29) subcutaneous injection protocol to a 32 mg/m^2 loading dose/16 mg/m^2 maintenance dose protocol. No serious adverse events were seen in 22 patients. Trends of improvement were found in clinical variables such as painful and swollen joint counts as well as in biochemical markers such as erythrocyte sedimentation rate and C reactive protein in all treated patients grouped together, but there was no clear dose response among the treatment groups.[166] The hope that TNFα receptor constructs may become valid therapeutic tools in RA was further supported by reports of phase II studies using TNFR75-IgG[193] and TNFR55-IgG.[167,168] Chronic administration of these TNFα receptor fusion proteins in dose ranges up to about 100 mg per month in single or repetitive dosing regimens have been remarkably well tolerated. The formation of antibodies has been reported. Interestingly, single intravenous administration of TNFR75-IgG resulted in antibody formation in 15%

of the patients,[190] whereas long-term treatment with repeated subcutaneous administration of the same compound was reported not to lead to antibody formation.[166,169,193] On the basis of successful phase III results, TNFR75-IgG has been approved recently for treatment of RA by the United States Food and Drug Administration. The complexity of the role of TNF in disease was further demonstrated in a TNFR55-IgG1 phase II study in MS patients. Broad, if circumstantial, evidence from clinical laboratory studies, and extensive preclinical data, suggest that TNF has a role in MS pathology. Yet, in this study of safety and efficacy of TNF neutralizing treatment in 167 patients with relapsing-remitting and secondary progressing MS, magnetic resonance imaging analyses did not reveal statistically significant differences between active treatment and placebo groups. However, the number of patients with exacerbations and the total number of exacerbations were higher, and the time to first relapse was shorter, in the active treatment groups.[194]

9.4 CONCLUDING REMARKS

Man, or animals, confronted with infection, injury, or chronic focal inflammation, respond with a host defense that is based on molecular mechanisms which are being increasingly unraveled. A highly complex network of stress and acute phase reactants, cytokines, hormones, complement, and coagulation cascades integrate into the host defense, but when elicited in excess can flip over to lethal systemic toxicity or progression of chronic inflammation and tissue injury. The elements that determine whether the host response will protect an individual or contribute to toxicity are unknown. TNFα is one of the central mediators in host defense, and there is increasing evidence that targeting but this one cytokine in the mediator network may become a valid therapeutic strategy in quite diverse disease conditions involving immune/inflammatory cell communication.

Modern molecular biology has provided tools to construct fusion proteins, combining elements of diverse genes and approaching a rational design of desired activities. The first fusion proteins combining receptor and immunoglobulin sequences were constructed from CD4 extracellular domain and immunoglobulin IgG and IgM heavy or light chain sequences.[195–197] Elements to consider in the design are molecular size—and the related pharmacokinetic properties—of the fusion construct as compared to the isolated receptor moiety, and secondary functions introduced by the immunoglobulin heavy chain sequences such as complement and Fc receptor binding, and potential oligomerization. The Fc moiety of fusion proteins can be deliberately used to target Fc receptors.[195–197]

To achieve a high ligand binding activity, one critically important element is the structural configuration of the receptor–ligand interaction. TNFα receptor-IgG constructs are potent TNFα binding compounds as a consequence of the stereochemical mode of TNFα–TNFα receptor interaction; the unique trimeric structure of the TNFα (and LTα) molecule allows the binding of the two receptor moieties of one TNFR-IgG molecule to one TNFα trimer, resulting in a much higher TNFα neutralizing activity of TNFR-IgG compared to monomeric soluble receptors. Extrapolating this design principle to other ligand–receptor pairs, and fusing extracellular receptor domains to heavy chain fragments in the absence of supportive information on the stereochemistry of ligand–receptor interaction, will not necessarily lead to similarly optimized binding activity. In general, it cannot be excluded that depending on the molecular architecture a dimeric receptor-IgG fusion protein might even have undesirable ligand binding properties. The design of a receptor-IgG fusion protein therefore must be carefully reasoned for each individual receptor–ligand pair in view of its desired properties.

REFERENCES

1. Coley, W. B. 1893. The treatment of malignant tumors by repeated inocculations of erysipelas. With a report of ten original cases. *Amer. J. Med. Sci.* 105: 487–511.
2. Coley, W. B. 1906. Late results of the treatment of inoperable sarcoma by the mixed toxins of erysipelas and Bacillus prodigiosus. *Am. J. Med. Sci.* 131: 375–430.
3. O'Malley, W. E., B. Achinstein, M. J. Shear. 1962. Action of bacterial polysaccharide on tumors. II. Damage of sarcoma 37 by serum of mice treated with Serratia marcescens polysaccharide, and induced tolerance. *J. Natl. Cancer Inst.* 29: 1169–1175.
4. Carswell, E. A., L. J. Old, R. L. Kassel, S. Green, N. Fiore, B. Williamson. 1975. An endotoxin induced serum factor that causes necrosis of tumors. *Proc. Natl. Acad. Sci. USA* 72: 3666–3670.
5. Beutler, B., A. Cerami. 1989. The biology of cachectin/TNF—a primary mediator of the host response. Ann. Rev. Immunol. 7: 625–655.
6. Vassalli, P. 1992. The pathophysiology of tumor necrosis factors. *Ann. Rev. Immunol.* 10: 411–452.
7. Sidhu, R. S., A. P. Bollon. 1993. Tumor necrosis factor activities and cancer therapy—a perspective. *Pharmac. Ther.* 57: 79–128.
8. Tagouchi, T., Y. Sohmura. 1991. Clinical studies with TNF. *Biotherapy.* 3: 177–186.
9. Lienard, D., P. Ewalenko, J. J. Delmotte, N. Renard, F. J. Lejeune. 1992. High-dose recombinant tumor necrosis factor alpha in combination with interferon gamma and melphalan in isolation perfusion of the limbs for melanoma and sarcoma. *J. Clin. Oncol.* 10: 52–60.
10. Beutler, B., D. Greenwald, J. D. Hulmes, M. Chang, Y. C. E. Pan, J. Mathison, R. Ulevitch, A. Cerami. 1985. Identity of tumour necrosis factor and the macrophage-secreted factor cachectin. *Nature* 316: 552–554.
11. Fong, Y., M. A. Marano, L. L. Moldawer, H. Wei, S. E. Calvano, J. S. Kenney, A. C. Allison, A. Cerami, G. T. Shires, S. F. Lowry. 1990. The acute splanchnic and peripheral tissue metabolic response to endotoxin in humans. *J. Clin. Invest.* 85: 1896–1904.

12. Hesse, D. G., K. J. Tracey, Y. Fong, K. R. Manogue, M. A. Palladino, A. Cerami, G. T. Shires, S. F. Lowry. 1988. Cytokine appearance in human endotoxemia and primate bacteremia. *Surg. Gynecol. Obstet.* 166: 147–153.

13. Michie, H. R., K. R. Manogue, D. R. Spriggs, A. Revhaug, S. O'Dwyer, C. A. Dinarello, A. Cerami, S. M. Wolff, D. W. Willmore. 1988. Detection of circulating tumor necrosis factor after endotoxin administration. *N. Engl. J. Med.* 318: 1481–1486.

14. van Deventer, S. J. H., H. R. Buller, J. W. ten Cate, L. A. Aarden, C. E. Hack, A. Sturk. 1990. Experimental endotoxemia in humans; analysis of cytokine release and coagulation, fibrinolytic and complement pathways. *Blood* 76: 2520–2526.

15. van Zee, K., L. deForge, E. Fischer, M. A. Marano, J. S. Kenney, D. G. Remick, S. F. Lowry, L. L. Moldawer. 1991. IL-8 in septic shock, endotoxemia and following IL-1 administration. *J. Immunol.* 146: 3478–3482.

16. Tracey, K., B. R. Beutler, S. F. Lowry, J. Merryweather, S. Wolpe, I. W. Milsark, R. J. Hariri, T. J. Fahey, A. Zentella, J. D. Albert, G. T. Shires, A. Cerami. 1986. Shock and tissue injury induced by recombinant human cachectin. *Science* 234: 470–474.

17. Tracey, K., S. F. Lowry, T. J. Fahey, J. D. Albert, Y. Fong, D. Hesse, B. R. Beutler, K. R. Manogue, S. Calvano, H. Wei, A. Cerami, G. T. Shires. 1987. Cachectin/tumor necrosis factor induces lethal shock and stress hormone responses in the dog. *Surg. Gynecol. Obstet.* 164: 415–422.

18. Redl, H., G. Schlag, M. Ceska, J. Davies, W. A. Buurman. 1993. Interleukin-8 release in baboon septicemia is partially dependent on tumor necrosis factor. *J. Infect. Dis.* 167: 1464–1466.

19. Redl, H., G. Schlag, S. Bahrami, R. Kargl, W. Hartter, W. Woloszczuk, J. Davies. 1994. Big-endothelin release in baboon bacteremia is partially TNF dependent. *J. Lab. Clin. Med.* 124: 796–801.

20. Redl, H., G. Schlag, A. Schiesser, J. Davies. 1995. Thrombomodulin release in baboon sepsis: Its dependence on the dose of Escherichia coli and the presence of tumor necrosis factor. *J. Infect. Dis.* 171: 1522–1527.

21. Tracey, K., Y. Fong, D. G. Hesse, K. R. Manogue, A. T. Lee, G. C. Kuo, S. F. Lowry, A. Cerami. 1987. Anti-cachectin/TNF monoclonal antibodies prevent septic shock during lethal bacteraemia. *Nature* 330: 662–664.

22. Hinshaw, L. B., P. Tekamp-Olsen, A. C. Chang, P. A. Lee, F. B. Taylor, C. K. Murray, G. T. Peer, T. E. Emerson, R. B. Passey, G. C. Kuo. 1990. Survival of primates in LD100 septic shock following therapy with antibody to tumor necrosis factor (TNF alpha). *Circ. Shock.* 30: 279–292.

23. Lesslauer, W., H. Tabuchi, R. Gentz, M. Brockhaus, E. J. Schlaeger, G. Grau, P. F. Piguet, P. Pointaire, P. Vassalli, H. R. Loetscher. 1991. Recombinant soluble TNF receptor proteins protect mice from LPS-induced lethality. *Eur. J. Immunol.* 21: 2883–2886.

24. Ashkenazi, A., S. A. Marsters, D. J. Capon, S. M. Chamow, I. S. Figari, P. Pennica, D. V. Goeddel, M. A. Palladino, D. H. Smith. 1991. Protection against endotoxic shock by a tumor necrosis factor receptor immunoadhesin. *Proc. Natl. Acad. Sci. USA* 88: 10535–10539.

25. Peppel, K., D. Crawford, B. Beutler. 1991. A tumor necrosis factor receptor IgG

heavy chain chimeric protein as a bivalent antagonist of TNF activity. *J. Exp. Med.* 174: 1483–1489.

26. van Zee, K. J., T. Kohno, E. Fischer, C. S. Rock, L. L. Moldawer, S. F. Lowry. 1992. Tumor necrosis factor soluble receptors circulate during experimental and clinical inflammation and can protect against excessive tumor necrosis factor a *in vitro* and *in vivo*. *Proc. Natl. Acad. Sci. USA* 89: 4845–4849.
27. Mohler, K. M., D. S. Torrance, C. A. Smith, R. E. Goodwin, K. E. Stremler, V. P. Fung, H. Madani, M. B. Widmer. 1993. Soluble tumor necrosis factor receptors are effective therapeutic agents in lethal endotoxemia and function simultaneously as both TNF carriers and TNF antagonists. *J. Immunol.* 151: 1548–1561.
28. Evans, T. J., D. Moyes, A. Carpenter, R. Martin, H. Loetscher, W. Lesslauer, J. Cohen. 1994. Protective effect of 55 kD but not 75 kD soluble tumor necrosis factor receptor eceptor-immunoglobulin G fusion proteins in an animal model of Gram-negative sepsis. *J. Exp. Med.* 180: 2173–2179.
29. Nakane, A., T. Minagawa, K. Kato. 1988. Endogenous tumor necrosis factor (cachectin) is essential to host resistance against Listeria monocytogenes infection. *Infect. Immun.* 56: 2563–2569.
30. Havell, E. 1989. Evidence that tumor necrosis factor has an important role in antibacterial resistance. *J. Immunol.* 143: 2894–2899.
31. Czuprynski, C. J., J. F. Brown, K. M. Young, A. J. Cooley, R. S. Kurtz. 1988. Effects of murine recombinant interleukin 1α on the host response to bacterial infection. *J. Immunol.* 140: 962–968.
32. Nakano, Y., K. Onozuka, Y. Terada, H. Shinomiya, M. Nakano. 1990. Protective effect of recombinant tumor necrosis fatcor α in murine salmonellosis. *J. Immunol.* 144: 1935–1941.
33. Blanchard, D. K., J. Y. Djeu, T. W. Klein, H. Friedman, W. E. I. Stewart. 1988. Protective effects of tumor necrosis factor in experimental Legionella pneumophilia infections of mice via activation of PMN function. *J. Leukocyte Biol.* 43: 429–435.
34. Echtenacher, B., W. Falk, D. Mannel, P. H. Krammer. 1990. Requirement of endogenous tumor necrosis factor/cachectin for recovery from experimental peritonitis. *J. Immunol.* 145: 3762–3766.
35. Havell, E. A. 1992. Role of TNF in resistance to bacteria. In *Tumor Necrosis Factors. Structure, Function and Mechanisms of Action*, ed. B. B. Aggarwal, J. Vilcek. New York: Marcel Dekker, pp. 341–361.
36. Pfeffer, K., T. Matsuyama, T. M. Kuendig, A. Shahinian, K. Wiegmann, P. S. Ohashi, M. Kronke, T. W. Mak. 1993. Mice deficient for the 55 kd tumor necrosis factor receptor are resistant to endotoxic shock, yet succumb to *L. monocytogenes infection*. *Cell* 73: 457–467.
37. Rothe, J., W. Lesslauer, H. R. Loetscher, Y. Lang, P. Koebel, F. Koentgen, A. Althage, R. Zinkernagel, M. Steinmetz, H. Bluethmann. 1993. Mice lacking the tumor necrosis factor 1 are resistant to TNF-mediated toxicity but highly susceptible to infection by Listeria monocytogenes. *Nature* 364: 798–802.
38. Kindler, V., A. P. Sappino, G. E. Grau, P. F. Piguet, P. Vassalli. 1989. The inducing role of tumor necrosis factor in the development of bactericidal granulomas during BCG infection. *Cell* 56: 731–740.

39. Titus, R., B. Sherry, A. Cerami. 1989. Tumor necrosis factor plays a protective role in experimental murine cutaneous leishmaniasis. *J. Exp. Med.* 170: 2097–2104.
40. Liew, F. Y., C. Parkinson, S. Millot, A. Severn, M. Carrier. 1990. Tumor necrosis factor (TNF) in leishmaniasis. *Immunology* 69: 570–573.
41. Garcia, I., Y. Miyazaki, K. Araki, R. Lucas, G. E. Grau, G. Milon, Y. Belkaid, C. Montixi, W. Lesslauer, P. Vassalli. 1995. Transgenic mice expressing high levels of soluble TNF-R1 fusion protein are protected from lethal septic shock and cerebral malaria, and are highly sensitive to *Listeria monocytogenes* and *Leishmania major* infections. *Eur. J. Immunol.* 25: 2401–2407.
42. Garcia, I., Y. Miyazaki, G. Marchanl, W. Lesslauer, P. Vassalli. 1997. High sensitivity of transgenic mice expressing soluble TNFR1 fusion protein to mycobacterial infections: synergistic action of TNF and IFNγ in the differentiation of protective granulomas. *Eur. J. Immunol.* 27: 3182–3190.
43. Bruce, A. J., W. Boling, M. S. Kindy, J. Peschon, P. J. Kraemer, M. K. Carpenter, F. W. Holtsberg, M. P. Mattson. 1996. Altered neuronal and microglial responses to excitotoxic and ischemic brain injury in mice lacking TNF receptors. *Nature Medicine* 2: 788–794.
44. Smith, C. A., T. Farrah, R. G. Goodwin. 1994. The TNF receptor superfamily of cellular and viral proteins: activation, costimulation, and death. *Cell* 76: 959–962.
45. Beutler, B., C. van Huffel. 1994. Unraveling function in the TNF ligand and receptor families. *Science* 264: 667–668.
46. Kriegler, M., C. Perez, K. DeFay, I. Albert, S. D. Lu. 1988. A novel form of TNF/cachectin is a cell surface cytotoxic transmembrane protein : ramifications for the complex physiology of TNF. *Cell* 53: 45–53.
47. Black, R. A., C. T. Rauch, C. J. Kozlosky, J. J. Peschon, J. L. Slack, M. F. Wolfson, B. J. Castner, K. L. Stocking, P. Reddy, S. Srinivasan, N. Nelson, N. Boiani, K. A. Schooley, M. Gerhart, R. Dacis, J. N. Fitzner, R. S. Johnson, R. J. Paxton, C. J. March, D. P. Cerretti. 1997. A metalloproteinase disintegrin that releases tumor-necrosis-factor-α from cells. *Nature* 385: 729–733.
48. Moss, M. L., S.-L. C. Jin, M. E. Milla, W. Burkhart, H. L. Carter, W. J. Chen, W. C. Clay, J. R. Didsbury, D. Hassler, C. R. Hoffman, T. A. Kost, M. H. Lambert, M. A. Leesnitzer, P. McCauley, G. McGeehan, J. Mitchell, M. Moyer, G. Pahel, W. Rocque, L. K. Overton, F. Schoenen, T. Seaton, J. L. W. Su, J., D. Willard, J. D. Becherer. 1997. Cloning of a disintegrin metalloproteinase that processes precursor tumor-necrosis-factor-α. *Nature* 385: 733–736.
49. Decker, T., M. L. Lohmann-Matthes, G. E. Gifford. 1987. Cell-associated tumor necrosis factor (TNF) as a killing mechanism of activated cytotoxic macrophages. *J. Immunol.* 138: 957–962.
50. Karp, S. E., P. Hwu, A. Farber, N. P. Restifo, M. Kriegler, J. J. Mule, S. A. Rosenberg. 1992. In vivo activity of tumor necrosis factor (TNF) mutants. *J. Immunol.* 149: 2076–2081.
51. Grell, M., E. Douni, H. Wajant, M. Lohden, M. Clauss, B. Maxeiner, S. Georgopoulos, W. Lesslauer, G. Kollias, K. Pfizenmaier, P. Scheurich. 1995. The transmembrane form of tumor necrosis factor (TNF) is the prime activating ligand of the 80kDa TNF receptor. *Cell* 83: 793–802.
52. Browning, J. L., A. Ngam-ek, P. Lawton, J. DeMarinis, R. Tizard, E. P. Chow, C. Hession, B. O'Brine-Greco, S. F. Foley, C. F. Ware. 1993. Lymphotoxin β, a novel

member of the TNF family that form a heteromeric complex with lymphotoxin on the cell surface. *Cell* 72: 847–856.

53. Browning, J. L., I. Dougas, A. Ngam-ek, P. R. Bourdon, B. N. Ehrenfels, K. Miatkowski, M. Zafari, A. M. Yampaglia, P. Lawton, W. Meier, C. P. Benjamin, C. Hession. 1995. Characterization of surface lymphotoxin forms. *J. Immunol.* 154: 33–46.
54. Dayer, J.-M., B. Beutler, A. Cerami. 1985. Cachectin/tumor necrosis factor stimulates collagenase and prostaglandin E_2 production by human synovial cells and dermal fibroblasts. *J. Exp. Med.* 162: 2163–2168.
55. Saklatvala, J. 1986. Tumor necrosis factor a stimulates resorption and inhibits synthesis of proteoglycan in cartilage. *Nature* 322: 547–549.
56. Shinmei, M., K. Masuda, T. Kikuchi, Y. Shimomura. 1989. The role of cytokines in chondrocyte mediated cartilage degradation. *J. Rheumatol. Suppl.* 18: 32–34.
57. Pratta, M. A., T. M. di Meo, D. M. Ruhl, E. C. Arner. 1989. Effect of interleukin-1β and tumor necrosis factor a on cartilage proteoglycan metabolism in vitro. *Agents Actions* 27: 250–253.
58. Kolibas, L. M., R. L. Goldberg. 1989. Effect of cytokines and anti-arthritic drugs on glycosaminoglycan synthesis by bovine articular chondrocytes. *Agents Actions* 27: 245–249.
59. Bertolini, D. R., G. E. Nedwin, T. S. Bringman, D. D. Smith, G. R. Mundy. 1986. Stimulation of bone resorption and inhibition of bone formation in vitro by human tumor necrosis factor. *Nature* 319: 516–518.
60. Hopkins, S. J., A. Meager. 1988. Cytokines in synovial fluid: II The presence of tumor necrosis factor and interferon. *Clin. Exp. Immunol.* 73: 88–92.
61. Saxne, T., M. A. Palladino, D. Heinegard, N. Tala, F. A. Wollheim. 1988. Detection of tumor necrosis factor a but not tumor necrosis factor b in rheumatoid arthritis synovial fluid and serum. *Arthritis Rheum.* 31: 1041–1045.
62. di Giovine, F. S., G. Nuki, G. Duff. 1988. Tumor necrosis factor in synovial exudates. *Ann. Rheum. Dis.* 47: 768–772.
63. Westacott, C. I., J. T. Whicher, I. C. Barnes, D. Thompson, A. J. Swan, P. A. Dieppe. 1990. Synovial fluid concentration of five different cytokines in rheumatoid diseases. *Ann. Rheum. Dis.* 49: 676–681.
64. Tetta, C., G. Camussi, V. Modena, C. di Vittorio, C. Baglioni. 1990. Tumor necrosis factor in serum and synovial fluid of patients with active and severe rheumatoid arthritis. *Ann. Rheum. Dis.* 49: 665–667.
65. Chu, C. Q., M. Field, M. Feldmann, R. N. Maini. 1991. Localisation of tumor necrosis factor a in synovial tissue and at the cartilage-pannus junction in patients with rheumatoid arthritis. *Arthritis Rheum.* 34: 1125–1132.
66. Brennan, F. M., D. L. Gibbons, T. Mitchell, A. P. Cope, R. N. Maini, M. Feldmann. 1992. Enhanced expression of tumor necrosis factor receptor mRNA and protein in mononuclear cells isolated from rheumatoid arthritis synovial joints. *Eur. J. Immunol.* 22: 1907–1912.
67. Wooley, P. H., J. Dutcher, M. B. Widmer, S. Gillis. 1993. Influence of a recombinant human soluble tumor necrosis factor receptor FC fusion protein on type II collagen induced arthritis in mice. *J. Immunol.* 151: 6602–6607.
68. Williams, R. O., M. Feldmann, R. Maini. 1992. Anti-tumor necrosis factor

ameliorates joint disease in murine collagen-induced arthritis. *Proc. Natl. Acad. Sci. USA* 89: 9784–9788.

69. Thorbecke, G. J., R. Shah, C. H. Leu, A. P. Kuruvilla, A. M. Hardison, M. A. Palladino. 1992. Involvement of endogenous tumor necrosis factor a and transforming growth factor b during induction of collagen type II arthritis in mice. *Proc. Natl. Acad. Sci. USA* 89: 7375–7379.
70. Piguet, P. F., G. E. Grau, C. Vesin, H. Loetscher, R. Gentz, W. Lesslauer. 1992. Evolution of collagen arthritis in mice is arrested by treatment with anti-tumor necrosis factor (TNF) antibody or a recombinant soluble TNF receptor. *Immunology* 77: 510–514.
71. Mori, L., S. Iselin, G. de Libero, W. Lesslauer. 1996. Attenuation of collagen-induced arthritis in TNFR1-IgG1-treated and TNFR1-deficient mice. *J. Immunol.* 157: 3178–3182.
72. Cheng, J., K. Turksen, Q. C. Yu, H. Schreiber, M. Teng, E. Fuchs. 1992. Cachexia and graft-vs.-host-disease-type skin changes in keratin promoter driven TNFα transgenic mice. *Genes & Development.* 6: 1444–1456.
73. Probert, L., D. Plows, G. Kontogeorgos, G. Kollias. 1995. The type I interleukin-1 receptor acts in series with tumor necrosis factor (TNF) to induce arthritis in TNF-transgenic mice. *Eur. J. Immunol.* 25: 1794–1797.
74. Keffer, J., L. Probert, H. Cazlaris, S. Georgopoulos, E. Kaslaris, D. Kioussis, G. Kollias. 1991. Transgenic mice expressing human tumor necrosis factor: a predictive genetic model of arthritis. *EMBO J.* 10: 4025–4031.
75. Hofman, F. M., D. R. Hinton, K. Johnson, J. E. Merrill. 1989. Tumor necrosis factor identified in multiple necrosis brain. *J. Exp. Med.* 170: 607–612.
76. Selmaj, K., C. S. Raine, B. Cannella, C. F. Brosnan. 1991. Identification of lymphotoxin and tumor necrosis factor in multiple sclerosis lesions. *J. Clin. Invest.* 87: 949–954.
77. Woodroofe, M. N., M. L. Cuzner. 1993. Cytokine mRNA expression in inflammatory multiple sclerosis lesions: detection by non-radioactive in situ hybridisation. *Cytokine* 5: 583–588.
78. Selmaj, K. W., C. S. Raine. 1988. Tumor necrosis factor mediates myelin and oligodendrocyte damage in vitro. *Ann. Neurol.* 23: 339–346.
79. Selmaj, K. W., M. Farooq, W. T. Norton, C. S. Raine, C. F. Brosnan. 1990. Proliferation of astrocytes in vitro in response to cytokines A primary role for tumor necrosis factor. *J. Immunol.* 144: 129–135.
80. Sharief, M. K., R. Hentges. 1991. Association between tumor necrosis factor a and disease progression in patients with multiple sclerosis. *New Engl. J. Med.* 325: 467–472.
81. Franciotta, D. M., L. M. E. Grimaldi, G. V. Martino, G. Piccolo, G. V. Melzi d'Eril. 1989. Tumor necrosis factor in serum and cerebrospinal fluid of patients with multiple sclerosis. *Ann. Neurol.* 26: 787–789.
82. Hauser, S. L., T. H. Doolittle, R. Lincoln, R. H. Brown, C. A. Dinarello. 1990. Cytokine accumulations in CSF of multiple sclerosis patients: frequent detection of interleukin-1 and tumor necrosis factor but not interleukin-6. *Neurology* 40: 1735–1739.
83. Maimone, D., S. Gregory, B. G. W. Arnason, A. T. Reder. 1991. Cytokine levels in

the cerebrospinal fluid and serum of patients with multiple sclerosis. *J. Neuroimmunol.* 32: 67–74.

84. Peter, J. B., F. N. Boctor, W. W. Tourtellotte. 1991. Serum and CSF levels of IL-2, sIL-2R, TNF-α, and IL-1β in chronic progressive multiple sclerosis: expected lack of clinical utility. *Neurology* 41: 121–123.

85. Rieckmann, P., M. Albrecht, B. Kitze, T. Weber, H. Tumani, A. Broocks, W. Luer, A. Helwig, S. Poser. 1995. Tumor necrosis factor-α messenger RNA expression in patients with relapsing-remitting multiple sclerosis is associated with disease activity. *Ann. Neurol.* 37: 82–88.

86. Rivers, T. M., D. H. Sprunt, G. P. Berry. 1933. Observations on attempts to produce acute disseminated encephalomyelitis in monkeys. *J. Exp. Med.* 58: 39–53.

87. Zamvil, S. S., L. Steinman. 1990. The T lymphocyte in experimental allergic encephalomyelitis. *Annu. Rev. Immunol.* 8: 579–621.

88. Wekerle, H., K. Kojima, J. Lannes-Vieira, H. Lassmann, C. Linington. 1994. Animal models. *Ann. Neurol.* 36: S47–S53.

89. Martin, R., H. F. McFarland. 1997. Immunology of multiple sclerosis and experimental allergic encephalomyelitis. In *Multiple Sclerosis*, ed. C. S. Raine, H. F. McFarland, W. W. Tourtelotte. London: Chapman & Hall Medical, pp. 221–242.

90. Powell, M. B., D. Mitchell, J. Lederman, J. Buckmeier, S. S. Zamvil, M. Graham, N. H. Ruddle, L. Steinman. 1990. Lymphotoxin and tumor necrosis factor production by myelin basic protein-specific T cell clones correlates with encephalitogenicity. *Intern. Immunol.* 2: 539–544.

91. Huitinga, I., N. van Rooijen, C. J. A. deGroot, B. M. J. Uitdehaag, C. D. Dijkstra. 1990. Suppression of experimental allergic encephalomyelitis in Lewis rats after elimination of macrophages. *J. Exp. Med.* 172: 1025–1033.

92. Baker, D., D. Butler, B. J. Scallon, J. K. O'Neill, J. L. Turk, M. Feldmann. 1994. Control of established experimental allergic encephalomyelitis by inhibition of tumor necrosis factor (TNF) activity within the central nervous system using monoclonal antibodies and TNF receptor-immunoglobulin fusion proteins. *Eur. J. Immunol.* 24: 2040–2048.

93. Klinkert, W. E. F., K. Kojima, W. Lesslauer, W. Rinner, H. Lassmann, H. Wekerle. 1997. TNF-α receptor fusion protein prevents experimental auto-immune encephalomyelitis and demyelination in Lewis rats: an overview. *J. Neuroimmunol.* 72: 163–168.

94. Ruddle, N. H., C. M. Bergman, K. M. McGrath, E. G. Lingenheld, M. L. Grunnet, S. J. Padula, R. B. Clark. 1990. An antibody to lymphotoxin and tumor necrosis factor prevents transfer of experimental allergic encephalomyelitis. *J. Exp. Med.* 172: 1193–1200.

95. Selmaj, K., C. S. Raine, A. H. Cross. 1991. Anti-tumor necrosis factor therapy abrogates autoimmune demyelination. *Ann. Neurol.* 30: 694–700.

96. Selmaj, K., W. Papierz, A. Glabinski, T. Kohno. 1995. Prevention of chronic relapsing experimental autoimmuneencephalomyelitis by soluble TNF receptor I. *J. Neuroimmunol.* 56: 135–141.

97. Martin, D., S. L. Near, A. Bendele, D. A. Russell. 1995. Inhibition of tumor

necrosis factor is protective against neurologic dysfunction after active immunisation of Lewis rats with myelin basic protein. *Exp. Neurol.* 131: 221–228.

98. Korner, H., A. L. Goodsall, F. A. Lemckert, B. J. Scallon, J. Ghrayeb, A. L. Ford, J. D. Sedgwick. 1995. Unimpaired autoreactive T-cell traffic within the central nervous system during tumor necrosis factor receptor-mediated inhibition of experimental autoimmune encephalomyelitis. *Proc. Natl. Acad. Sci. USA* 92: 11066–11070.
99. Dick, A. D., P. G. McMenamin, H. Koerner, B. J. Scallon, J. Ghrayeb, J. V. Forrester, J. D. Sedgwick. 1996. Inhibition of tumor necrosis factor activity minimizes target organ damage in experimental autoimmune uveoretinitis despite quantitatively normal activated T cell traffic to the retina. *Eur. J. Immunol.* 26: 1018–1025.
100. Akassoglou, K., L. Probert, G. Kontogeorgos, G. Kollias. 1997. Astrocyte, but not neuron-specific transmembrane TNF triggers inflammation and degeneration in the central nervous system of transgenic mice. *J. Immunol.* 158: 438–445.
101. Probert, L., K. Akassoglou, M. Pasparakis, G. Kontogeorgos, G. Kollias. 1995. Spontaneous inflammatory demyelinating disease in transgenic mice showing central nervous system-specific expression of tumor necrosis factor alpha. *Proc. Natl. Acad. Sci. USA* 92: 11294–11298.
102. Aggarwal, B. B., T. E. Eessalu, P. E. Hass. 1985. Characterization of receptors for human tumor necrosis factor and their regulation by gamma-interferon. *Nature* 318: 665–667.
103. Loetscher, H., M. Steinmetz, W. Lesslauer. 1991. Tumor necrosis factor: receptors and inhibitors. *Cancer Cells* 3: 221–226.
104. Hohmann, H. P., R. Remy, M. Brockhaus, A. P. G. M. van Loon. 1989. Two different cell types have different major receptors for human tumor necrosis factor (TNFα). *J. Biol. Chem.* 264: 14927–14934.
105. Smith, R. A., M. Kirstein, W. Fiers, C. Baglioni. 1986. Species specificity of human and murine tumor necrosis factor. *J. Biol. Chem.* 261: 14871–14874.
106. Lewis, M., L. A. Tartaglia, A. Lee, G. L. Bennett, G. C. Rice, G. H. W. Wong, E. Y. Chen, D. V. Goeddel. 1991. Cloning and expression of cDNAs for two distinct murine tumor necrosis factor receptors demonstrate one receptor is species specific. *Proc. Natl. Acad. Sci. USA* 88: 2830–2834.
107. Brockhaus, M., H. J. Schoenfeld, E. J. Schlaeger, W. Hunziker, W. Lesslauer, H. Loetscher. 1990. Identification of two types of tumor necrosis factor receptors on human cell lines by monoclonal antibodies. *Proc. Natl. Acad. Sci. USA* 87: 3127–3131.
108. Loetscher, H., Y.-C. E. Pan, H.-W. Lahm, R. Gentz, M. Brockhaus, H. Tabuchi, W. Lesslauer. 1990. Molecular cloning and expression of the human 55 kd tumor necrosis factor receptor. *Cell* 61: 351–359.
109. Dembic, Z., H. Loetscher, U. Gubler, Y.-C. E. Pan, H.-W. Lahm, R. Gentz, M. Brockhaus, W. Lesslauer. 1990. Two human TNF receptors have similar extracellular, but distinct intracellular, domain sequences. *Cytokine* 2: 231–237.
110. Smith, C. A., T. Davis, D. Anderson, M. Solam, M. P. Beckmann, R. Jerzy, S. K. Dower, D. Cosman, R. G. Goodwin. 1990. A receptor for tumor necrosis factor defines an unusual family of cellular and viral proteins. *Science* 248: 1019–1023.

111. Peetre, C., H. Thysell, A. Grubb, I. Olsson. 1988. A tumor necrosis factor binding protein is present in human biological fluids. *Eur. J. Haematol.* 41: 414–419.

112. Seckinger, P., S. Isaaz, J. M. Dayer. 1988. A human inhibitor of tumor necrosis factor. *J. Exp. Med.* 167: 1511–1516.

113. Engelmann, H., D. Aderka, M. Rubinstein, D. Rotman, D. Wallach. 1989. A tumor necrosis factor-binding protein purified to homogeneity from human urine protects cells from tumor necrosis factor toxicity. *J. Biol. Chem.* 264: 11974–11980.

114. Olsson, I., M. Lantz, E. Nilsson, C. Peetre, H. Thysell, A. Grubb, G. Adolf. 1989. Isolation and characterisation of a tumor necrosis factor binding protein from urine. *Eur. J. Haematol.* 42: 270–275.

115. Engelmann, H., D. Novick, D. Wallach. 1990. Two tumor necrosis factor-binding proteins purified from human urine. Evidence for immunological cross-reactivity with cell surface tumor necrosis factor receptors. *J. Biol. Chem.* 265: 1531–1536.

116. Seckinger, P., E. Vey, G. Turcatti, P. Wingfield, J. M. Dayer. 1990. Tumor necrosis factor inhibitor: purification, NH2-terminal amino acid sequence and evidence for anti-inflammatory and immunomodulatory activities. *Eur. J. Immunol.* 20: 1167–1174.

117. Seckinger, P., J. H. Zhang, B. Hauptmann, J. M. Dayer. 1990. Characterisation of a tumor necrosis factor α (TNF-α) inhibitor: Evidence of immunological cross-reactivity with the TNF receptor. *Proc. Natl. Acad. Sci. USA* 87: 5188–5192.

118. Engelmann, H., H. Holtmann, C. Brakebusch, Y. S. Avni, I. Sarov, Y. Nophar, E. Hadas, O. Leitner, D. Wallach. 1990. Antibodies to a soluble form of a tumor necrosis factor (TNF) receptor have TNF-like activity. *J. Biol. Chem.* 265: 14497–14504.

119. Espevik, T., M. Brockhaus, H. Loetscher, U. Nonstad, R. Shalaby. 1990. Characterization of binding and biological effects of monoclonal antibodies against a human tumor necrosis factor receptor. *J. Exp. Med.* 171: 415–426.

120. Schall, T. J., M. Lewis, K. J. Koller, A. Lee, G. C. Rice, G. H. W. Wong, T. Gatanaga, G. A. Granger, R. Lentz, H. Raab, W. J. Kohr, D. V. Goeddel. 1990. Molecular cloning and expression of a receptor for human tumor necrosis factor. *Cell* 61: 361–370.

121. Gatanaga, T., C. Hwang, W. Kohr, F. Capuccini, J. A. Lucci, E. W. B. Jeffes, R. Lentz, J. Tomich, R. S. Yamamoto, G. A. Granger. 1990. Purification and characterisation of an inhibitor (soluble tumor necrosis factor receptor) for tumor necrosis factor and lymphotoxin obtained from the serum ultrafiltrates of human cancer patients. *Proc. Natl. Acad. Sci. USA* 87: 8781–8784.

122. Nophar, Y., O. Kemper, C. Brakebusch, H. Engelmann, R. Zwang, D. Aderka, H. Holtmann, D. Wallach. 1990. Soluble forms of tumor necrosis factor receptors (TNF-Rs). The cDNA for the type I TNF-R, cloned using amino acid sequence data of its soluble form, encodes both the cell surface and a soluble form of the receptor. *EMBO J.* 9: 3269–3278.

123. Jones, E. Y., D. I. Stuart, N. P. C. Walker. 1989. Structure of tumor necrosis factor. *Nature* 338: 225–228.

124. Eck, M. J., S. R. Sprang. 1989. The structure of tumor necrosis factor alpha at 2.6 Å resolution: implications for receptor binding. *J. Biol. Chem.* 264: 17595–17605.

125. Eck, M. J., M. Ultsch, E. Rinderknecht, A. M. de Vos, S. R. Sprang. 1992. The

structure of human lymphotoxin (tumor necrosis factor-β) at 1.9-Å resolution. *J. Biol. Chem.* 267: 2119–2122.

126. Banner, D. W., A. D'Arcy, W. Janes, R. Gentz, H.-J. Schoenfeld, C. Broger, H. Loetscher, W. Lesslauer. 1993. Crystal structure of the soluble human 55 kd TNF receptor-human TNFβ complex: implications for TNF receptor activation. *Cell* 73: 431–445.
127. Naismith, J. H., T. Q. Devine, B. J. Brandhuber, S. R. Sprang. 1995. Crystallographic evidence for dimerisation of unliganded tumor necrosis factor receptor. *J. Biol. Chem.* 270: 13303–13307.
128. Naismith, J. H., T. Q. Devine, T. Kohno, S. R. Sprang. 1996. Structures of the extracellular domain of the type I tumor necrosis factor receptor. *Structure* 4: 1251–1262.
129. Gehr, G., R. Gentz, M. Brockhaus, H. R. Loetscher, W. Lesslauer. 1992. Both tumor necrosis factor receptor types mediate proliferative signals in human mononuclear cell activation. *J. Immunol.* 149: 911–917.
130. Mackay, F., H. R. Loetscher, D. Stueber, G. Gehr, W. Lesslauer. 1993. TNFα-induced cell adhesion to human endothelial cells is under dominant control of one TNF receptor type, TNF-R55. *J. Exp. Med.* 177: 1277–1286.
131. Tartaglia, L. A., D. Pennica, D. V. Goeddel. 1993. Ligand passing: the 75-kDa tumor necrosis factor (TNF) receptor recruits TNF for signaling by the 55-kDa TNF receptor. *J. Biol. Chem.* 268: 18542–18548.
132. Tartaglia, L. A., D. V. Goeddel. 1992. Tumor necrosis factor receptor signaling: a dominant negative mutation suppresses the activation of the 55kDa tumor necrosis factor receptor. *J. Biol. Chem.* 267: 4304–4307.
133. Brakebusch, C., Y. Nophar, O. Kemper, H. Engelmann, D. Wallach. 1992. Cytoplasmic truncation of the p55 tumour necrosis factor (TNF) receptor abolishes signalling, but not induced shedding of the receptor. *EMBO J.* 11: 943–950.
134. Kalb, A., H. Bluethmann, M. W. Moore, W. Lesslauer. 1996. Both TNF receptors in TNFR1 and TNFR2 deficient mice are signaling competent and activate the MAPK pathway with differential kinetics. *J. Biol. Chem.* 271: 28097–28104.
135. Shakov, A. N., M. A. Collart, P. Vassalli, S. A. Nedospasow, C. V. Jongeneel. 1990. KappaB-type enhancers are involved in lipopolysaccharide-mediated transcriptional activation of the tumor necrosis factor a gene in primary macrophages. *J. Exp. Med.* 171: 35–47.
136. Han, J., T. Brown, B. Beutler. 1990. Endotoxin responsive sequences control cachectin/tumor necrosis factor biosynthesis at the translational level. *J. Exp. Med.* 171: 465–475.
137. Lee, J. C., J. T. Laydon, P. C. McDonnell, T. F. Gallagher, S. Kumar, D. Green, D. McNultey, M. J. Blumenthal, J. R. Heys, S. W. Landvatter, J. E. Strickler, M. M. McLaughlin, I. R. Siemens, S. M. Fisher, G. P. Livi, J. R. White, J. L. Adams, P. R. Young. 1994. A protein kinase involved in the regulation of inflammatory cytokine biosynthesis. *Nature* 372: 739–746.
138. Mohler, K. M., P. R. Sleath, J. N. Fitzner, D. P. Ceretti, M. Alderson, S. S. Kerwar, T. S. Torrance, C. Otten-Evans, T. Greenstreet, K. Weerawarna, S. R. Kronheim, M. Peterson, M. Gerhart, C. J. Kozlosky, C. J. March, R. A. Black. 1994. Protection against a lethal dose of endotoxin by an inhibitor of tumor necrosis

factor processing. *Nature* 370: 218–220.

139. Gearing, A. J. H., P. Beckett, M. Christodoulou, M. Churchill, J. Clements, A. H. Davidson, A. H. Drummond, W. A. Galloway, R. Gilbert, J. L. Gordon, T. M. Leber, M. Mangan, K. Miller, P. Nayee, K. Owen, S. Patel, W. Thomas, G. Wells, L. M. Wood, K. Woolley. 1994. Processing of tumor necrosis factor α precursor by metalloproteinases. *Nature* 370: 555–557.
140. McGeehan, G. M., J. D. Becherer, R. C. Bast, C. M. Boyer, B. Champion, K. M. Conolly, J. G. Conway, P. Furdon, S. Karp, S. Kidao, A. B. McElroy, J. Nichols, K. M. Pryzwansky, F. Schoenen, L. Sekut, A. Truesdale, M. Verghese, J. Warner, J. P. Ways. 1994. Regulation of tumor necrosis factor α processing by a metalloproteinase inhibitor. *Nature* 370: 558–561.
141. Bagby, G. J., K. J. Plessala, L. A. Wilson, J. J. Thompson, S. Nelson. 1991. Divergent efficacy of antibody to tumor necrosis factor-alpha in intravascular and peritonitis models of sepsis. *J. Infect. Dis.* 163: 83–88.
142. Eskandari, M. K., G. Bolgos, C. Miller, D. T. Nguyen, L. E. DeForge, D. G. Remick. 1992. Anti-tumor necrosis factor antibody therapy fails to prevent lethality after cecal ligation and puncture or endotoxemia. *J. Immunol.* 148: 2724–2730.
143. Hinshaw, L. B., T. E. Emerson, F. B. Taylor, A. C. Chang, M. Duerr, G. T. Peer, D. J. Flournoy, G. L. White, S. D. Kosanke, C. K. Murray, R. Xu, R. B. Passey, M. A. Fournel. 1992. Lethal Staphylococcus aureus-induced shock in baboons: prevention of death with anti-TNF antibody. *J. Trauma* 33: 568–573.
144. Emerson, T. E., D. C. Lindsey, G. J. Jesmok, M. L. Duerr, M. A. Fournel. 1992. Efficacy of monoclonal antibody against tumor necrosis factor alpha in an endotoxemic baboon model. *Circ. Shock.* 38: 75–84.
145. Fiedler, V. B., I. Loof, E. Sander, V. Voehringer, C. Galanos, M. A. Fournel. 1992. Monoclonal antibody to tumor necrosis factor-alpha prevents lethal endotoxin sepsis in adult rhesus monkeys. *J. Lab. Clin. Med.* 120: 574–588.
146. Walsh, C. J., H. J. Sugerman, P. G. Mullen, P. D. Carey, S. K. Leeper-Woodford, G. J. Jesmok, E. F. Ellis, A. A. Fowler. 1992. Monoclonal antibody to tumor necrosis factor alpha attenuates cardiopulmonary dysfunction in porcine gram-negative sepsis. *Arch. Surg.* 127: 138–144.
147. Fisher, C. J., S. M. Opal, J. F. Dhainaut, S. Stephens, J. L. Zimmerman, P. Nightingale, S. J. Harris, R. M. Schein, E. A. Panacek, J. L. Vincent, et al. 1993. Influence of an anti-tumor necrosis factor monoclonal antibody on cytokine levels in patients with sepsis. The CB0006 sepsis syndrome study group. *Crit. Care Med.* 21: 318–327.
148. Boekstegers, P., S. Weidenhofer, R. Zell, G. Pilz, E. Holler, W. Ertel, T. Kapsner, H. Redl, G. Schlag, M. Kaul, et al. 1994. Repeated administration of a F(ab)2 fragment of an anti-tumor necrosis factor alpha monoclonal antibody in patients with severe sepsis: effects on the cardiovascular system and cytokine levels. *Shock* 1: 237–245.
149. Boekstegers, P., S. Weidenhofer, R. Zell, E. Holler, T. Kapsner, H. Redl, G. Schlag, M. Kaul, J. Kempeni, K. Werdan. 1994. Changes in skeletal muscle pO2 after administration of anti-TNF alpha-antibody in patients with severe sepsis: comparison to interleukin-6 serum levels, APACHE II and Elebute scores. *Shock* 1: 246–253.

150. Ertel, W., M. H. Morrison, A. Ayala, I. H. Chaudry. 1994. Biological significance of elevated TNF levels: in vivo administration of monoclonal antibody against TNF following hemorrhagic shock increases the capacity of macrophages to release TNF while restoring immunoresponsiveness. *Cytokine* 6: 624–632.

151. Mullen, P. G., B. J. Fisher, C. J. Walsh, B. M. Susskind, S. K. Leeper-Woodford, G. J. Jesmok, A. A. Fowler, H. J. Sugerman. 1994. Monoclonal antibody to tumor necrosis factor-alpha attenuates plasma interleukin-6 levels in porcine gram-negative sepsis. *J. Surg. Res.* 57: 625–631.

152. Maini, R. N., M. Elliott, F. M. Brennan, R. O. Williams, M. Feldmann. 1994. Targeting TNF alpha for the therapy of rheumatoid arthritis. *Clin. Exp. Rheumatol.* 12: S63–S66.

153. Elliott, M. J., R. N. Maini, M. Feldmann, J. R. Kalden, C. Antoni, J. S. Smolen, B. Leeb, F. C. Breedveld, J. D. Macfarlane, H. Bijl, et al. 1994. Randomised double-blind comparison of chimeric monoclonal antibody to tumor necrosis factor alpha (cA2) versus placebo in rheumatoid arthritis. *Lancet* 344 (8930): 1105–1110.

154. Elliott, M. J., R. N. Maini. 1994. New directions for biological therapy in rheumatoid arthritis. *Int. Arch. Allergy Immunol.* 104: 112–125.

155. Stack, A. M., R. A. Saladino, C. Thompson, F. Sattler, D. L. Weiner, J. Parsonnet, H. Nariuchi, G. R. Siber, G. R. Fleisher. 1995. Failure of prophylactic and therapeutic use of a murine anti-tumor necrosis factor monoclonal antibody in Escherichia coli sepsis in the rabbit. *Crit. Care Med.* 23: 1512–1518.

156. Dhainaut, J. F., J. L. Vincent, C. Richard, P. Lejeune, C. Martin, L. Fierobe, S. Stephens, U. M. Ney, M. Sopwith. 1995. CDP571, a humanised antibody to human tumor necrosis factor—alpha: safety, pharmacokinetics, immune response, and influence of the antibody on cytokine concentrations in patients with septic shock. CDP571 study group. *Crit. Care Med.* 23: 1461–1469.

157. Junger, W. G., D. B. Hoyt, H. Redl, F. C. Liu, W. H. Loomis, J. Davies, G. Schlag. 1995. Tumor necrosis factor antibody treatment of septic baboons reduces the production of sustained T-cell suppressive factors. *Shock* 3: 173–178.

158. Abraham, E., R. Wunderink, H. Silverman, T. M. Perl, S. Nasraway, H. Levy, R. Bone, R. P. Wenzel, R. Balk, R. Allred, J. E. Pennington, J. C. Wherry, T.-A. S. S. Group. 1995. Efficacy and safety of monoclonal antibody to human tumor necrosis factor alpha in patients with sepsis syndrome. A randomized, controlled, double-blind, multicenter clinical trial. TNF alpha MAb sepsis study group. *JAMA* 273: 934–941.

159. Scallon, B. J., M. A. Moore, H. Trinh, D. M. Knight, J. Ghrayeb. 1995. Chimeric anti-TNF-alpha antibody cA2 binds recombinant transmembrane TNF-alpha and activates immune effector functions. *Cytokine* 7: 251–259.

160. Cohen, J., J. Carlet, I. S. group. 1996. INTERSEPT: an international, multicenter, placebo-controlled trial of monoclonal antibody to human tumor necrosis factor-alpha in patients with sepsis. International sepsis trial study group. *Crit. Care Med.* 24: 1431–1440.

161. Reinhart, K., C. Wiegand-Lohnert, F. Grimminger, M. Kaul, S. Withington, D. Treacher, J. Eckart, S. Willatts, C. Bouza, D. Krausch, F. Stockenhuber, J. Eiselstein, L. Daum, J. Kempeni. 1996. Assessment of the safety and efficacy of the monoclonal anti-tumor necrosis factor antibody-fragment, MAK 195F, in patients

with severe sepsis and septic shock: a multicenter, randomized, placebo-controlled, dose-ranging study. *Crit. Care Med.* 24: 733–742.

162. Paleolog, E. M., M. Hunt, M. J. Elliott, M. Feldmann, R. N. Maini, J. N. Woody. 1996. Deactivation of vascular endothelium by monoclonal anti-tumor necrosis factor alpha antibody in rheumatoid arthritis. *Arthritis Rheum.* 39: 1082–1091.
163. Lorenz, H. M., C. Antoni, T. Valerius, R. Repp, M. Grunke, N. Schwerdtner, H. Nusslein, J. Woody, J. R. Kalden, B. Manger. 1996. In vivo blockade of TNF alpha by intravenous infusion of a chimeric monoclonal TNF-alpha antibody in patients with rheumatoid arthritis. Short term cellular and molecular effects. *J. Immunol.* 156: 1646–1653.
164. Abraham, E., M. P. Glauser, T. Butler, J. Garbino, D. Gelmont, P. F. Laterre, K. Kudsk, H. A. Bruining, C. Otto, E. Tobin, C. Zwingelstein, W. Lesslauer, A. Leighton. 1997. p55 tumor necrosis factor receptor fusion protein in the treatment of patients with severe sepsis and septic shock: a randomized controlled multicenter trial. *JAMA* 277: 1531–1538.
165. Lesslauer, W., H. Tabuchi, R. Gentz, E. J. Schlaeger, M. Brockhaus, G. Grau, P. F. Piguet, P. Pointaire, P. Vassalli, H. R. Loetscher. 1991. Bioactivity of recombinant human TNF receptor fragments. *J. Cell. Biochem.* Suppl 15F: 115 (abstr).
166. Moreland, L. W., G. Margolies, L. W. Heck, A. Saway, C. Blosch, R. Hanna, W. J. Koopman. 1996. Recombinant soluble tumor necrosis factor receptor (p80) fusion protein: Toxicity and dose finding trial in refractory rheumatoid arthritis. *J. Rheumatol.* 23: 1849–1855.
167. Sander, O., R. Rau, P. van Riel, L. van de Putte, F. Hasler, M. Baudin, E. Ludin, T. McAuliffe, S. Dickinson, M. R. Kahny, W. Lesslauer, P. van der Auwera. 1996. Neutralisation of TNF by lenercept (TNFR55-IgG1, Ro45-2081) in patients with rheumatoid arthritis treated for 3 months: results of a European phase II trail. *Arthritis Rheum.* 39: S242.
168. Furst, D., M. Weisman, H. Paulus, K. Bulpitt, M. Weinblatt, R. Polisson, P. St. Clair, P. Milnarik, M. Baudin, E. Ludin, T. McAuliffe, M. R. Kahny, W. Lesslauer, P. van der Auwera. 1996. Neutralisation of TNF by lenercept (TNFR55-IgG1, Ro45-2081) in patients with rheumatoid arthritis treated for 3 months: results of an US phase II trail. *Arthritis Rheum.* 39: S243.
169. Baumgartner, S., L. W. Moreland, M. H. Schiff, E. Tindall, R. M. Fleischmann, A. Weaver, R. E. Ettlinger, B. L. Gruber, R. S. Katz, J. L. Skosey, R. B. Lies, A. Robison, C. M. Blosch. 1996. Double-blind, placebo-controlled trial of tumor necrosis factor receptor (p80) fusion protein (TNFR:Fc) in active rheumatoid arthritis. *Arthritis Rheum.* 39: S74.
170. Aderka, D., H. Engelmann, Y. Maor, C. Brakebush, D. Wallach. 1992. Stabilisation of the bioactivity of tumor necrosis factor by its soluble receptors. *J. Exp. Med.* 175: 323–329.
171. Digel, W., F. Porzsolt, M. Schmid, F. Herrmann, W. Lesslauer, M. Brockhaus. 1992. High levels of circulating soluble receptors for tumor necrosis factor in chronic B-lymphocytic leucemia and hairy cell leucemia. *J. Clin. Invest.* 89: 1690–1693.
172. Loetscher, H. R., P. Anghern, E. J. Schlaeger, R. Gentz, W. Lesslauer. 1993. Efficacy of a chimeric TNFR-IgG fusion protein to inhibit TNF activity in animal models of septic shock. In *Bacterial Endotoxin: Recognition and Effector Mechan-*

isms, ed. J. Levin, C. R. Alving, R. S. Munford, P. L. Stutz, Endotoxin Research Series, vol. 2. Amsterdam: Excerpta Medica, pp. 455–462.

173. Loetscher, H. R., R. Gentz, M. Zulauf, A. Lustig, H. Tabuchi, E. J. Schlaeger, M. Brockhaus, H. Gallati, M. Manneberg, W. Lesslauer. 1991. Recombinant 55kDa TNF receptor. Stoichiometry of binding to TNFalpha and TNFbeta and inhibition of TNF activity. *J. Biol. Chem.* 266: 18324–18329.

174. Scallon, B. J., H. Trinh, M. Nedelman, F. M. Brennan, M. Feldman, J. Ghrayeb. 1995. Functional comparisons of different tumor necrosis factor receptor/IgG fusion proteins. *Cytokine* 7: 759–770.

175. Smith, C. A., L. Lauffer, T. Davis, M. Widmer, S. Dower, C. Jacobs, K. Mohler, R. G. Goodwin. 1991. Chimeric TNF receptors of viral and cellular origin. *Cytokine* 3: 478 (abstr).

176. Hohmann, H. P., R. Remy, B. Pöschl, A. P. G. M. van Loon. 1990. Tumor necrosis factors-α and -β bind to the same two types of tumor necrosis factor receptors and maximally activate the transcription factor NF-κB at low receptor occupancy and within minutes after receptor binding. *J. Biol. Chem.* 265: 15183–15188.

177. Loetscher, H., D. Belluoccio, H. Kurt, M. Dellenbach, W. Lesslauer. 1993. Binding of TNF alpha to rsTNFR-p55-hgamma1 (Ro 45-2081/000) and to an anti-TNF alpha antibody in comparison to protective efficacy from TNF cytotoxicity. *Roche Research Report # B-161 245*, unpublished data.

178. Jin, H., R. Yang, S. A. Marsters, S. A. Bunting, F. M. Wurm, S. M. Chamow, A. Ashekenazi. 1994. Protection against rat endotoxic shock by p55 tumor necrosis factor (TNF) receptor immunoadhesin: comparison with anti-TNF monoclonal antibody. *J. Infect. Dis.* 170: 1323–1326.

179. Haak-Frendscho, M., S. A. Marsters, J. Mordenti, S. Brady, N. A. Gillett, S. A. Chen, A. Ashkenazi. 1994. Inhibition of TNF by a TNF receptor immunoadhesin. Comparison to an anti-TNF monoclonal antibody. *J. Immunol.* 152: 1347–1353.

180. Beutler, B., I. W. Milsark, A. C. Cerami. 1985. Passive immunisation against cachectin/tumor necrosis factor protects mice from lethal effect of endotoxin. *Science* 229: 869–871.

181. van Zee, K. J., L. L. Moldawer, H. S. A. Oldenburg, W. A. Thompson, S. A. Stackpole, W. J. Montegut, M. A. Rogy, C. Meschter, H. Gallati, C. Schiller, W. F. Richter, H. R. Loetscher, A. Ashkenazi, S. Chamow, F. Wurm, S. E. Calvano, S. F. Lowry, W. Lesslauer. 1996. Protection against lethal E. coli bacteraemia in baboons (Papio anubis) by pretreatment with a 55kDa TNF receptor (CD120a)-immunoglobulin fusion protein, Ro 45-2081. *J. Immunol.* 156: 2221–2230.

182. Deleuran, B. W., C. Q. Chu, M. Field, F. M. Brennan, T. Mitchell, M. Feldmann, R. N. Maini. 1992. Localisation of tumor necrosis factor receptors in the synovial tissue and cartilage-pannus junction in patients with rheumatoid arthritis. Implications for local actions of tumor necrosis factor alpha. *Arthritis Rheum.* 35: 1170–1178.

183. Courtenay, J. S., M. J. Dallman, A. D. Dayan, A. Martin, B. Mosedale. 1980. Immunisation against heterologous type II collagen induces arthritis in mice. *Nature* (*Lond.*) 283: 666–668.

184. Wooley, P. H., H. S. Luthra, J. M. Stuart, C. S. David. 1981. Type II collagen-induced arthritis in mice. I. Major hystocompatibility complex (I region) linkage and antibody correlates. *J. Exp. Med.* 154: 688–700.

185. Wooley, P. H., H. S. Luthra, M. M. Griffiths, J. M. Stuart, A. Huse, C. S. David. 1985. Type II collagen-induced arthritis in mice. IV. Variations in immunogenetic regulation provide evidence for multiple arthritogenic epitopes on the collagen molecule. *J. Immunol.* 135: 2443–2451.

186. Kennedy, M. K., D. S. Torrance, K. S. Picha, K. M. Mohler. 1992. Analysis of cytokine mRNA expression in the central nervous system of mice with experimental autoimmune encephalomyelitis reveals that IL-10 mRNA expression correlates with recovery. *J. Immunol.* 149: 2496–2505.

187. Kojima, K., T. Berger, H. Lassmann, D. Hinze-Selch, Y. Zhang, J. Gehrmann, K. Reske, H. Wekerle, C. Linington. 1994. Experimental autoimmune panecephalitis transferred to the Lewis rat by T lymphocytes specific for the S100β molecule, a calcium binding protein of astroglia. *J. Exp. Med.* 180: 817–829.

188. Renzetti, L. M., P. M. Paciorek, S. A. Tannu, N. C. Rinaldi, J. E. Tocker, M. A. Wassermann, P. R. Gater. 1996. Pharmacological evidence for tumor necrosis factor as a mediator of allergic inflammation in the airways. *J. Pharmacol. Exp. Ther.* 278: 847–853.

189. Gater, P. R., M. A. Wassermann, P. M. Paciorek, L. M. Renzetti. 1996. Inhibiton of Sephadex-induced lung injury in the rat by Ro45-2081, a tumor necrosis factor receptor fusion protein. *Am. J. Respir. Cell Mol. Biol.* 14: 454–460.

190. Fisher, C. J., J. M. Agosti, S. M. Opal, S. F. Lowry, R. A. Balk, J. C. Sadoff, E. Abraham, R. M. Schein, E. Benjamin. 1996. Treatment of septic shock with the tumor necrosis factor receptor:Fc fusion protein. The soluble TNF receptor sepsis study group. *N. Engl. J. Med.* 334: 1697–1702.

191. Silva, A. T., K. F. Bayston, J. Cohen. 1990. Prophylactic and therapeutic effects of a monoclonal antibody to tumor necrosis factor-alpha in experimental Gram-negative shock. *J. Infect. Dis.* 162: 421–427.

192. Suffredini, A. E., D. Reda, S. M. Banks, M. Tropea, J. M. Agosti, R. Miller. 1995. Effects of recombinant dimeric TNF receptor on human inflammatory responses following intravenous endotoxin administration. *J. Immunol.* 155: 5038–5045.

193. Moreland, L. W., S. W. Baumgartner, M. H. Schiff, E. A. Tindall, R. M. Fleischmann, A. L. Weaver, R. E. Ettlinger, S. Cohen, W. J. Koopman, K. Mohler, M. B. Widmer, C. M. Blosch. 1997. Treatment of rheumatoid arthritis with a recombinant human tumor necrosis factor receptor (p75)-Fc fusion protein. *N. Engl. J. Med.* 337: 141–147.

194. Paty, D. W., The Lenercept Multiple Sclerosis Study Group, The UBC. MS/MRI Analysis Group. 1998. TNF neutralisation induces an increase in relapses in patients with multiple sclerosis. *Canadian Congr. Neurol. Sci.*, June 16–20.

195. Capon, D. J., S. M. Chamow, J. Mordenti, S. A. Marsters, T. Gregory, H. Mitsuya, R. A. Byrn, C. Lucas, F. M. Wurm, J. E. Groopman, S. Broder, D. H. Smith. 1989. Designing CD4 immunoadhesins for AIDS therapy. *Nature* 337: 525–531.

196. Traunecker, A., J. Schneider, H. Kiefer, K. Karjalainen. 1989. Highly efficient neutralisation of HIV with recombinant CD4-immunoglobulin molecules. *Nature* 339: 68–70.

197. Byrn, R. A., J. Mordenti, C. Lucas, D. Smith, S. A. Marsters, J. S. Johnson, P. Cossum, S. M. Chamow, F. M. Wurm, T. Gregory, J. E. Groopman, D. J. Capon. 1990. Biological properties of a CD4 immunoadhesin. *Nature* 344: 667–670.

10

OPTIMIZING PRODUCTION AND RECOVERY OF IMMUNOADHESINS

FLORIAN W. WURM
Swiss Federal Institute of Technology, Lausanne, Switzerland

AVI ASHKENAZI
Genentech, Inc., South San Francisco, CA 94080

STEVEN M. CHAMOW
Scios, Inc., Mountain View, CA 94043

10.1 BACKGROUND

Immunoadhesins are chimeric, antibody-like molecules that combine the functional domain of a binding protein (usually a receptor or a cell-adhesion molecule) with the constant domains (usually the hinge and Fc regions) of an immunoglobulin. Immunoadhesins have many applications as research tools, for example in studies on receptor–ligand interactions. In addition, human immunoadhesins have potential applications in the clinic and provide an alternative to human monoclonal antibodies as agents that recognize and neutralize foreign or human proteins that possess pathologic functions.

The name *immunoadhesin* derives from its hybrid polypeptide structure. These molecules contain two parts: one part is derived from an immunoglobulin (Ig) molecule, and the other, from a protein that possesses an adhesion or a binding function. The Ig part of an immunoadhesin confers Ig-like properties, and the "adhesin" part determines target specificity of the chimeric molecule.

Antibody Fusion Proteins, Edited by Steven M. Chamow and Avi Ashkenazi
ISBN 0471-18358-X

Initial attempts to create this type of molecule employed fusions of the Ig-like V regions of the T cell receptor α and β chains with the constant region of IgG2a.[1] The Vα-heavy chain fusion was assembled with light chain and secreted, when cotransfected with a normal λ light chain. In contrast, the Vβ-heavy chain fusion was not assembled with λ light chain and was not secreted. Attempts to reconstruct the entire T-cell receptor by coexpressing the Vα-heavy chain fusion with a Vβ-κ light chain fusion were unsuccessful.

Subsequent attempts to create immunoadhesins were based on CD4, the primary cellular receptor of human immunodeficiency virus (HIV).[2] Because CD4 contains four Ig-like domains in its extracellular region, it appeared possible that part of CD4 could be substituted for the V regions of Ig without compromising the structure of either portion of the chimeric molecule. Various chimeric constructions were tested, and efficient expression and secretion of the resulting immunoadhesins by transfected mammalian cells was obtained with molecules that contained two or four V-like domains of human CD4 linked to the hinge, CH2, and CH3 domains of human IgG heavy-chain (reviewed in Ref. 3). In this chapter, we use this immunoadhesin structure as a prototype. Many other variations on this theme are possible, including molecules that contain light chains,[4] and molecules based on other Ig subtypes, such as IgM,[5] or IgA.[6] In addition, many studies show that the adhesin portion of the molecule need not be derived from a protein that contains Ig-like domains. This was demonstrated first with the lymphocyte adhesion molecule L-selectin,[7,8] and later with other molecules, including receptors for the cytokines tumor necrosis factor (TNF)[9–11] and interferon (IFN) γ,[12,13] and natriuretic peptide receptors.[14] Further, the adhesin part of the molecule need not be derived from a cell surface protein such as a receptor, but may be derived from a soluble protein, e.g., the cytokines interleukin (IL) 2 or IL-10.[15] Finally, the antigen- binding site of an antibody can be retained in the chimeric molecule, and the Fc region or some portion of it can be replaced with a protein possessing some desired biological function. This type of molecule has been called an antigen binding fusion protein.[16] Examples of such chimeras include fusions with insulin-like growth factor 1 (IGF-1),[17] TNF,[18] and IL-1.[19] These types of molecular constructs are detailed in Section I of this book.

In this chapter, we outline methods for the construction, expression, purification, and characterization of immunoadhesins. Additional information on immunoadhesins can be found in review articles[3,16,20–23] and in Chapters 8 and 9 of this volume.

10.2 MOLECULAR DESIGN

The simplest immunoadhesin design combines the binding region of the adhesin protein with the hinge and Fc regions of an IgG heavy chain (Fig. 10.1). The resulting glycoprotein is a covalent homodimer, which resembles an

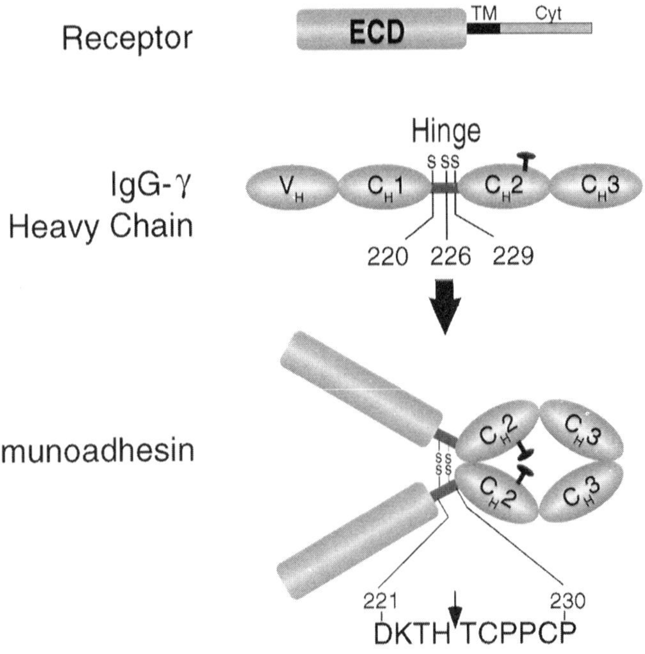

Figure 10.1 Schematic structure of a prototypic immunoadhesin. An immunoadhesin is derived from the parental ligand binding protein, in this case a type I transmembrane receptor, and an IgG1 heavy chain molecule. ECD, TM and Cyt refer to the extracellular, transmembrane and cytoplasmic domains of the receptor, respectively. The variable (V_H) and constant (C_H1, hinge, C_H2, and C_H3) regions of IgG heavy chain are shown. The domains are delimited by the following amino acid residues, according to the IgG1 Eu index[36]: C_H1, 118–215; hinge, 216–230, C_H2, 231–344; C_H3, 345–447. The N-linked carbohydrate chain at Asn-298 within the C_H2 domain is shown. The three cysteine residues within the hinge region are involved in interchain disulfide bonds: Cys-220, H-L, Cys-226, H-H, and Cys-229, H-H. To construct the immunoadhesin, the receptor ECD is fused to the IgG hinge beginning at Asp-221, so that only Cys-226 and Cys-229 are present in the chimeric molecule (the hinge region sequence of the immunoadhesin is shown). Cys-220 is deleted, since the chimeric construct lacks a light chain. Thus, because the two hinge cysteines produce H-H chain disulfide bonds, the immunoadhesin is secreted from transfected mammalian cells as a homodimer. It can be purified using protein A affinity chromatography.

IgG antibody, but lacks CH1 domains and light chains. Deletion of the CH1 domain improves the association and secretion of heavy chains in the absence of light chains, probably because secretion is inhibited in the absence of light chain by interaction of the hydrophobic face of CH1 with the H chain-binding protein.[5,7,24,25] In addition, joining the fusion partners at the flexible hinge usually facilitates proper folding and function of both parts of the immunoadhesin.

10.2.1 Choice of Ig Fusion Partner

For human immunoadhesins, human IgG1 has been used most frequently, although the IgG3 isotype,[26] or other Ig subtypes[5,6] have been used as well. A major advantage of using IgG1 is that $\gamma 1$ immunoadhesins can be purified efficiently on immobilized *Staphylococcus aureus* protein A (see Section 10.6.1 and Box 10.1). In contrast, purification of $\gamma 3$ requires *Streptococcus* protein G, a less versatile medium. However, other structural and functional properties of immunoglobulins should be considered when choosing the Ig fusion partner for a particular immunoadhesin construction. For example, the $\gamma 3$ hinge is longer and more flexible, so it can accommodate larger adhesin domains that may not fold or function properly when fused to $\gamma 1$. Another consideration may be valency: IgG immunoadhesins are bivalent homodimers, whereas the Ig subtypes IgA and IgM give rise to dimeric or pentameric structures, respectively, of the basic Ig homodimer unit.[5,6] For immunoadhesins designed for in vivo application, the pharmacokinetic properties and the effector functions specified by the Fc region are important as well. Although IgG1, IgG2, and IgG4 all have in vivo half-lives of 21 days, their relative potencies at activating the complement system are different. IgG4 does not activate complement, and IgG2 is significantly weaker at complement activation than IgG1. Moreover, unlike IgG1, IgG2 does not bind to Fc receptors on mononuclear cells or neutrophils. While IgG3 is optimal for complement activation, its in vivo half-life is approximately one-third that of the other IgG isotypes. Another important consideration for immunoadhesins designed to be used as human therapeutics is the number of allotypic variants of the particular isotype. In general, IgG isotypes with fewer serologically defined allotypes are preferred. For example, IgG1 has only four such allotypic sites, two of which (G1m1 and 2) are located in the Fc region; one of these sites, G1m1, is nonimmunogenic. In contrast, there are 12 serologically defined allotypes in IgG3, all of which are in Fc; only three of these sites (G3m5, 11, and 21) have one allotype which is nonimmunogenic. Thus, the potential allotypic immunogenicity of a $\gamma 3$ immunoadhesin is much greater than that of a $\gamma 1$ immunoadhesin.

Human immunoadhesins are immunogenic in mice, and antibodies to the administered immunoadhesin can appear within 5–7 days. Therefore, if one's purpose is to carry out long-term studies in experimental animals, it is advisable to use fusion partners that are derived from the autologous species. For example, mouse IgG2a[12] and IgG1 (A. Ashkenazi, S. M. Chamow, D. Peers, and S. Marsters, unpublished results) have been used for fusion with murine adhesin partners in order to conduct long-term studies in mice. An alternative approach is to introduce the heterologous species immunoadhesin as a transgene.[27,28]

10.2.2 Choice of Adhesin Fusion Partner

When a binding protein that interacts with the ligand of interest is known, this binding protein can be used as the adhesin fusion partner. In such cases, the

Box 10.1 Structure of Protein A.

Staphylococcal protein A (SpA) is a pathogenicity factor bound to the cell wall of the bacterium *Staphylococcus aureus*. It exhibits tight binding to IgG, IgA, and IgM from several different mammalian species including human.[96] First exploited as an immunological reagent in 1978,[97] SpA has been used extensively during the past 20 years, both in analytical and preparative applications. Its biochemical properties lend itself well to use as an antibody-binding ligand in affinity chromatography. SpA is a 524 amino acid protein with a molecular weight of 57 kDa.[98] The extracellular portion of SpA contains a tandem repeat of five highly homologous IgG-binding domains[99] designated (from the N-terminus) E, D, A, B, and C. Each domain consists of approximately 58 amino acid residues, arranged into three α-helices that overlay one another. The C-terminal portion of the sequence contains cell wall-binding and transmembrane domains, designated X_r and X_c, respectively. Domains E, D, A, B, and C all bind to the Fc fragment of IgG,[100] whereas domains D and E bind to Fab fragments at a site distinct from the antigen combining site.[100,101] SpA binds to IgG multivalently—SpA and rabbit IgG have been shown to form soluble complexes in which the stoichiometry is IgG_4-SpA_2.[102] SpA contains no cysteines. Due to its tandem domains, predominantly α-helical structure, and its lack of disulfide bonds, SpA is structurally resiliant. Its activity is recovered quantitatively after exposure to extremes in pH, making it an excellent ligand for affinity chromatography. Moreover, its capacity for IgG is high—typcally 20–30 mg/ml for human IgG. Recombinant SpA, truncated to contain only the five Ig binding domains, has been produced in *E. coli* and is used for several commercially available affinity supports. Immunoglobulins as well as Fc-containing fusion proteins can be produced in high purity and yield using SpA.

portion of the binding protein that contains the ligand-binding site (e.g., the ectodomain of a cell-surface receptor or subdomains thereof) can be used as the adhesin region. In cases where only one member of a binding pair is known, an immunoadhesin based on the known member can be used to identify the unknown binding partner.[7,8,29–35] The adhesin domain may be derived also from a soluble ligand. For example, immunoadhesins based on the cytokines IL-2 and IL-10 have been reported.[15] For most applications, if the adhesin contains its own natural hydrophobic leader sequence, it is advisable to include that sequence in the immunoadhesin to achieve efficient secretion and processing of the recombinant polypeptide.

10.2.3 Design of the Fusion Junction

The design of the fusion junction depends on the particular fusion partners. If the adhesin is a Type I cell surface receptor, joining the extracellular sequence just upstream of the hydrophobic transmembrane region is usually suitable (Fig. 10.1). If the adhesin has a known domain structure, one can attempt to delete domains that are not required for binding; in such cases, it is important to place the junction at residues located between domains, to avoid misfolding. With respect to the parental Ig, a useful joining point is just upstream of the hinge cysteines that form the disulfide bonds between the two heavy chains. For IgG1 there are two such cysteines, located at positions 226 and 229.[36] A junction design that we use frequently is to place the codon for the C-terminal residue of the adhesin part of the molecule directly upstream of the codons for the sequence DKTHTCPPCP of the γ1 hinge.[9,13,24]

10.3 EXPRESSION

To date, all recombinant proteins from mammalian cells used for pharmaceutical purposes are based on expression vectors with constitutive promoters. Constitutive overexpression of complex recombinant proteins in mammalian cells does not pose a major metabolic burden on the cell hosts and does not significantly affect growth and other activities of genetically engineered cells. This is a major difference from products made in *E. coli*, which are mostly derived from expression vectors with inducible promoters.

Vector constructions are similar for many recombinant glycoproteins expressed in mammalian cells. SV40 promoter-based vectors have been used for immunoadhesins such as CD4-IgG[24] or p55 TNF receptor-IgG (TNFR-IgG).[9] In our work on CD4-IgG and TNFR-IgG, both of which are Type I transmembrane glycoproteins, cDNA constructs were used in which the Fc-segment of the heavy chain of an IgG1-gene-cDNA was directly fused with the amino-terminal ectodomain of the respective receptor (Fig. 10.1). We found that proper folding of the fusion protein occurs if the molecule is constructed such that the two cysteines encoded within the hinge region of the Ig heavy chain (C226 and C229) serve to form two interheavy chain disulfide bonds (Fig. 10.1). Significantly, the heavy chain cysteine involved in H-L chain bonding (C220) is deleted from the fusion polypeptide.

10.3.1 Vectors for Transient Expression

Transient expression in mammalian cells is the method of choice for initial expression of a recombinant immunoadhesin. Very popular for transient expression are human embryonic kidney 293 cells which have been immortalized by transfection with adenovirus DNA.[37] It has been shown that these cells express intracellularly the adenovirus-derived EIA protein. Subpopulations of these cells have been transfected stably with DNA sequences coding for EBNA protein from Epstein-Barr virus or the large T-antigen of SV40. These cell lines are reported to have superior characteristics as hosts for transient transfection.[38] In transient transfections with human embryonic kidney 293 cells, using calcium phosphate, 1–10 μg/ml of secreted recombinant protein has been observed in cell supernatants.[39,40] More recently, transient expression in mammalian cells has been used to produce larger quantities of desired proteins (Section 10.4.6).

10.3.2 Vectors for Stable Expression

The choice of a vector for stable expression in mammalian cells is driven mainly by the choice of the host system. The most successful and frequently used host system for mammalian expression is Chinese hamster ovary (CHO) cells; other cell lines have been used as well, including baby hamster kidney (BHK) cells and the myeloma cell line SP2/0. In this chapter, we focus on the CHO-DHFR expression system, since this is what we have used for all our immunoadhesin proteins; moreover, for many groups, it continues to be the preferred system for large-scale production of recombinant proteins.

A mutagenized cell line lacking enzyme activity for dihydrofolate reductase (DHFR), developed by Urlaub and Chasin[41] for metabolic and genetic studies, can be propagated only in media supplemented with glycine, hypoxanthine, and thymidine (GHT). When these cells are transfected with a functional expression vector for DHFR, cells that have acquired a functional DHFR gene no longer require the presence of GHT, and so they can be selected and expanded in media lacking GHT. A second expression cassette for the protein of interest can be included in the DHFR plasmid or cotransfected as part of a second vector. Screening a number of clones in GHT-selective media leads to the identification of cells that coexpress the product of interest with DHFR.

Expression in CHO cells can be augmented by exposing recombinant cells to methotrexate (MTX), a folate derivative that blocks DHFR activity completely and irreversibly. Exposure to low concentrations of MTX selects cells that express DHFR at an elevated level. Frequently, the elevated DHFR expression is the result of an increased copy number of the recombinant sequence. Stepwise selection with elevated concentrations of MTX can be repeated many times, and may result in the isolation of cells that contain dramatically increased copy numbers of the transferred genes. The same principles as those discussed previously for the CHO system apply also to other systems with stable integration of exogenous DNA.

One can choose from two vector strategies to generate recombinant CHO cells via the DHFR selection route. To introduce DNA sequences for both the protein of interest and the protein that is the selective marker, either (1) a single vector, or (2) a set of two or more vectors, is assembled. A two vector system offers increased flexibility to improve expression of oligomeric proteins (e.g., immunoglobulins). With a two-vector system, one can optimize the molar ratio of vectors for individual polypeptide chains within the transfection cocktail, assuring that appropriate amounts of each polypeptide are produced.

10.4 LARGE-SCALE PRODUCTION

Immunoadhesins are usually designed to be very effective antagonists, inhibiting the interaction of a receptor with its ligand. When considered for in vivo applications, they are needed in relatively large doses—the annual requirement for production can be in excess of one kilogram. Large-scale production requires an efficient process. Productivity from mammalian cell-based processes for recombinant proteins has changed dramatically during the last decade. While a concentration of about 50 mg of recombinant human tissue plasminogen activator[39] per liter of cell culture supernatant was regarded as good productivity in 1985, today's cell culture processes have been shown to provide 500 mg/L or more of recombinant antibody in a batch process. This improvement in volumetric productivity is the result of several advances in cell culture technology: (1) a more profound understanding of the physiology of cells, (2) better control of process parameters, and (3) richer and more sophisticated media formulations to support higher culture densities in bioreactors. As an example, $1–2 \times 10^6$ cells per ml of culture medium were considered peak densities for CHO cells in 1985. Now, with some CHO cell lines in optimized processes for batch cultures, cell densities 4–5 times higher have been obtained.

10.4.1 Laboratory Scale Cell Culture

Here we focus on process development, using "deep tank technology," for manufacture of human therapeutic proteins. During the early phase of development, progress depends on regular production of small quantities of the new protein to supply efforts in recovery and assay development.

One strategy to speed early progress is to use parallel cell culture systems—one for production of R&D material and the other for cell culture process development. For example, early in the development of CD4-IgG, an extended perfusion culture with cells in suspension was operated in a 2 L bioreactor for a period of 3 months. With very high perfusion rates (up to 6 reactor volumes/day), supporting very high cell densities (up to 1.4×10^8 cells/ml, corresponding to 35% packed cell volume), this culture provided gram quantities of CD4-IgG. With this material—much more abundant than might have been expected during this early phase of development—useful insights were gained, both of the biology of the molecule and of available recovery options. In cell culture, meanwhile, we were able to assess different media compositions and the effect of these on product quality and cell productivity.

10.4.2 Transfection, Amplification, and Clonal Selection

The most frequently used method for introduction of DNA into mammalian cells is the calcium phosphate DNA coprecipitation method. This method was developed about 20 years ago by Graham and van der Eb.[42] It is a reliable method for stable and for transient expression of recombinant proteins from mammalian cells. Recent insights in the physicochemical parameters for the generation of calcium-phosphate DNA coprecipitates have improved its reliability and efficiency.[40,43]

Following transfection of the DHFR-deficient CHO host cell line with a vector or vector combination containing a functional DHFR cassette, cell lines were established from colonies emerging under selective conditions. We used an approach that provided an increased frequency of high expressor cell lines among all emerging cell colonies (Fig. 10.2). Two days after transfection using the calcium-phosphate method, cells were plated into media lacking glycine, hypoxanthine, and thymidine (GHT medium). For the GHT selection, only a portion of the transfected cell population was used. The majority of transfected cells was plated into media containing various concentrations of MTX (see Section 10.3.2). In most cases, concentrations of 30, 100, and 300 nM were used. We preferred to isolate, expand, and analyze cell colonies that emerged under selection from the highest concentration of MTX, since these colonies often show the highest level of coexpression for the protein of interest. This approach allowed us to reduce the number of cell lines for expansion and analysis to not more than 100 (24 clones from GHT medium, and 24 each at 30, 100, and 300 nM MTX).

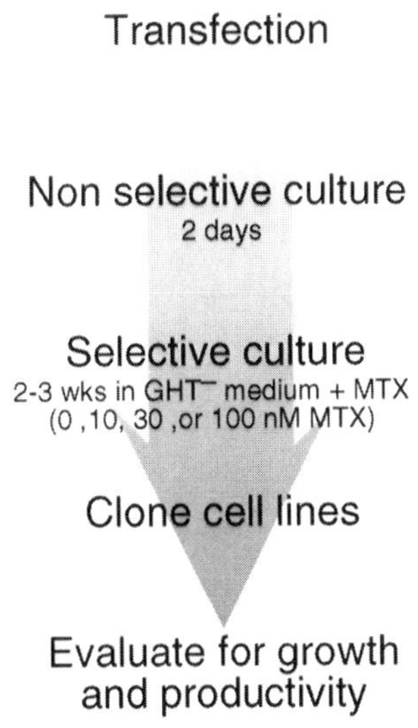

Figure 10.2 Strategy for transfection of DHFR vector cocktails into CHO cells and identification of high producer cell lines.

As many as 10 cell lines were carried through to the final evaluation phase. Growth behavior, adaptability to suspension conditions, robustness for serum-free culture, and scalability were the criteria used for evaluation of cell lines. Also, responsiveness to medium additives that enhance productivity and the impact of MTX removal were assessed. Growth for extended periods of time in non-selective medium and assessment of specific productivity allowed us to develop confidence in the long-term stability of candidate cell lines as good production hosts.

Two to three cell lines were finally considered acceptable for scale up. These candidate cell lines were transferred into good manufacturing practice (GMP) facilities for cell banking, while further process development continued. We chose to use robust methods for the final process. As a robust process engineering method, a batch cell culture process was developed in which product derived from a fixed-period production phase was harvested in one step. In processes developed more recently, extended batch approaches with feeding of culture additives have been used.

10.4.3 Cell Banking

Cells contained in a master cell bank are stored frozen in liquid nitrogen. The freezing of cells for banking was done according to standard cell culture techniques,[44] utilizing 10–20% dimethyl sulfoxide (DMSO) in fresh cell culture medium, into which cells were taken up, after gentle centrifugation and removal of spent medium. In most cases, cryovials with 1 ml of cell suspension were immersed directly into the vapor phase of liquid nitrogen storage tanks.

Selection of a cell line from which to establish the master cell bank is a critical decision. The generation of a master cell bank, usually 50–500 vials containing highly concentrated suspensions of mammalian cells, requires the use of facilities meeting GMP requirements. Cell genetics, cytogenetics, production stability, growth pattern, viability status, and absence of adventitious agents all must be evaluated. We minimized the risk inherent in this process by generating multiple prebanks from which vials were thawed and cultivated. Cells derived from these individual seed trains were used both for comparison in process development efforts and for stable productivity assessment in a standardized assay.

It is important to note that cells derived from a prebank and master-bank are not necessarily identical, since they may have been treated differently. For example, cells that eventually go into production have to endure two freezing and thawing cycles, whereas cells under study for long-term stability experience only one freeze-thaw cycle. This is significant because, unlike prokaryotic cells, mammalian cell populations have to be considered heterogenetic. Exposure to variations in culture conditions—the most drastic being freezing and thawing of cells in specialized media formulation—can often result in a change in overall performance of the cell line under study.

10.4.4 Growth of the Seed Train

To begin development of a seed train, ampules containing viable cells are thawed and the cells are expanded. The quantity of cells in the vial may vary, according to procedure. If, for example, the cell line had been previously adapted to growth in suspension culture, the number of viable cells should be large enough to inoculate a small spinner flask. To seed a spinner at a minimum volume of 50 ml with a seed density of 3×10^5 cells/ml, the thawed vial should contain at least 15×10^6 cells.

Adaptation to growth in suspension could take anywhere from 3 weeks to 3 months, depending on the particular cell line. Several subcultivations of cells grown on plastic cell culture substrates (splits) were performed, to produce a homogeneous population in 850 cm^2 roller bottles. Confluent cultures of these cells in roller bottles were trypsinized, and the cells were suspended in media suitable for growth in spinner flasks. One roller bottle would contain enough cells to prepare about 200 ml of a suspension culture. Cell suspensions established in this manner were then processed as described above, until a robust cell seed train was established.

Growth of the seed train can be accomplished in medium with or without added fetal bovine serum (FBS). Most cell lines required 5% or less FBS and, during the last few years, FBS-free culture media have been developed and have proved to be reliable.

After an initial lag phase of 24–48 h, cells began to grow in spinners, reaching a final cell density of $0.8–2 \times 10^6$ cells/ml within 4–5 days. However after 5–6 days, a subcultivation was initiated, even when growth was poor. The protocol for the subcultivation was adapted to the performance of the cell line. In the case of poor growth, e.g., a cell density of $0.8–1 x 10^6$ cells/ml, the medium was removed by gentle centrifugation and replaced with fresh, prewarmed medium, utilizing a volume of medium that adjusted cell density to about 0.3×10^6 cells/ml. In the case of excellent growth, e.g., a density of $1.5–2 \times 10^6$ cells/ml, a subcultivation method was used which simply diluted out the medium, adjusting the cell density again to about 0.3×10^6 cells/ml. Thus, the seed train was kept at a constant volume, by repetitive subcultivation of spinners or bioreactors with a smaller working volume.

In summary, the goal of the seed train development is to establish a robust procedure and a reliable growth performance of the cells under study. Critical for success is the availability of media formulations that have been specifically developed for this purpose.

10.4.5 Cell Culture Production

We have preferred to do process development work using bioreactors of nominal 2 L volume, with a maximal working volume of 1.5 L. To optimize process parameters, a battery of bioreactors was used. To compensate for reactor variability, duplicates or triplicates of the same condition were employed.

One key parameter was the need to create sufficient cell mass (cell number) for the inoculation of the production vessel at large scale. Therefore, factors such as media composition, stirring rate, oxygenation, pH- and osmolarity control—all of which allow for high growth rates and a high final density of the cell suspension prior to transfer to a larger tank—were optimized. If the cell line had not been adapted to growth in serum-free media, a reduction of the required serum to less than 5% was another development goal. Finally, since high cell viability at the start of production phase was found to be essential for high productivity, much effort went into developing an optimum procedure for transferring cells between vessels.

Production phase is defined as the final phase of the cell culture process, which is optimized for product accumulation in a large-volume bioreactor. For the inoculum train, a complete exchange of the culture medium prior to the start of production phase can result in improved performance and productivity. We evaluated various methods for this purpose. Initially, exchange of medium by centrifugation was used; however, this method was not suitable for large-scale operation. Therefore, other methods, such as tangential flow filtration, were evaluated. Tangential flow filtration[45] was found to be an excellent means by which to achieve medium exchange, insofar as cell viability remained high during the operation. Consequently, this method of medium exchange was adopted for large-scale use.

We found that achievement of five objectives was necessary for optimal productivity during production phase: (1) rapid attainment of high cell density, (2) maintenance of high cell density and viability for the entire production run, (3) manipulation of cellular metabolism to promote high specific productivity, (4) maintenance of consistent product quality, and (5) selection of an efficient cell harvest technology capable of sustaining cell integrity and product quality.

10.4.6 Rapid Manufacturing of Non-clinical Grade Proteins for Evaluation and Analysis

Transient gene expression in animal cells, when performed at the 1–10 L scale, represents a fast approach for the synthesis of tens to hundreds of milligrams of recombinant proteins. In addition to virus-mediated nucleic acid transfer, calcium-phosphate, polyethylenimine, cationic lipids, and electroporation have been reported to be useful in promoting efficient DNA transfer. Efforts to increase the scale of such production processes to 100 L are underway in many laboratories, since the classical approach for recombinant protein production in animal cells using stable integration of exogeneous DNA is costly, very time consuming, and requires substantial know-how in many areas of process development and process engineering.

Human embryonic kidney 293 cells seem to be the most popular host system for these efforts. Attempts to use CHO cells in transient expression approaches have not been successful. Expression levels for secreted proteins like CD4-IgG, TNFR-IgG, or rtPA are only 1-10% of those observed in human embryonic kidney 293 cells (F.M. Wurm, unpublished). For virus mediated transfer of genetic information to animal cells, Sindbis virus-based RNA transfer to BHK, CHO, or Raji cells[46,47] has become popular recently. Lastly, baculovirus-mediated DNA transfer to insect cells is used successfully in many laboratories (for example, see Ref. 48). Following is a brief discussion of the various transient expression technologies that are currently in use.

Calcium phosphate-mediated DNA transfer. Jordan et al.[49] reported the use of calcium phosphate-mediated DNA transfer to suspended 293 cells, grown in a bioreactor or in spinners. A most critical factor in these transfections is the removal of DNA-calcium phosphate precipitates 3–6 h after transfer of the DNA to the reactor. Persistant precipitates are toxic to mammalian cells, most pronounced in the absence of serum and in cultures with low cell densities. In small-scale experiments (e.g., cell culture plates), DNA-calcium phosphate coprecipitates are removed or solubilized by a medium exchange; in reactors, solubilization is mediated by a deliberate change of the culture pH, coupled with the addition of fresh, low-calcium growth medium.

Up to 3 mg of an rtPA variant has been obtained within 5 days upon transfection in 2 L mini-bioreactors with 1 mg of purified plasmid DNA transferred.[49]

Polyethyleneimine as carrier for DNA transfer. Using cationic polyethyleneimine[50] or polyethyleneimine in combination with dioleoyl-melittin as carrier complexes for plasmid DNA, Schlaeger and coworkers achieved transient production of up to 1.4 mg/L of TNFR55-IgG with human embryonic kidney 293 (EBNA) cells grown in stirred tanks at 10 L scale.[51]

Lipid- and liposome-based DNA transfer systems. A number of lipid- and liposome-based DNA transfer systems have been developed over the last 5 years and are commercially available. Among the most popular ones are LipofectAMINE (GIBCO-BRL), DOTAP and DOSPER (Boehringer Mannheim), Transfectam (Promega), and CLONfectin (Clontech). These are intended for small-scale experiments and usually for attached cells. They work well in multiwell plates or in single 100 mm dishes, and they provide similar product concentrations for secreted proteins as the ones seen with calcium phosphate. None of these have been reported to be used at a scale beyond 100 ml of cell culture medium—most likely because they are very expensive. So far, the only nonviral systems that may allow scale-up beyond the 1 L scale are calcium phosphate and polyethyleneimine.

Electroporation for DNA transfer. Blasey and coworkers[52] have pioneered the use of electroporation at 10 L scale. Using a commercial electroporator, COS cells, derived from a CellCube for expansion of cell mass, were electroporated and transferred subsequently to a 10 L bioreactor. Other cell lines, like CHO and human embryonic kidney 293 cells were tested as well. Milligram quantities of product were harvested from the culture 7–12 days later. This technique appears to require some further improvement in order to achieve robustness and reproducibility.

Virus-mediated protein expression. Baculovirus mediated protein expression is popular in many laboratories.[53] The possibility that susceptible insect cells (*Spodoptera frugiperda* Sf9 cells, *Drosophila* Schneider cells) can be grown in suspension in stirred bioreactors or spinners allows for scale-up, so far reported up to the 100 L scale.[54,55]

Using mammalian cells, semliki forest virus (SFV) has been used for rapid production of recombinant protein at 15 L scale with BHK cells.[56] The powerful promoter of this virus system allows for very high protein synthesis by infected cells. However, this results in increased lysis of cells,—a disadvantage for recovery. In CHO cells, high levels of the serotonin 5-HT receptor were shown following SFV-mediated gene transfer.[57]

10.5 IMMUNOADHESIN STRUCTURE: IMPLICATIONS FOR PURIFICATION

The prototypic IgG1-based immunoadhesin has a disulfide-bonded homodimeric structure that resembles an antibody but lacks CH1 domains and light chains (Fig. 10.1). The subunit structure of the immunoadhesin can be analyzed by sodium dodecylsulfate polyacrylamide gel electrophoresis (SDS-PAGE). One technique is to label the transfected cells with ^{35}S-methionine and/or ^{35}S-cysteine and to precipitate the secreted immunoadhesin with protein A or protein G; then, the labeled immunoadhesin can be analyzed by SDS-PAGE and autoradiography.[58] Alternatively, serum-free supernatants from transfected cells can be subjected directly to SDS-PAGE, electroblotted onto nitrocellulose, and the immunoadhesin can be visualized on the blot with labeled protein A or G, with antibodies, or with labeled ligand.

Fig. 10.3 shows an example of this type of analysis; in this case, with a protein A-purified preparation of human interferon γ receptor immunoadhesin (IFNγR-IgG). The purified immunoadhesin was visualized directly by staining with coomassie blue (*A*), or transferred onto nitrocellulose and visualized with anti-Fc antibody (*B*), which confirms the presence of the Fc region, or with labeled ligand (*C*), which confirms the presence of the adhesin region in functional form. Under nonreducing conditions, the protein migrates with an apparent molecular weight approximately twice that of the reduced protein, confirming the molecule's disulfide-linked homodimeric structure.

Results with many such immunoadhesins show that fusion of the adhesin and the Fc regions usually does not perturb the folding of each domain. Comparisons of the binding function of the adhesin portion with those of the soluble or cell-surface counterparts generally show similar properties (e.g., CD4-IgG,[59] IFNγR-IgG,[12,13] or IL-1R-IgG[60]). The Fc region also seems to

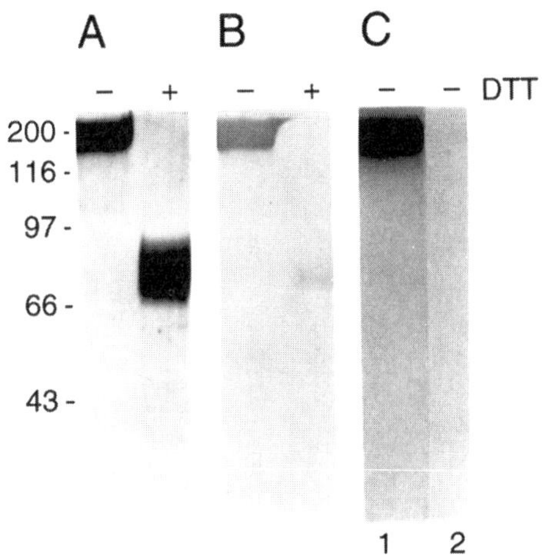

Figure 10.3 Subunit structure of an immunoadhesin as analyzed by SDS-PAGE and immunoblotting and ligand blotting techniques. Human embryonic kidney 293 cells were transfected with a vector directing transient expression of human interferon-γ receptor immunoadhesin (hIFN-γR-IgG). The protein was recovered from culture supernatants and purified on protein A agarose. SDS-PAGE was carried out without (–) or with (+) reduction by 10 mM dithiothreitol (DTT). The protein was stained with coomassie blue (*A*) electroblotted onto nitrocellulose and incubated with (*B*) antibodies to IgG-Fc or with (*C*) ^{125}I-labeled hIFN-γ (10 mM) alone (lane 1), or with 1 mM unlabeled hIFN-γ (lane 2). Blots were developed with horseradish peroxidase-conjugated second antibody and 4-chloronapthol (*B*) or by autoradiography (*C*). (Reprinted with permission from A. Ashkenazi and S. M. Chamow, Immunoadhesins: An alternative to human monoclonal antibodies, *Methods: A Companion to Meth. Enzymol.* vol. 8, 104–115. Copyright © 1995 by Academic Press, Inc.).

fold correctly, because it is recognized by antibodies and by protein A or protein G, and because it retains many of the effector functions characteristic of antibodies, such as binding to Fc receptors.[24]

IgG1-based immunoadhesins are produced in mammalian cells as glycoproteins, due to the presence of a single N-linked glycosylation site within the CH2 domain (see Fig. 10.1). Additional glycosylation sites (both N- and O-linked) can be present within the adhesin portion of the molecule, as in the TNF and IFNγ receptor immunoadhesins.[9,13] Heterogeneity of oligosaccharide sequences (particularly sialic acid residues) can produce overall molecular charge heterogeneity.[61,62] In immunoadhesins as in antibodies, the mature heavy-chain polypeptide contains no C-terminal lysine, notwithstanding a C-terminal lysine predicted by the DNA coding sequence.[63] Rather, the C-terminal residue in the mature polypeptide is the penultimate glycine.[36] Presumably, the C-terminal lysine is removed during cellular processing of the nascent polypeptide.[64]

One unique feature of immunoadhesins is the potential to display a charge dipole, resulting from their chimeric nature. A charge dipole will exist if the pI's of the adhesin domain and the Fc domain are widely different. Chimeras like this will behave non-ideally in separation techniques based on ionic charge. CD4-IgG is an example of this type of immunoadhesin molecule. Because the pI of the CD4 and Fc domains differ by more than 2 pH units, the behavior of this molecule in ion exchange chromatographic systems is difficult to predict.

10.6 PURIFICATION

10.6.1 Affinity Chromatography

A significant advantage of expressing adhesins as immunoadhesins is that they can be purified easily by affinity chromatography. The suitability of protein A as an affinity ligand depends on the species and isotype of the immunoglobulin Fc domain used in the chimera. Protein A can be used to purify immunoadhesins that are based on human $\gamma 1$, $\gamma 2$, or $\gamma 4$ heavy chains.[65] Protein G is recommended for mouse $\gamma 1$ and for human $\gamma 3$.[66] The matrix to which the affinity ligand is attached is most often agarose, but other matrices, are available. Mechanically stable matrices such as controlled pore glass[67] or poly(styrene-divinylbenzene)[68] allow faster flow rates and shorter processing times than can be achieved with agarose.[69]

The conditions for binding an immunoadhesin to the protein A or G affinity column are dictated entirely by the characteristics of the Fc domain; that is, its species and isotype. Generally, when the proper ligand is chosen, efficient binding occurs directly from unconditioned culture fluid. One distinguishing feature of immunoadhesins is that, for human $\gamma 1$-containing molecules, the binding capacity for protein A is somewhat diminished relative to an antibody of the same Fc type.[70] Bound immunoadhesin can be efficiently eluted either at acidic pH ($\leqslant 3.0$), or in a neutral pH buffer containing a mildly chaotropic salt. For cases in which the adhesin portion of the molecule is sensitive to treatment with acid, the latter method of elution is preferred.[14,71] This affinity chromatography step can result in an immunoadhesin preparation that is $>95\%$ pure.

10.6.2 Other Chromatographic Methods

Other methods can be used in place of, or in addition to, affinity chromatography on protein A or G (see, for example, Refs. 72, 73). In our experience, immunoadhesins behave similarly to antibodies in thiophilic gel chromatography[74] and immobilized metal affinity chromatography.[75,76] In addition, partitioning via aqueous two-phase extraction[77] or chromatographic recovery in expanded beds[78] have both been applied to antibodies and should be directly applicable to immunoadhesins. In contrast to antibodies, however, the behavior of immunoadhesins on ion exchange columns is dictated not only by their isoelectric points, but also by a potential charge dipole in the molecules (see Section 10.5). Microheterogeneity of charge can also be a factor for immunoadhesins in which the adhesin portion of the molecule is glycosylated and contains sialic acid.

We designed our recovery process for CD4-IgG so that it was easily scalable. In general, we sought to link successive operations in a logical flow, combining tangential flow filtration with affinity, ion exchange and hydrophobic interaction chromatography operations in series (for example, see Ref. 79).

10.6.3 Large-Scale Recovery

Several immunoadhesins have been advanced as candidates for human therapy (Table 10.1). In the case of CD4-IgG, which was advanced to clinical trials as an HIV antiviral, we designed a recovery process that was intended to produce 0.5–1 kg of purified product per batch, using cell culture operation in a 12 kL bioreactor. Initially the process was developed at laboratory scale, using 2 L feedstocks provided by the cell culture development lab. As larger fluid volumes of feedstocks became available, we scaled the recovery process stepwise in three 10–20 fold size increments, checking performance at each step as new production equipment was introduced.

Guidelines for scaling column operations were relatively simple (Table 10.2). In making the transition from laboratory to pilot to manufacturing scale, parameters such as sample load volume, volumetric flow rate, and media volume were increased, while the column bed height, linear flow rate, sample concentration, and ratio of sample to gel were kept constant. Since a gradient was used in operation of one of the columns, the ratio of gradient volume to gel volume was held constant. Thus, the time required for the gradient to develop was approximately the same on the larger column.

10.6.4 Nonprotein Contaminants

Pharmaceutical production requires removal of nonprotein as well as protein contaminants from the product. Among the nonprotein contaminants that must be removed are host cell DNA, bacterial cells (e.g., bioburden) and components such as endotoxin and viruses. DNA can be present in the feedstock as a consequence of cell lysis during production. As much as 1–5 mg/ml of DNA can be present in harvested cell culture fluid that must be removed. Removal of DNA can be obtained in many types of chromatography operations, but since it is a polyanion, removal is best achieved using anion exchange methods. The potential for bioburden in a process always exists; we have dealt with this by using filtration (0.22 μm) of process buffers and pools. Efficient removal of bacterial endotoxin can occur in anion exchange and hydrophobic interaction chromatography steps.

TABLE 10.1 Immunoadhesins Studies for Clinical Application

	CD4-IgG	TFN Receptor (p55)-IgG	TFN Receptor (p75)-IgG	LFA3-IgG	CTLA4-IgG
IgG subclass	IgG1	IgG1	IgG1	IgG1	IgG1
Target ligand	HIV gp 120	TNFα, β	TNFα, β	CD2	B7
Product designation		lenercept Tenefuse	TNFR:Fc Enbrel	LFA3TIP	
Company	Genentech, Inc. (S. San Francisco, CA)	Hoffmann-LaRoche Basel, Switzerland/ Nutley, NJ) Genentech (S. San Francisco, CA)	Immunex (Seattle, WA) Wyeth-Ayerst Laboratories (Philadelphia, PA)	Biogen (Cambridge, MA)	Bristol-Myers Squibb, New York, NY)
Main clinical indication	HIV infection	Severe sepsis	Rheumatoid	Psoriasis	Psoriasis
Reference	24, 90	9, 10, 91	92, 93	61, 94	30, 95

*see also chapter 9.
†see also chapter 8.

TABLE 10.2 Guidelines for Scaling Chromatographic Operations from Laboratory to Plant

Maintain	Type of chromatographic media (including particle size) Column bed height Linear flow rate (cm/h) Sample load concentration Gradient volume: media volume
Increase	Sample load volume Volumetric flow rate (L/min) Column diameter
Check System Factors:	Distribution system Wall effects Piping (linear flow in whole system)

Source: Modified from Ref. 86.

As outlined previously, all rodent cells that are used for protein expression contain viruslike C-particles. Although these particles are noninfectious, they warrant concern and should be removed. Since our production method utilized cultured hamster (CHO) cells, we had to validate that any viral particles present in the feedstock were removed or inactivated during the recovery process. Typically the log titer reduction (LTR) values (LTR = log of challenge titer to recovery titer ratio) for affinity, ion exchange, hydrophobic and size exclusion chromatography range between 2 and 4.[80] This means that a purification scheme that uses four chromatography steps can be expected to yield an LTR value between 8 and 16. The most effective methods for removal of retroviruses were affinity, anion exchange, and hydrophobic interaction chromatography. In order to increase process robustness, we sought to combine removal steps with additional steps in which virus inactivation occurred. We used several methods for inactivation of viruses, including exposure to low pH, mild chaotropic agents, and nonionic detergents.[80]

10.6.5 Protein Quality, Purity and Consistency

The required purity of a product is, of course, dictated by its final use. For recombinant protein products, purity can be compromised by the presence of contaminating host cell proteins, but also by variant forms of the desired product. Aggregated product is often of concern with immunoadhesins, since these proteins are nonnatural and so it is difficult to predict their solution behavior. If the recovery process includes affinity chromatography on protein A, then care must be taken that leached protein A is not detectable in the product. The recovery process should be designed to remove both of these potential contaminants.

Protein consistency can be an issue of concern with immunoadhesins, since these are intended as recombinant glycoproteins with long in vivo half-lives. Often the adhesin portion of the molecule contains sialylated complex oligosaccharides through which clearance of the glycoprotein can occur.[81–83] In situations where batch-to-batch consistency varies, the serum stability of the product can be affected due to the variable glycosylation in these oligosaccharides.[62,84,85]

10.6.6 Large-Scale Process Integration

To develop a robust process, integration of the cell culture and recovery processes must begin early in the development effort. The goal of all activities is to develop a reliable process, capable of delivering a protein of consistent quality (at high yield and low cost).

In our experience, three issues are important for successful integration of cell culture and recovery development in scale-up.[79,86] First, equipment used for development and production will likely be different. Piping, vessels, pumps, and other equipment that come in contact with product streams can have an impact on product quality and recoverability. Second, changes in the composition of the production medium can have a major influence on recovery steps, sometimes even at downstream steps in the recovery train. Finally, differences in handling cell culture fluid streams during production and harvest can have a major impact on the amount and type of contaminants (proteins, nucleic acids, particulate cell debris).

10.7 CONCLUSIONS

Immunoadhesins have an established practical utility as research tools, and are beginning to show efficacy as human therapeutic agents. As research tools, they offer many of the advantages of monoclonal antibodies and are useful especially in studies on ligand–receptor interactions and in identification or isolation of unknown ligands for known receptors. An immunoadhesin approach, if applicable, also should be considered when it is difficult to obtain antibodies to a given antigen, for example, when the antigen is available in limited quantity or is poorly immunogenic. As human therapeutics, immunoadhesins share many of the properties of monoclonal antibodies, including target specificity and Fc effector functions. As compared with chimeric antibodies, which contain nonhuman sequences, immunoadhesins are potentially less immunogenic in humans. In addition, immunoadhesins can be easier to obtain than "humanized" antibodies, because they do not require CDR grafting. Immunoadhesins also may provide a useful alternative to human monoclonal antibodies, for which successful production has lagged far behind that of mouse antibodies, due to several factors including human tolerance to human antigens, ethical limitations to experimental immunization of humans, and instability of human antibody-producing hybrid cell lines. These obstacles are just now being overcome, as advances in genetic engineering techniques[87] for producing human antibodies from combinatorial libraries[88] or by genetic manipulation of laboratory animals[89] continue to develop. Finally, as compared with soluble receptors, immunoadhesins can have advantages in areas such as ease of purification, increased binding affinity or avidity, added effector functions, and extended in vivo half-life. Immunoadhesins generally are straightforward to produce at large scale. To date, several immunoadhesins have been produced using mammalian cell culture at large scale for human therapeutic use.

ACKNOWLEDGMENTS

We wish to acknowledge all of the many dedicated members of the CD4-IgG and TNFR-IgG project teams at Genentech, whose contributions were so important to development of this technology. In particular, we wish to acknowledge Dan Capon, Joanne Beck, Adriana Johnson, Scot Marsters, David Peers, Robert Pitti, Hardat Prashad, Douglas Smith, and Rebecca Ward for their critical contribution to these efforts.

REFERENCES.

1. Gascoigne, N. R. J., C. C. Goodnow, K. I. Dudzik, V. T. Oi, M. M. Davis. 1987. Secretion of a chimeric T-cell receptor-immunoglobulin protein. *Proc. Natl. Acad. Sci. USA* 84: 2936–2940.
2. Capon, D. J., S. M. Chamow, J. Mordenti, S. A. Marsters, T. Gregory, H. Mitsuya, R. A. Byrn, C. Lucas, F. M. Wurm, J. E. Groopman, S. Broder, D. H. Smith. 1989. Designing CD4 immunoadhesins for AIDS therapy. *Nature* 337: 525–531.
3. Ashkenazi, A., D. J. Capon, R. H. Ward. 1993. Immunoadhesins. *Intl. Rev. Immunol.* 10: 219–227.
4. Berg, J., E. Lotscher, K. S. Steimer, D. J. Capon, J. Baenziger, H. Jack, M. Wabl. 1991. Bispecific antibodies that mediate killing of cells infected with HIV of any strain. *Proc. Natl. Acad. Sci. USA* 88: 4723–4727.
5. Traunecker, A., J. Schneider, H. Kiefer, K. Karjalainen. 1989. Highly efficient neutralization of HIV with recombinant CD4–immunoglobulin molecules. *Nature* 339: 68–70.
6. Martin, S., J. M. Casasnovas, D. E. Staunton, T. A. Springer. 1993. Efficient neutralization and disruption of rhinovirus by chimeric ICAM-1/immunoglobulin molecules. *J. Virol.* 67: 3561–3568.
7. Watson, S. R., Y. Imai, C. Fennie, J. S. Geoffroy, S. D. Rosen, L. A. Lasky. 1990. A homing receptor-IgG chimera as a probe for adhesive ligands of lymph node high endothelial venules. *J. Cell Biol.* 110: 2221–2229.

8. Watson, S. R., C. Fennie, L. A. Lasky. 1991. Neutrophil influx into an inflammatory site inhibited by a soluble homing receptor-IgG chimera. *Nature* 349: 164–167.
9. Ashkenazi, A., S. A. Marsters, D. J. Capon, S. M. Chamow, I. S. Figari, D. Pennica, D. V. Goeddel, M. A. Palladino, D. H. Smith. 1991. Protection against endotoxic shock by a tumor necrosis factor receptor immunoadhesin. *Proc. Natl. Acad. Sci. USA* 88: 10535–10539.
10. Lesslauer, W., H. Tabuchi, R. Gentz, M. Brockhaus, E. J. Schlaeger, G. Grau, P. F. Piguet, P. Pointaire, P. Vassalli, H. Loetscher. 1991. Recombinant soluble tumor necrosis factor receptor proteins protect mice from lipopolysaccharide-induced lethality. *Eur. J. Immunol.* 21: 2883–2886.
11. Peppel, K., D. Crawford, B. Beutler. 1991. A tumor necrosis factor (TNF) receptor-IgG heavy chain chimeric protein as a bivalent antagonist of TNF activity. *J. Exp. Med.* 174: 1483–1489.
12. Kurschner, C., G. Garotta, Z. Dembic. 1992. Construction, purification, and characterization of new interferon gamma inhibitor proteins. *J. Biol. Chem.* 267: 9354–9360.
13. Haak-Frendscho, M., S. A. Marsters, S. M. Chamow, D. H. Peers, N. J. Simpson, A. Ashkenazi. 1993. Inhibition of interferon-gamma by an interferon-gamma receptor. *Immunology* 79: 594–599.
14. Bennett, B. D., G. L. Bennett, R. V. Vitangcol, J. R. Jewett, J. Burnier, W. Henzel, D. G. Lowe. 1991. Extracellular domain-IgG fusion proteins for three human natriuretic peptide receptors. Hormone pharmacology and application to solid phase screening of synthetic peptide antisera. *J. Biol. Chem.* 266: 23060–23067.
15. Steele, A. W., X. X. Zheng, T. B. Strom. 1993. Fourth IBC Intl. Conf. on Antibody Engineering, Coronado, CA. International Business Communications, Southborough, MA.
16. Mayforth, R. D. 1993. Designing Antibodies. San Diego, CA: Academic Press.
17. Shin, S. U., S. L. Morisson. 1990. Expression and characterization of an antibody binding specificity joined to insulin-like growth factor 1: Potential applications for cellular targeting. *Proc. Natl. Acad. Sci. USA* 87: 5322–5326.
18. Hoogenboom, H. R., J. C. Raus, G. Volckaret. 1991. Targeting of tumor necrosis factor to tumor cells: Secretion by myeloma cells of a genetically engineering antibody-tumor necrosis factor hybrid molecule. *Biochim. Biophys. Acta.* 1096: 345–354.
19. Fell, H. P., M. A. Gayle, L. Grosmaire, J. A. Ledbetter. 1991. Genetic construction and characterization of a fusion protein consisting of a chimeric F(ab') with specificity for carcinomas and human IL-2. *J. Immunol.* 146: 2446–2452.
20. Hollenbaugh, D., N. J. Chalupny, A. Aruffo. 1992. Recombinant globulins: Novel research tools and possible pharmaceuticals. Curr. Opin. Immunol. 4: 216–219.
21. Ashkenazi, A., S. M. Chamow. 1995. Immunoadhesins: An alternative to human monoclonal antibodies. Methods: A Companion to *Meth. Enzymol.* 8: 104–115.
22. Chamow, S. M., A. Ashkenazi. 1996. Immunoadhesins. *Trends Biotechnol.* 14: 52–60.
23. Ashkenazi, A., S. M. Chamow. 1997. Immunoadhesins as research tools and therapeutic agents. *Curr. Opin. Immunol.* 9: 195–200.
24. Byrn, R. A., J. Mordenti, C. Lucas, D. Smith, S. A. Marsters, J. S. Johnson, P.

Cossum, S. M. Chamow, F. M. Wurm, T. Gregory, J. E. Groopman, D. J. Capon. 1990. Biological properties of a CD4 immunoadhesin. *Nature* 344: 667–670.

25. Zettlmeissl, G., J.-P. Gregersen, J. M. Duport, S. Mehdi, G. Reiner, B. Seed. 1990. Expression and characterization of human CD4:immunoglobulin fusion proteins. *DNA and Cell Biol.* 9: 347–353.
26. Loetscher, H., R. Gentz, M. Zulaug, A. Lustig, H. Tabuchi, E. J. Schlaeger, M. Brockhous, H. Gallati, M. Manneberg, W. Lesslauer. 1991. Recombinant 55–kDa tumor necrosis factor receptor: Stoichiometry of binding to TNF and inhibition of TNF activity. *J. Biol. Chem.* 266: 18324–18329.
27. Garcia, I., Y. Miyazaki, K. Araki, M. Araki, R. Lucas, G. E. Grau, G. Milon, Y. Belkaid, C. Montixi, W. Lesslauer, P. Vassalli. 1995. Transgenic mice expressing high levels of soluble TNFR1 fusion protein are protected from lethal septic shock and cerebral malaria, and are highly sensitive to *Listeria monocytogenes* and *Leishmania major* infections. *Eur. J. Immunol.* 25: 2401–2407.
28. Mori, L., S. Iselin, G. De Libero, W. Lesslauer. 1996. Attenuation of collagen-induced arthritis in 55–kDa TNF receptor type 1–IgG treated and TNFR1–deficient mice. *J. Immunol.* 157: 3178–3182.
29. Linsley, P. S., W. Brady, L. Grosmaire, A. Aruffo, N. K. Damle, J. A. Ledbetter. 1991. Binding of the B cell activation antigen B7 to CD28 costimulates T cell proliferation and interleukin 2 mRNA accumulation. *J. Exp. Med.* 173: 721–730.
30. Linsley, P. S., W. Brady, M. Urnes, L. S. Grosmaire, N. K. Damle, J. A. Ledbetter. 1991. CTLA-4 is a second receptor for the B cell activation antigen B7. *J. Exp. Med.* 174: 561–569.
31. Stamenkovic, I., D. Sgroi, A. Aruffo, M. S. Sy, T. Anderson. 1991. The B lymphocyte adhesion molecule CD22 interacts with leukocyte common antigen CD45RO on T cells and a2–6 sialyltransferase, CD75, on B cells. *Cell* 66: 1133–1144.
32. Chalupny, N. J., R. Peach, D. Hollenbaugh, J. A. Ledbetter, A. G. Farr, A. Aruffo. 1992. T-cell activation molecule 4–1BB binds to extracellular matrix proteins. *Proc. Natl. Acad. Sci. USA* 89: 10360–10364.
33. Armitage, R. J., T. A. Sato, B. M. Macduff, K. N. Clifford, A. R. Alpert, C. A. Smith, W. C. Fanslow. 1992. Identification of a source of biologically active CD40 ligand. *Eur. J. Immunol.* 22: 2071–2076.
34. Smith, C. A., H.-J. Gruss, T. Davis, D. Anderson, T. Farrah, E. Baker, G. R. Sutherland, C. I. Brannan, N. G. Copeland, N. A. Jenkins, K. H. Grabstein, B. Gliniak, I. B. McAlister, W. Fanslow, M. Alderson, B. Falk, S. Gimpel, S. Gillis, W. S. Din, R. G. Goodwin, R. J. Armitage. 1993. CD30 antigen, a marker for Hodgkin's lymphoma, is a receptor whose ligand defines an emerging family of cytokines with homology to TNF. *Cell.* 73: 1349–1360.
35. Suda, T., T. Takahashi, P. Golstein, S. Nagata. 1993. Molecular cloning and expression of the fas ligand, a novel member of the tumor necrosis factor family. *Cell* 75: 1169–1178.
36. Kabat, E. A., T. T. Wu, H. M. Perry, K. S. Gottesman, C. Foeller. 1991. . In Sequences of Proteins of Immunological Interest. Washington, D.C.: U.S. Department of Health and Human Services, U.S. Government Printing Office.
37. Gorman, C. M., D. R. Gies, G. McCray. 1990. Transient production of proteins using an adenovirus transformed cell line. *DNA Prot. Eng. Tech.* 2: 3–10.

38. Boie, Y., N. Sawyer, D. M. Slipetz, K. M. Metters, M. Abramovitz. 1995. Molecular cloning and characterization of the human prostanoid DP receptor. *J. Biol. Chem.* 270: 18910–18916.

39. Bennett, W. F., N. Paoni, D. Botstein, A. J. S. Jones, B. Keyt, L. Presta, F. M. Wurm, M. Zoller. 1991. Functional properties of a collection of charged-to-alanine substitution variants of tissue-type plasminogen activator. *J. Biol. Chem.* 266: 5191–5201.

40. Jordan, M., F. M. Wurm. 1995. High-level transient expression in mammalian cells: Identification and optimization of physicochemical parameters of the calcium phosphate transfection method, In: Animal Cell Technology: Developments towards the 21st Century, eds. Beuvery, Griffiths, Zeijlemaker, Kluwer Academic Publishers, England, pp. 49–56.

41. Urlaub, G., L. A. Chasin. 1980. Isolation of Chinese hamster cell mutants deficient in dihydrofolate reductase activity. *Proc. Natl. Acad. Sci. USA* 77: 4216–4220.

42. Graham, F. L., A. J. van der Eb. 1973. A new technique for the assay of infectivity of human adenovirus 5 DNA. *Virology* 52: 456–467.

43. Jordan, M., A. Schallhorn, F. M. Wurm. 1996. Transfecting mammalian cells: Optimization of critical parameters affecting calcium-phosphate precipitate formation. *Nucl. Acids Res.* 24: 596–601.

44. Porterfield, J. S., M. J. Ashwood-Smith. 1962. Preservation of cells in tissue culture by glycerol and dimethyl sulfoxide. *Nature* 193: 548–550.

45. Michaels, S. L., C. Antoniou, V. Goel, P. Keating, R. Kuriyel, A. S. Michaels, S. R. Pearl, G. de los Reyes, E. Rudolph, M. Siwak. 1995. Tagential flow filtration. In Separations Technology: Pharmaceutical and Biotechnology Applications, ed. W. P. Olson, Buffalo Grove, IL: Interpharm Press, pp. 57–194.

46. Liljestrom, P., H. Garoff. 1991. A new generation of animal cell expression vectors based on the semliki forest virus replicon. *Bio/Technol.* 9: 1356–1361.

47. Berglund, P., M. Sjoberg, H. Garoff, G. J. Atkins, B. J. Sheahan, P. Liljestrom. 1993. Semliki forest virus expression system: Production of conditionally infectious recombinant particles. *Bio/Technol.* 11: 916–920.

48. Kost, T. A., D. M. Ignar, W. C. Clay, J. Andrews, J. D. Leray, L. Overton, C. R. Hoffman, K. E. Kilpatrick, B. Ellis, D. L. Emerson. 1997. Production of a urokinase plasminogen activator-IgG fusion protein (uPA-IgG) in the baculovirus expression system. *Gene* 190: 139–144.

49. Jordan, M., C. Kohne, F. M. Wurm. 1998. Calcium-phosphate mediated DNA transfer into HEK 293 cells in suspension: Control of physicochemical parameters allows transfection in stirred media. *Cytotechnol.* 26: 39–47.

50. Abdallah, B., A. Hassan, C. Benoist, D. Goula, J.-P. Behr, B. Demeneix. 1996. A powerful nonviral vector for in vivo gene transfer into the adult mammalian brain: polythyleneimine. *Hum. Gene Ther.* 7: 1947–1954.

51. Schlaeger, E. J., J. Y. Legendre, A. Trzeciak, E. A. Kitas, K. Christensen, U. Deuschle, A. Supersaxo. 1997. Transient transfection in mammalian cells: A basic study for an efficient and cost-effective scale-up process. In *Transient Gene Expression in Animal Cells*, Jersey, UK.

52. Blasey, H. D., L. Rey, L. Garcia, A. Bernard. 1997. Transient expression by transfection: Potential, scale-up, limitations. In *Transient Gene Expression in*

Animal Cells, Jersey, UK.

53. Luckow, V. A., M. D. Summers. 1988. Trends in the development of baculovirus expression vectors. *Bio/Technol.* 6: 47–55.
54. Barkhem, T., B. Carlsson, A. Danielsson, U. Norinder, H. Frieberg, L. Ohman. 1992. Production in a 100–liter stirred tank reactor of functional, full-length, human thyroid receptor beta 1 in Sf9 insect cells using a recombinant baculovirus. In *Baculovirus and Recombinant Protein Production Processes*, ed. J. M. Vlak, E.-J. Schlaeger, A. R. Bernard, Editiones Roche, Basel, pp. 235–246.
55. Massie, B., R. Tom, A. W. Caron. 1992. Scale-up of a baculovirus expression system: Production of recombinant proteins in perfused high density Sf9 cell cultures. In *Baculovirus and Recombinant Protein Production Processes*, eds. J. M. Vlak, E.-J. Schlaeger, A. R. Bernard, Editiones Roche, Basel, pp. 234–???.
56. Blasey, H. D., B. Brethon, K. Lundstrom, R. Hovius, A. R. Bernard. 1997. Recombinant protein production using the Semliki Forest virus expression system. *Cytotechnology* 24(1) 65–72.
57. Lundstrom, K., A. Michel, H. Blasey, A. Bernard, R. Hovius, H. Vogel, A. Surprenant. 1997. Expression of ligand-goted ion channels with the Semliki Forest virus expression system. *J. Recog. Signal Transduction Res* 17(1-3) 115–126.
58. Marsters, S. A., A. D. Frutkin, N. J. Simpson, B. M. Fendly, A. Ashkenazi. 1992. *J. Biol. Chem.* 267: 5747–5750.
59. Chamow, S. M., D. H. Peers, R. A. Byrn, M. G. Mulkerrin, R. J. Harris, W.-C. Wang, P. J. Bjorkman, D. J. Capon, A. Ashkenazi. 1990. Enzymatic cleavage of a CD4 immunoadhesin generates crystallizable, biologically active Fd-like fragments. *Biochemistry* 29: 9885–9891.
60. Pitti, R. M., S. A. Marsters, M. Haak-Frendscho, G. C. Osaka, J. Mordenti, S. M. Chamow, A. Ashkenazi. 1994. Molecular and biological properties of an interleukin-1 receptor immunoadhesin. *Molec. Immunol.* 31: 1345–1351.
61. Meier, W., A. Gill, M. Rogge, R. Dabora, G. R. Majeau, F. B. Oleson, W. E. Jones, D. Frazier, K. Maitkowski, P. S. Hochman. 1995. Immunomodulation by LFA3TIP, an LFA3/IgG1 fusion protein: cell line dependent glycosylation effects on pharmacokinetics and pharmacodynamic markers. *Therap. Immunol.* 2: 159–171.
62. Flesher, A. R., J. Marzowski, W.-C. Wang, H. V. Raff. 1995. Fluorophore-labeled carbohydrate analysis of immunoglobulin fusion proteins: Correlation of oligosaccharide content with in vivo clearance profile. *Biotechnol. Bioeng.* 46: 399–407.
63. Harris, R. J., K. L. Wagner, M. W. Spellman. 1990. Structural characterization of a recombinant CD4–IgG hybrid molecule. *Eur. J. Biochem.* 194: 611–620.
64. Ellison, J. W., B. J. Berson, L. E. Hood. 1982. The nucleotide sequence of a human immunoglobulin C-gamma1 gene. *Nucl. Acids Res.* 10: 4071–4079.
65. Lindmark, R., K. Thoren-Tolling, J. Sjoquist. 1983. Binding of immunoglobulins to protein A and immunoglobulin levels in mammalian sera. *J. Immunol. Meth.* 62: 1–13.
66. Guss, B., M. Eliasson, A. Olsson, M. Uhlen, A. K. Frej, H. Jornvall, J. I. Flock, M. Lindberg. 1986. Structure of the IgG-binding regions of streptococcal protein G. *EMBO J.* 5: 1567–1575.
67. Phillips, T. M., W. D. Queen, N. S. More, A. M. Thompson. 1985. *J. Chromatogr.*

544: 267–279.

68. Afeyan, N. B., S. P. Fulton, F. E. Regnier. 1991. Perfusion chromatography packing materials for proteins and peptides. *J. Chromatogr.* 544: 267–279.
69. Lee, S. M., M. Gustafson, D. Pickle, M. Flickinger, G. Muschik, A. Morgan, Jr. 1986. Large-scale purification of a murine anti-melanoma monoclonal antibody. J. *Biotechnol.* 4: 189–204.
70. Chamow, S. M., D. H. Peers, G. S. Blank, R. D. Hershberg. 1996. Large-scale protein A affinity chromatography of immunoadhesins, Recovery of Biological Products VIII, Tucson, AZ Engineering Foundation and the American Chemical Society, New York, NY.
71. Chamow, S. M., D. Z. Zhang, X. Y. Tan, S. M. Mhatre, S. M. Marsters, D. H. Peers, R. A. Byrn, A. Ashkenazi, R. P. Junghans. 1994. A humanized, bispecific immunoadhesin-antibody that retargets CD3+ effectors to kill HIV1–infected cells. *J. Immunol.* 153: 4268–4280.
72. Nau, D. R. 1989. Chromatographic methods for antibody purification and analysis. *BioChromatography* 4: 4–18.
73. Anspach, F. B., D. Petsch, W.-D. Deckwer. 1996. Purification of murine IgG1 on group specific affinity sorbents. *Bioseparation* 6: 165–184.
74. Hutchens, T. W., J. Porath. 1986. Thiophilic adsorption of immunoglobulins—analysis of conditions optimal for selective immobilization and purification. *Anal. Biochem.* 159: 217–226.
75. Al-Mashikhi, S. A., S. Nakai. 1988. *J. Dairy Sci.* 71: 1756–1763.
76. Hale, J. E., D. E. Beidler. 1994. Purification of humanized murine and murine monoclonal antibodies using immobilized metal-affinity chromatography. *Anal. Biochem.* 222: 29–33.
77. Andrews, B. A., S. Nielsen, J. A. Asenjo. 1996. Partitioning and purification of monoclonal antibodies in aqueous two-phase systems. *Bioseparation* 6: 303–313.
78. Chase, H. A. 1994. Purification of proteins by adsorption chromatography in expanded beds. *Trends Biotechnol.* 12: 296–303.
79. Wheelwright, S. M. 1991. *Protein Purification: Design and Scale Up of Downstream Processing*. New York, NY: Wiley.
80. White, E. M., J. B. Grun, C.-S. Sun, A. F. Sito. 1991. Process validation for virus removal and inactivation. *Biopharm.* 4: 34.
81. Fukuda, M. N., H. Sasaki, L. Lopez, M. Fukuda. 1989. Survival of recombinant erythropoietin in the circulation: The role of carbohydrates. *Blood* 73: 84–89.
82. Thotakura, N. R., R. K. Desai, L. G. Bates, E. S. Cole, B. M. Pratt, B. D. Weintraub. 1991. Biological activity and metabolic clearance of a recombinant human thyrotropin produced in CHO cells. *Endocrinology* 128: 341–348.
83. Bishop, L. A., T. V. Nguyen, P. R. Schofield. 1995. Both of the b-subunit carbohydrate residues of follicle stimulating hormone determine the metabolic clearance rate and in vivo potency. *Endocrinology* 136: 2635–2640.
84. Hooker, A. D., M. H. Goldman, N. H. Markham, D. C. James, A. P. Ison, A. T. Bull, P. G. Strange, I. Salmon, A. J. Baines, N. Jenkins. 1995. N-glycans of recombinant human interferon-g change during batch culture of CHO cells. *Biotechnol. Bioeng.* 48: 639–648.
85. Jenkins, N. 1996. Role of physiology in the determination of protein heterogeneity.

Curr. Opin. Biotechnol. 7: 205–209.

86. Sofer, G. K., L. E. Nystrom. 1989. *Process Chromatography: A Practical Guide*. San Dicgo: Academic Press.
87. Hayden, M. S., L. K. Gilliland, J. A. Ledbetter. 1997. Antibody engineering. *Curr. Opin. Immunol.* 9: 201–212.
88. Vaughn, T. J., A. J. Williams, K. Pritchard, J. K. Osbourn, A. R. Pope, J. C. Earnshaw, J. McCafferty, R. A. Hodits, J. Wilton, K. S. Johnson. 1996. Human antibodies with sub-nanomolar affinities isolated from a large non-immunized phage display library. *Nature Biotechnol.* 14: 309–314.
89. Jakobovits, A., A. L. Moore, L. L. Green, G. J. Vergara, C. E. Maynard-Currie, H. A. Austin, S. Klapholz. 1993. Germ-line transmission and expression of a human-derived yeast artificial chromosome. *Nature* 362: 255–258.
90. Hodges, T. L., J. O. Kahn, L. D. Kaplan, J. E. Groopman, P. A. Volberding, A. J. Ammann, C. J. Arri, L. M. Bouvier, J. Mordenti, A. E. Izu, J. D. Allan. 1991. Phase I study of recombinant CD4–IgG therapy of patients with AIDS and AIDS-related complex. *Antimicrobiol. Agents Chemother.* 35: 2580–2586.
91. Abraham, E., M. P. Glauser, T. Butler, J. Garbino, D. Gelmont, P. F. Laterre, K. Kudsk, H. A. Bruining, C. Otto, E. Tobin, C. Zwingelstein, W. Lesslauer, A. Leighton. 1997. p55 Tumor necrosis factor receptor fusion protein in the treatment of patients with severe sepsis and septic shock. JAMA 277: 1531–1538.
92. Wooley, P. H., J. Dutcher, M. B. Widmer, S. Gillis. 1993. Influence of a recombinant human soluble tumor necrosis factor receptor Fc fusion protein on type II collagen-induced arthritis in mice. *J. Immunol.* 151: 6602–6607.
93. Moreland, L. W., S. W. Baumgartner, M. H. Schiff, E. A. Tindall, R. M. Fleischmann, A. L. Weaver, R. E. Ettlinger, S. Cohen, W. J. Koopman, K. Mohler, M. B. Widmer, C. M. Blosch. 1997. Treatment of rheumatoid arthritis with a recombinant human tumor necrosis factor receptor (p75)-Fc fusion protein. *New Engl. J. Med.* 337: 141–147.
94. Kaplon, R. J., P. S. Hochman, R. E. Michler, P. A. Kwiatkowski, N. M. Edwards, C. L. Berger, H. Xu, W. Meier, B. P. Wallner, P. Chisholm, C. C. Marboe. 1996. Short course single agent therapy with an LFA3–IgG1 fusion protein prolongs primate cardiac allograft survival. *Transplantation* 61: 356–363.
95. Larsen, C. P., E. T. Elwood, D. Z. Alexander, S. C. Ritchie, R. Hendrix, C. Tucker-Burden, H.-R. Cho, A. Aruffo, D. Hollenbaugh, P. S. Linsley, K. J. Winn, T. C. Pearson. 1996. Long-term acceptance of skin and cardiac allografts after blocking CD40 and CD28 pathways. *Nature* 381: 434–438.
96. Lindmark, R., K. Thoren-Tolling, J. Sjoquist. 1983. Binding of immunoglobulins to protein A and immunoglobulin levels in mammalian sera. *J. Immunol. Meth.* 62: 1–13.
97. Goding, J. W. 1978. Use of staphylococcal protein A as an immunological reagent. *J. Immunol. Meth.* 20: 241–253.
98. Uhlen, M., B. Guss, B. Nilsson, S. Gatenbeck, L. Philipson, M. Lindberg. 1984. Complete sequence of the staphylococcal gene encoding protein A. A gene evolved through multiple duplications. *J. Biol. Chem.* 259: 1695–1702.
99. Moks, T., L. Abrahamsen, B. Nilsson, U. Hellman, J. Sjoquist, M. Uhlen. 1986. Staphylococcal protein A consists of five IgG-binding domains. *Eur. J. Biochem.*

156: 637–643.

100. Ljungberg, U. K., B. Jansson, U. Niss, R. Nilsson, B. E. B. Sanberg, B. Nilsson. 1993. The interaction between different domains of staphylococcal protein A and human polyclonal IgG, IgA, IgM and $F(ab')_2$: separation of affinity from specificity. *Molec. Immunol.* 30: 1279–1285.

101. Inganas, M., S. G. O. Johnsson, H. H. Bennich. 1980. Interaction of human polyclonal IgE and IgG from different species with protein A from Staphylococcus aureus: Demonstration of protein A reactive sites located in the $F(ab')_2$ fragment of human IgG. *Scand. J. Immunol.* 12: 23–31.

102. Hanson, D. C., V. N. Schumaker. 1984. A model for the formation and interconversion protein A-immunoglobulin G soluble complexes. *J. Immunol.* 132: 1397–1409.

INDEX